ELEPHANTS AND WHALES

ELEPHANTS AND WHALES

RESOURCES FOR WHOM?

Edited by

Milton M.R. Freeman

University of Alberta
Canada

and

Urs P. Kreuter

Texas A&M University
USA

Gordon and Breach Science Publishers

Australia • Austria • Belgium • France • Germany • India • Japan • Malaysia • Netherlands
Russia • Singapore • Switzerland • Thailand • United Kingdom • United States

Gordon and Breach Science Publishers S.A.

Postfach
4004 Basel
Switzerland

Front cover: Elephants photograph — courtesy of Stephen J. Thomas
Whales photograph — courtesy of Ole Wich

British Library Cataloguing in Publication Data

Elephants and Whales: Resources for Whom?
I. Freeman, Milton M.R.   II. Kreuter,
Urs P.
639.9795

ISBN 2-88449-010-8 (hardcover)
ISBN 2-88449-011-6 (softcover)

CONTENTS

DEDICATION

To John H. Peterson Jr.

In memory of his friendship, enthusiasm and energetic work to foster human welfare and the rights of communal people to manage their own resources. We thank him for initiating and then playing a major role in the Elephants and Whales Symposium held at the Third Annual Meeting of the International Association for the Study of Common Property. Dr. Peterson's untimely death has left a void in the effort to promote the concept of wildlife-based community development. In gratitude for his example and contributions, royalties arising from the publication of this book will be donated to the Choctaw Cultural Resources Center (Mississippi Band of Choctaw Indians) in memory of Dr. John Peterson.

ACKNOWLEDGEMENTS

The editors wish to express appreciation to the late John Peterson for his inspiration and work in organizing the symposium, the papers of which provide the basis for this book.

Financial contributions toward publication of this book were provided through an Earhart Foundation Research Fellowship to Urs Kreuter, the Thomas M. O'Conner Professorship by Dr. J.R. Conner and the Social Sciences and Humanities Research Council of Canada (Grant 410-911539). Thanks also go to Jerry Stuth, Richard Conner and Wayne Hamilton (Texas A&M University) for allowing Urs Kreuter relief from his official duties to edit the manuscript.

Richard Hoare, Urs Kreuter and Steve Thomas kindly donated the photographs on elephants; Richard Caulfield, Bjorn Tore Forberg, Kate Sanderson and Einar Tallaksen provided other photographs used in this book.

Sylvia Dudash and Kary Yourman (Texas A&M University), Elaine Maloney, Gail Mathew and Marlys Rudiak (University of Alberta) provided prompt and careful typing of early drafts of chapters.

We wish to acknowledge permission provided by the Arctic Institute of North America to allow reprinting of the chapters by Richard Caulfield and Mats Ris that first appeared in *Arctic*; the Institute of Cetacean Research for permission to print a re-edited paper by Nancy Doubleday; *Maritime Anthropological Studies* for permission to reprint an edited version of Arne Kalland's paper that was first published in that journal; the editor of *Pachyderm* for permission to publish a re-edited version of Russell Taylor's paper; National Academy Press for permission to publish an edited version of Emizet Kisangani's paper from the *Proceedings of the Conference on Common Property Resource Management*; and Alfred A. Knopf, Inc. for allowing us to publish an edited abstract from Ray Bonner's book *At the Hand of Man: Peril and Hope for Africa's Wildlife*.

Finally, thanks are due to the authors for their contributions to the book and their enthusiasm for the project.

CONTRIBUTORS

Raymond Bonner has been a foreign correspondent for the *New York Times* and a staff writer at *The New Yorker*. His earlier books, *Weakness and Deceit: U.S. Foreign Policy and El Salvador* and *Waltzing with a Dictator: The Marcoses and the Making of American Policy* won the Robert F. Kennedy, Overseas Press Club, and Hillman Foundation awards. A graduate of MacMurray College and Stanford Law School, Mr. Bonner practised public interest law for several years before turning to journalism. From 1988 through 1992 he lived in Nairobi and he now lives in Warsaw.

Harold Beyer Broch is assistant professor in the Institute of Social Anthropology at the University of Oslo. He has conducted fieldwork in northern Canada and in coastal societies in Indonesia, as well as in east Africa as an advisor to the Royal Northern Ministry of Development Cooperation. He is the author of *Woodland Trappers — Hare Indians of Northwestern Canada* and *Growing Up Agreeably — Bonerate Childhood Observed*. Dr Broch's research interests included cultural ecology, psychological anthropology and the study of socialization processes.

Richard A. Caulfied teaches in the Department of Rural Development, University of Alaska Fairbanks, USA. His doctorate, at the School of Development Studies, University of East Anglia, UK, focused on Greenlandic aboriginal whaling. His professional interests include natural resources anthropology, the political economy of Greenland and Alaska, and cooperative renewable resources management regimes in the circumpolar north. He has worked extensively with indigenous peoples and communities in the circumpolar north.

Nancy Doubleday is trained as a biologist and lawyer, and has worked extensively on environmental and wildlife issues at the community, regional, national and international levels. She has also worked as a naturalist, outdoor educator and wilderness guide. Currently she is conducting research on arctic contaminants. Ms. Doubleday also writes on international law, traditional knowledge and nature.

Milton Freeman is Henry Marshall Tory Professor of Anthropology at the University of Alberta. His interests are in sustainable and equitable use of common property resources, traditional ecological knowledge and management systems, and environmental and socio-economic impact assessment. Current research involves traditional ecological knowledge studies of several economically-important marine and freshwater species in the Canadian western and eastern arctic regions, and the cultural ecology of community-based coastal whaling in the northern circumpolar regions.

Richard Hasler has a Ph.D in Anthropology from Michigan State University. His on-going research focuses on the social aspects of natural resource management in Zimbabwe's Communal Areas. The cultural and political dynamics associated with wildlife resource use and the potential of community-based tourism are two of his main interests.

Arne Kalland is Senior Research Fellow and Deputy Director at the Nordic Institute of Asian Studies, Copenhagen. Dr. Kalland is a social anthropologist having conducted fieldwork on fishing and agricultural societies in Japan, Norway and Thailand. He has published a number of books and articles on Japanese society and culture, including *Japanese Whaling — End of an Era?* (with B. Moeran, 1992) and he recently initiated and led an international team of researchers investigating the importance of minke whaling in North Norwegian coastal culture (*Norwegian Small Type Whaling in Cultural Perspectives*, 1992). Current research includes people's perceptions of whales and Japanese perceptions of nature.

Stephen R. Kellert is Professor of Social Ecology at the Yale University School of Forestry and Environmental Studies. He has published over 100 articles, books and monographs, mainly in the areas of natural resource policy and conservation of biological diversity. He has conducted extensive research on human values relating to nature and its protection. His most recent books include *The Biophilia Hypothesis* (1993, with E.O. Wilson) and *Ecology, Economics and Ethics: The Broken Circle* (1991, with F.H. Bormann).

Emizet Kisangani (Information not available)

Urs P. Kreuter was born in Tanzania and raised in Zimbabwe and now lives in Texas. He holds degrees in range science and economics and obtained his Ph.D from Utah State University in 1992. His first professional appointment was with the South African Department of Agriculture where he worked as a research officer in range and pasture management. Dr. Kreuter has also worked as a lecturer at the University of Natal, and is currently employed as a Research Associate at Texas A & M University. He is also a Research Associate of the Competitive Enterprise Institute, Washington, D.C.. Aside from developing decision support software for range management, Dr. Kreuter has a keen interest in international wildlife policy and has published newspaper articles on the ivory-trade ban in several major U.S. newspapers.

John H. Peterson, Jr. was born in Tifton, Georgia and died in 1992. He received his Ph.D degree from the University of Georgia in 1970. He was a social ecologist, specializing in natural resource development and a professor of anthropology at Mississippi State University. He was involved for many years in the development of communal lands including heading the planning council of the Choctaw Tribal Government in the U.S. From 1980–1984 Dr. Peterson served on the Grazing Lands Committee of the U.S. Man and the Biosphere Directorate. He visited Zimbabwe from 1989 to 1991 where he was a Fulbright Fellow in the Tropical Resource Ecology Program of the University of Zimbabwe, and in 1991 he became senior lecturer in the Centre for Applied Social Sciences. His research in Zimbabwe focused on the early implementation of the CAMPFIRE program.

Mats Ris is a doctoral candidate in the Department of Ethnology, University of Gothenburg and Visiting Researcher at the Canadian Circumpolar Institute, University of Alberta. He has conducted fieldwork on the Faroe Islands with special reference to the traditional pilot whale fishery, and in 1992 was a member of an International Study Group that reported on the traditional minke whale fishery in northern Norway. For the past five years Mr. Ris has attended the International Whaling Commission meetings representing IWGIA (International Work Group for Indigenous Affairs). His present research is on cultural perspectives concerning the international management of whaling.

Kate Sanderson studied Faroese and Icelandic in Copenhagen, Torshavn and Reykjavik from 1985–87 and served as Information Consultant to the Faroese Department of Fisheries from 1986–1993, in which capacity she has represented the Faroese on the Danish Delegation to the International Whaling Commission. In 1993 she was awarded a Master of Philosophy degree from the University of Sydney, Australia, for the thesis *Grindadrap — A Textual History of Whaling Traditions in the Faroes to 1900*. Since late 1993 Ms. Sanderson has been Secretary of the North Atlantic Marine Mammal Commission (NAMMCO) in Tromso, Norway.

Randy T. Simmons is director of the Institute of Political Economy and head of the Department of Political Science at Utah State University. As a political economist he specializes in applying the methods and assumptions of economics to political questions. He is co-author of *Markets, Policy and Welfare*, a primer for analysizing and understanding economic, environmental and social policies.

Ike C. Sugg is Alex C. Walker Fellow in Environmental Studies at the Competitive Enterprise Institute in Washington, D.C. The CEI is a non-partisan, non-profit, public interest group advocating free-market, individual-liberty alternatives to political problems. He researches and analyses U.S. and international wildlife policies and rural land-use issues. Prior to receiving his fellowship in 1992, Sugg served as CEI's Director of Development for two years. He studied philosophy and political science at the University of Texas at Austin, and has practical experience in land and wildlife management.

Russell D. Taylor is Zimbabwean born and holds a B.Sc (Agriculture) degree from the University of Natal, and a Ph.D from the University of Zimbabwe. He was first employed in 1971 as an agricultural extension officer before joining the Department of National Parks and Wild Life Management as a wildlife ecologist. He has worked in the northern Sebunge region of Zimbabwe for nearly twenty years, where he has been engaged in research, park management, land use planning and, more recently, the CAMPFIRE program. For the past five years he has been with the Multispecies Animal Production Systems Project of the World Wide Fund for Nature and presently heads WWF's support for the CAMPFIRE project in Harare.

Stephen J. Thomas was born and educated in England. In 1984 he resigned a career in commercial banking to take up the post of administrator with an agricultural co-operative development project in Zimbabwe. He currently manages the Institutional Development Unit of the Zimbabwe Trust, a non-governmental organisation which is assisting Zimbabwe's Department of National Parks and Wild Life Management in the implementation of the CAMPFIRE program.

1. INTRODUCTION

MILTON M.R. FREEMAN and URS P. KREUTER

Elephants and whales, arguably represent the best known examples of "charismatic megafauna". As such they provide sharpened focus to the international conflict over the appropriate use and management of wild animals, and the question of who should benefit from wildlife. Concerns about the impact of human activity on the environment, contemporary western affluence often juxtaposed with third-world poverty, the views of urban dwellers pitted against those living and working close to nature, and varied interpretations of the ethical responsibility of humans towards other life forms and future generations, have all fueled the argument. Associated with and exacerbating these concerns is the longstanding difficulty experienced by government agencies managing common pool resources, of which elephants and whales are prominent examples.

This book focuses attention on three elements of the resource management *problematique*: elephants and whales as species; the concept of resources, which implies both consumptive and non-consumptive human usage and economic and non-economic resource values; and the question of who has the right to use these animal resources. To a large extent the book emphasizes anthropological and related issues having relevance to the use and management of elephants and whales as common pool resources.

The book has its origins in the recent institutional responses to this growing interest in common property, starting with the Panel on Common Property Resource Management convened by the U.S. National Research Council. This led to creation of an international common property network, quickly followed by the formation of the International Association for the Study of Common Property (IASCP). The 2nd Annual Meeting of the IASCP, held in Winnipeg in 1991, included a symposium on community-based whaling. During the symposium, discussion broadened to consider cultural and political issues influencing the management of various charismatic animal species. This uncovered commonalities between issues involving elephants as well as whales (see Peterson 1993). To continue this discussion about these common management problems, it was decided to convene a session during the 3rd Annual Meeting of IASCP held in 1992 in Washington, D.C. The chapters in this book include some that evolved from papers presented during that 1992 meeting (Freeman, Kalland, Kreuter and Simmons, Sanderson, Thomas) as well as two relevant papers (Caulfield, Ris) presented at the whaling symposium held during the 1991 IASCP meeting.

ELEPHANTS AND WHALES

The central position of elephants and whales in this debate results from the question of whether these animals are to be treated in a fundamentally different manner to other wildlife species. In short, are they to be regarded as resource stocks to be utilized for the benefit of human society, or are they to enjoy a special status that precludes some uses that other animals are subject to? This question has emerged because some sectors of western society increasingly imbue elephants and whales (and other selected species, see Kellert 1986) with characteristics that requires them to be treated fundamentally differently from other species for management and conservation purposes.

The special status accorded to elephants and whales stems from their actual or perceived unique biological and behavioral qualities, though often these characteristics are imagined rather than having a secure scientific foundation. For instance, the reported intelligence, communication abilities, and social behaviour of these animals frequently invoke human reactions based more upon sentimentality, imagination, and anthropocentric illusions rather than scientific objectivity. Included among those known to suffer these lapses in critical thinking are some scientists and self-professed environmentalists. Widely reported statements by officials of governments opposed to the killing of whales, various public figures, and journalists, have misled many into believing that elephants and whales are highly intelligent and close to humans in many other qualities, everywhere endangered yet subject to needless slaughter and thus in urgent need of total protection.

However, as subsequent chapters will illustrate, whilst some individuals and governments consider the consumptive use of, and trade in commodities derived from, these animals as unnecessary or even frivolous, other societies still attach diverse (including economic) value to certain of these animals as resources. This difference in perspective is the basic cause of conflicts in international debates about the appropriate use of these large and mediagenic animals.

Biological Attributes

Attributes, such as being hairless mammals, able to communicate over great distances, being limbless and thus helpless on land (whales), and having a prehensile nose (elephants), are anomalies which have imbued elephants and whales with a "specialness" that has formed the basis of campaigns by groups seeking to protect them (Kalland 1993; Peterson 1993). The most imposing feature of elephants and whales is, however, their size: elephants are the largest living land animals (males can exceed 5,000 kg), and whales include the largest animals known to science (weighing up to 100,000 kg). Large body size in animals is generally associated with slow growth, delayed sexual maturity, long gestation periods (16 months in some whale species and 22 months in elephants), a single offspring at each birth, and longevity (60 years or more in elephants and up to 90 years in one species of whale).

Since large animals require more food and space than smaller animals they tend to occur in smaller numbers. One question that frequently dominates discussions about large animal conservation is whether the relatively small populations can be used sustainably or if human use will inevitably result in over-use. Many people believe that animals producing a single offspring at parturition are especially vulnerable to over-exploitation. Yet biologists recognize that any population can be perpetuated if each female produces only two to three surviving offspring during her lifetime; given the longevity of elephants and whales and the low adult mortality rates from natural causes, the probability that females attain, indeed surpass, this level of reproduction is high.

Elephants and whales, having few predators save humans, can exhibit steady rates of population increase if food supplies are adequate and predation rates remain low. Whale populations, for example, have demonstrated their ability to continue growing whilst being prudently harvested. North Atlantic pilot whales are currently estimated to number about 778,000 even though 1500-2000 animals from this stock have been harvested each year for several centuries. Antarctic minke whales number around 760,000 and yet have been commercially harvested since the 1930's when their population was much smaller than at present. Bowhead whales in the Chukchi-Beaufort seas, which were seriously depleted by commercial exploitation in the 19th century, currently exceed 8000 and are increasing by about 3% per year despite a current annual harvest of about 50 animals. Today the eastern stock of Pacific gray whale has fully recovered (to it's pre-exploitation number of around 20,000) from near extinction earlier this century in spite of a continuing annual harvest numbering about 170 whales over the past several years. Though most species of whale stocks can increase from 3–5% annually, some have exhibited higher rates of increase (e.g., 7% in southern right whales, Martin *et al.* 1990; see also Schmidt 1994 for a general observation on increasing whale populations).

Elephants have exhibited similar population increases when hunted, provided their nutritional requirements are met. For instance, the total number of elephants in Zimbabwe increased at an average annual rate of 4.9% from 1979 to 1989 (see Kreuter and Simmons, this volume). This increase occurred even though nearly 27,000 elephants were killed during the same ten-year period (Martin and Conybeare 1992). Such increases in elephant populations have been associated with their potentially high annual reproductive rate combined with a high degree of protection against predators that the herd provides to the young elephants. Thus maximum population growth rates under ideal conditions have been estimated to range from 4.73% to 6% per annum (Hanks 1979; Owen-Smith 1988).

Brain Size and Intelligence

In addition to reproductive potential, brain size is another characteristic that is correlated with large body-size. Due to a widely-held view that a large brain implies a high level of intelligence, this characteristic has led to exaggerated claims about the intelligence of elephants and whales and has contributed to the perceived "specialness" of these large-bodied animals. Brain size is, however, largely a function of body size, because the greater the muscle mass the greater the size of the neural network required for motor-coordination. In past years, scientists believed that large brain size was also a manifestation of intelligence, a belief that is still held by many non-scientists today. If this assumption is correct, then elephants and whales are certainly prime candidates in the search for animals with superior intelligence. However, what constitutes "intelligence" in animals is hard to define and impossible to measure. What is clear is that "trainability" of animal species, such as dolphins and porpoises, does not reflect intelligence in the sense that the term is commonly understood to mean when applied to humans (Gaskin 1982, Klinowska 1992).

One reason it is not too fruitful to speak about brain size in different species is that published figures are not always strictly comparable: when a brain is removed *post mortem*, varying amounts of associated tissue may be included or excluded, and in addition brain size is comparatively larger in younger, smaller, mammals than in fully mature or older specimens of the same species. In addition, some brains are preserved before being studied, and depending on the preserving fluids in use weight changes will occur. However, though aware that there is no basis for evaluating either the quality or comparability of published information on the subject, it is briefly addressed here due to conclusions others have drawn using these data.

Averaging 7.8 kg, the adult sperm whale brain is far larger than the human brain (ca. 1.5 kg) and somewhat larger than that of an African elephant (ca. 7.5 kg). However, since a sperm whale is about 500 times heavier than a human and more than seven times heavier than an African elephant, its larger brain size compared to humans and elephants is neither unexpected nor remarkable. What may be more unexpected is that the brain to body-size ratio in the large-brained sperm whale is only one-quarter that observed in domestic cattle (Table 1.1).

Table 1.1 Approximate Brain and Body Weight of Some Animals

	Brain Weight (kg)	Body Weight (kg)	Brain wt/ Body wt(%)
Sperm whale (male)	7.82	37,000	0.021
Cow	0.50	600	0.083
Killer whale	6.00	5,620	0.107
African elephant	7.50	5,000	0.150
Bottlenose dolphin	1.60	170	0.941
Human	1.50	70	2.143

After Klinowska 1992

It is now clear that brain to body size comparisons have little value for assessing animal intelligence (see Freeman 1990:112–113); indeed, within a species, individual brain and body weights can vary substantially even among fully developed adults. For instance, adult human brain weights range from 0.9 to 2.0 kg, and in some whales a 40% variation in individual body weight (but not brain weight) occurs at different seasons of the year (Klinowska 1992).

More recently attention has been focused on brain structure, rather than brain size, in order to gauge the mental capacity of whales compared to other animals. For example, the neocortex (the part of the brain thought to be associated with more advanced mental functions and which is most highly developed in primates and humans) is extensive in whales, dolphins and porpoises. However, despite its large area the cetacean neocortex is thin and anatomically exhibits a quite primitive structure (see Klinowska 1989, 1992, for references). Furthermore, the animal having the relatively greatest neocortical development is not a cetacean, elephant or even a primate, but rather *Echidna*, a primitive egg-laying mammal (Macphail 1982). Anomalies such as this make the search for "intelligence" within either brain size or structure a less than rewarding exercise.

It is possible that intelligence can best be defined in terms of an organism's behavioural, rather than anatomical characteristics, specifically its expressed ability to comprehend cause and effect relationships and to use such knowledge to enhance its probability of survival and wellbeing. Many animals appear to be aware of the effects of their actions, and thus, arguably, to exhibit intelligence. However, there is no evidence that any species other than *Homo sapiens* has the capacity for making such associations systematically and to consciously select, use and manipulate natural entities and to create abstracted symbols to achieve predetermined and long term goals (see Sugg and Kreuter, this volume). Entities used by humans in this fashion are known as resources.

RESOURCES

Humans, like all other living organisms, must use resources to survive and must often compete with various other organisms for certain of these resources. For instance, in Africa humans and elephants have historically competed directly for land resources through their preference for more fertile areas (Parker and Graham 1989a, 1989b). Some whales may also compete with humans for food resources by consuming fish, squid or crustacea (krill). Elephants raid African's gardens and destroy food stores, and coastal fishermen complain of damaged nets or the fish in traps when whales encounter these gear. In inshore whaling areas of Japan, fishermen observe that after the IWC whaling ban came into effect the fishery for certain small shoaling fish has been seriously damaged due to the increased nearshore presence and activity of minke whales (see also Kalland and Moeran 1992:175).

Since humans are generally omnivorous, use animals for purposes other than food, and are capable of controlling most non-human species with which they compete, competitor species may themselves become human resources. Yet if the competing species are not beneficial to people, they are likely to be increasingly at risk or even eliminated as human demand for the resources being competed for grows. In discussing elephants and whales it is imperative to understand this ecological relationship, including their diverse values, both positive and negative, as resources for human societies.

Elephants

Elephants can not only provide useful human resources (including ivory, hide, meat, fat, and revenues from tourism), they are also a pivotal agent for maintaining biodiversity and productivity in African savannas (Western 1993). They occur in at least 30 countries in Africa, and range throughout and, more importantly, beyond the continent's 500,000 km² of state protected areas (Adams and McShane 1992). Owen-Smith (1988) stated that elephants have a large home range (20–3,000 km² depending on water and forage availability), high daily food intake (ca. 1.0% of body mass), and destructive foraging behaviour (debarking trees or pushing them over to reach young foliage).

These traits have accentuated the competitive interaction between people and elephants and they have resulted in significant effects on ecosystem dynamics. Such effects may be highly destructive, particularly when elephants are prevented from migrating to other areas and population increases are not contained. However, herbivory by elephants can also have positive consequences by maintaining elephant-induced ecological processes that promote savanna biodiversity, which argues for conserving elephants over larger areas than just in national parks (Owen-Smith 1988; Western 1989). Yet, this would increase the potential for human-elephant interactions, a situation which will more readily be tolerated by Africans if they derive direct benefits from elephants (see Kreuter and Simmons, Hasler, Peterson, Taylor and Thomas, this volume).

Whales

The high economic value of whale products in the past resulted primarily from the global demand for whale oil, but many other industrial uses existed for products obtained from the whale carcass, including meat (for human and animal foods), baleen, gut, bone (for fertilizer), leather, and substances used in manufacturing various pharmaceuticals. Today however, the principal whaling activity is carried out by small-scale community-based whalers who hunt for food. In recent years whale watching has also become a fast-growing element of local tourism development in several regions of the world. However, though economically rewarding for some entrepreneurs, there have been few critical studies that assess the environmental and socio-economic impacts of this new activity (see, e.g. Ris, this volume; also Lanfant and Graburn 1992).

As many whales feed on commercially valuable fish, squid and crustaceans, the potential for human-whale competition is high, especially when considering the large body size and population numbers of some whales. For example, about one million sperm whales live in the North Pacific Ocean, and each whale consumes from 3 to 3.5% of it's body weight daily (Martin *et al.* 1990). This food includes mostly squid or octopus, though in northern waters substantial quantities of cod and redfish are also eaten. The total amount of food consumed by sperm whales in the North Pacific alone is about 50 million tonnes per year, and a similar amount is consumed by the approximately one million sperm whales in other oceans. To place this in a fishery context, the total world fisheries catch is presently about 100 million tonnes each year. Killer whales consume around 4% of their body weight daily, and their diet includes salmon, cod, herring and other commercially important fish (Martin *et al.* 1990). Approximately 7000 killer whales are reported to live in the waters around Iceland (a country where about 90% of the economy is based upon fishing), and they consume an estimated 45 million kg of fish annually. Some 778,000 pilot whales also occur in the North Atlantic, eating mainly squid, and consuming the equivalent of 5% of their body weight daily, or a total of about 150 million kg of food per year (Martin *et al.* 1990).

In the Southern Oceans, krill fishing is important to six nations. Krill is also important to various antarctic whale species, including 760,000 minke whales directly competing with these fishing fleets; this minke population likely increases in size by at least 30,000 whales each year.

For these reasons, many whalers, fishermen and fishery managers believe that maintaining the ban on commercial whaling is ecologically damaging and represents a great threat to the economic viability of certain coastal communities (e.g. Hoydal 1990, and Broch and Ris, this volume).

Other Resource Values

Besides these economic and environmental considerations, whales and elephants have non-material values that may be of greater importance to many people. For instance, high demand for African wildlife safaris and whale watching opportunities reflect the non-consumptive aesthetic, educational and recreational interest many people have in elephants and whales.

Similar non-material values are accorded to these animals by local communities. Given the size, strength, potential danger, and potential food source that these animals represent, it is not surprising that they feature prominently in the diets, arts, folklore, social relations and rituals of some societies (see Caulfield, Hasler, and Kisangani, this volume). It can truly be said that such animals are highly valued multidimensionally; for example, in a Japanese coastal whaling village, minke whale meat and blubber was considered important for about thirty different culturally-significant events (Braund *et al.* 1990).

Removing local people from highly-valued traditional relationships to these resources by coercive means can be expected to endanger societies and cultures
that have elaborated these important interdependencies often over many generations (e.g. Manderson and Hardacre 1989; Kalland and Moeran 1992; Lynge
1992).

Successfully taking these large animals requires not just courage and skill,
but considerable community cooperation. The hunting, carcass and product
processing and distribution, consumption and celebration phases of elephant
and whale utilization are thus often accorded high socio-cultural value. Whaling, for example, continues to be an important component in certain societies'
cultural identity (e.g. in Greenland: Lynge 1992; Iceland: Brydon 1990; Canadian Inuvialuit: Freeman *et al.* 1992; Japan: Kalland and Moeran 1992,
Manderson and Akatsu 1993), and this has strengthened the resolve of certain
societies who strongly identify with whaling to oppose the efforts of protectionists who would foreclose their right to hunt whales (see Broch, Caulfield,
Doubleday, Ris, and Sanderson, this volume).

FOR WHOM?

The domain in which natural resource use by humans is decided is the
noosphere or "thinking layer" (see Sugg and Kreuter, this volume). Mankind
alone has the discretionary power to decide how to use resources, whether to
use them consumptively or non-consumptively, whether to use them immediately or to reserve them for future use.

As Sugg and Kreuter state in this volume, the debate over the management
and use of these mediagenic animals is a debate over values. At centre it is a
conflict between users, managers and conservationists, who understand that
wild animals must have economic value in order to be conserved, and preservationists, many of whom argue that people do not have the right to consume,
let alone to derive monetary profit, from wild animals. Such conflicts are especially pronounced with regard to those charismatic animals having widespread
public recognition and high media profile.

The debate about values is generally driven by preservationists from affluent
western countries who use elephants and whales as "conservation symbols"
(Peterson 1993, see also, Bonner, Kalland, and Ris, this volume) with little regard for the values of the people who coexist with the animals or who have
traditionally used them as resources. In order to use images of these animals
as effective symbols, preservationists have frequently portrayed wild animals
in anthropomorphic terms to increase public identification with the species in
question. Indeed, many authors (see Kalland, this volume) writing about animals for non-scientists, frequently seek to establish a connection between humans and animals by using particular terms. For instance, the terms "matri-
arch" and "patriarch" are used to describe old elephants (e.g. Payne 1989),

while kin terms such as "uncles", "aunts", "nieces" and "nephews", used to describe individuals in some whale groupings (e.g. Corkeron 1988), even though such kin terms are scientifically meaningless in polygamous species and quite ambiguous in human societies unless carefully defined.

In drawing attention to this dubious but powerful use of language in discussing conservation issues (see Lee 1988), some scientists question whether "these mistaken substitutions. . . (are) manifestations of political correctness modulated by the view that there is no proper distinction between human and non-human animals?" (Williams *et al.* 1993). The question of "political correctness" is central to much of the debate about whether elephants and whales should be treated as consumable resources or as emotionally-charged icons for the benefit of certain preservationist groups, and several papers in this book speak to this issue (see e.g., Sugg and Kreuter, Bonner, Kalland, and Ris, this volume).

The current debate concerning the appropriateness of using these particular animals consumptively, arose because of unsustainable levels of exploitation of many populations of elephants and whales in the past. The debate is sustained by preservationists' claims, *inter alia,* that these "special" species represent part of the common heritage of mankind and are not merely resources for the exclusive use of certain countries or particular groups of people. These claims are used to justify international intervention in the management of resources occurring within national territories, often on the grounds that elephants and whales are open access resource stocks (see Sugg and Kreuter, and Kreuter and Simmons in this volume). Yet, as many conservation organizations now recognize (e.g, IUCN 1993; Makombe 1994; WWF 1993) these stocks will continue to be mismanaged as long as the people who directly affect their population dynamics are not provided with incentives to protect them by being given clearly defined, enforceable and divestible property rights and meaningful management responsibilities toward these resources.

The Common Property Debate

There are, however, problems with ascribing individual property rights to species, such as elephants and whales, that migrate across regions governed by multiple jurisdictions. Indeed, even with well defined user rights, they are likely to remain common property. While Garrett Hardin (1968), in his seminal article on the "Tragedy of the Commons", claimed that common property status inevitably leads to over exploitation, his analysis was incomplete. Current scholarship recognizes there are numerous community-based management systems throughout the world that have resulted in the sustainable use of a wide range of environmental resources upon which human societies depend, often over extended periods (see e.g., Williams and Hunn 1982, Ruddle and Akimichi 1984, NRC 1986, McCay and Acheson 1987, Berkes 1989b, Cordell 1989, Ruddle and Johannes 1990, Freeman *et al.* 1991).

In commenting on the failure of many resource managers to appreciate the realities of other cultural systems and societies' abilities to sustainably manage their resources, it has been perceptively observed that:

> "... to many whose worldviews are shaped by the urban-industrial society in which they live, with little intimate contact with neighbours and other members of society, the 'tragedy' may appear inevitable. By contrast, use of commons for long-term sustainable yields is relatively more likely in the case of people living in small groups with tight communal control over the resource base and over social behaviour. Not being familiar with such societies, many Western-trained, urban-based resource managers and scientists overlook possibilities for sustainable management of commons in such situations" (Madhav Gadgil, in Berkes 1989a).

A comprehensive understanding of what successful management requires has been thoroughly treated in several recent syntheses that present the current understanding of adaptive management arrangements (e.g. Holling 1978, Berkes *et al.* 1989, Pinkerton 1989, Feeny *et al.* 1990, Bromley 1991, Ostrom 1990). It is apparent that sustainable resource use may depend less upon gaining greater understanding of resource stock behaviour and mathematical modelling, and more upon creating or sustaining appropriate local-level management institutions.

The importance of community-based institutions

Community-based institutions are a frequent feature of successful common property management outcomes. As mentioned earlier, one principal criticism directed toward Hardin's "tragedy of the commons" thesis was his failure to consider communal management as an option for sustainable utilization, a failure that continues to be repeated in much of the discourse on wildlife and fisheries (e.g. Ludwig *et al.* 1993). In order to prevent the tragedy of the commons, it is necessary to transform open access to resources to defined, or limited, access, which ordinarily restricts resource use to members of a specified group (which may be the local community, or an identified group of customary or otherwise-sanctioned users). The danger posed to both the resource stock and the dependent human population when non-local elites usurp this right of customary local users' restricted access, is illustrated in the chapter by Kisangani.

The advantage of devolving management responsibility to the local level is that compliance and enforcement are greatly facilitated. The chapters describing the CAMPFIRE program in this book bear testimony to this reality (see, Peterson, Taylor, and Thomas, in this volume and also papers in Green and Smith 1986; Freeman and Carbyn 1988). As many understand, and numerous examples attest, the community as both manager and beneficiary has built-in incentives that ensure members remain prudent harvesters. While it is recognized that no system is infallible, such community-level involvement facilitates adaptive management by enabling rapid correction of erroneous decisions.

However, there exists resistance to the notion of empowering local communities by supporting communal management systems, for such action requires not just consultation and information sharing, but rather, handing over of actual authority. As the chapters by Peterson and Thomas demonstrate, even in well intentioned wildlife administrations, this is not easily accomplished. Several chapters in this book illustrate just how difficult effecting management for sustainable use of some common property resources is in practice. The history of whaling provides a classic illustration of a tragedy of the commons resulting from the activities of a management regime that did not work properly. The IWC in its attempts to address the open access problem, intensified the harvesting pressure on particular whale stocks whilst failing to address the compliance and enforcement provisions made even more necessary by their inappropriate actions. The failure of the global whaling regime to operate sustainable commercial whale fisheries appears to be continuing due to its collective inability to understand the critical differences between historic industrial whaling and today's community-based whaling (see Freeman, in this book and Freeman 1993) and the lack of knowledge about new management paradigms that recognize the importance accorded to equitability in ensuring sustainable resource utilization (IUCN/UNEP/WWF 1992; Young and Osherenko 1993).

Failures are also apparent in attempts by CITES members to manage elephants, where the strategy adopted has been to eliminate trade (resulting in the loss of economic benefits needed to effectively manage the resource) rather than to enshrine rights and a management regime that would facilitate controlled marketing of elephant products. Furthermore, at the present time the IWC and CITES have both abandoned the goal of sustainable resource use in favour of restricting use (and compromising equity considerations) to the greatest degree possible.

CONCLUSION

The objectives of the World Conservation Strategy are to maintain essential ecological processes, life support systems and biodiversity, and to ensure the sustainable use of species and ecosystems (IUCN/UNEP/WWF 1992). That widely-endorsed environmental manifesto also recognizes the essential role to be played in management by the users of the resources if sustainable outcomes are to be attained, a perspective supported implicitly at least by most contributions in this book. Moreover, many international agreements seek to advance the fundamental rights of all people to dignity, employment, and cultural freedom (see Doubleday, this volume). The activities by animal protectionists, and environmental organizations and governments supporting their cause, at the expense of human welfare should therefore be seriously questioned.

As human populations continue to grow, demand for resources (including competition for space) and environmental conflicts will inevitably increase. Since the elephant and whale conflict is especially intense, it provides a window on likely future conflict scenarios. Inter-governmental bodies, such as CITES and IWC were established to address wildlife and whale management problems in a manner perceived to be appropriate by member governments. They were established to conserve resources by controlling trade by considering objective criteria based upon the best scientific information available. These organizations have, however, been subverted by preservationist interests seeking to impose subjective sentimental ideas about elephants and whales onto the people who live with them or who have traditionally used them, and who plausibly claim an ecological understanding of the role of these species in the local environments. Thus CITES and IWC now appear to provide a prominent international stage for promoting preservationist ideologies whilst failing to provide a suitable forum for developing and adopting effective conservation strategies that respect the cultural and socio-economic needs of affected communities. For example, even though Zimbabwe has an over-abundance of elephants, at their meeting in 1992 the CITES members rejected a proposal by southern African countries to allow them to resume trade in elephant products. Similarly at the 1993 IWC meeting, the majority of IWC members voted against Norway's and Japan's applications to resume controlled harvesting of abundant stocks of the non-endangered minke whale. These results were mainly due to the influence of international preservationist organizations' upon the voting behaviour of several member nations unaffected by the negative impacts of these actions.

These restrictive measures do not enhance conservation of wildlife stocks for sustainable use, but rather, aim to place stocks outside of the range of consumable natural resources. Though such a goal justifies the existence of many groups working to advance an animal welfare or animal rights agenda, it nevertheless seems strange behaviour for intergovernmental resource management regimes established to facilitate conservation of biological resources based upon scientific principles. In accepting public sentiments toward certain animals as a reasonable basis for their anti-harvest positions, member governments should be mindful how poorly informed the public is about such issues (see the chapter by Freeman and Kellert). However, several chapters in the book indicate how dedicated and single-minded the animal preservation groups are in pressing their cause as well as some of the reasons their methods appear to succeed so well.

There is no doubt that one important reason these preservationist campaigns work is because of the continuing and widespread, yet erroneous, view held by the public and politicians that the problem of managing common property resources remains intractable. Yet, in closing, we state once more that the fate of elephants and whales is ultimately in the hands of people affected by the presence of these animals yet denied the economic incentive to conserve them. The critical importance of ensuring community wellbeing and ecological integration is well articulated in the World Conservation Strategy. The challenge for environmentalists is to ensure these principles are well understood by the public at large, by those who write about biodiversity and sustainable resource use, and by government decision makers, so that inter-governmental agencies will be effective in promoting sustainable and equitable resource use with due regard for human rights and cultural diversity.

REFERENCES

Adams, J.S. and T.O. McShane, 1992.
   The Myth of Wild Africa: Conservation without Illusion. W.W. Norton, New York.
Berkes, F., 1989a.
   Cooperation from the perspective of human ecology. In F. Berkes (ed), *Common Property Resources: Ecology and Community-Based Sustainable Development,* pp. 70–88. Belhaven, London.
Berkes, F. (ed), 1989b.
   Common Property Resources: Ecology and Community-Based Sustainable Development. Belhaven, London.
Berkes, F., D. Feeny, B.J. McCay and J.M. Acheson, 1989.
   The benefits of the commons. *Nature* 340, 13 July: 91–93.
Braund, S.R., J. Takahashi, J.A. Kruse and M.M.R. Freeman, 1990.
   Quantification of Local Need for Minke Whale Meat for the Ayukawa-Based Minke Whale Fishery. Document TC/42/SEST 8. International Whaling Commission, Cambridge.
Bromley, D.W., 1991.
   Environment and Economy: Property Rights and Public Policy. Blackwell, Oxford.
Brydon, A., 1990.
   Icelandic nationalism and the whaling issue. *North Atlantic Studies* 2(1–2):185–191.
Cordell, J. (ed), 1989.
   A Sea of Small Boats. Cultural Survival, Cambridge, Mass.
Corkeron, 1988.
   Social behaviour. In R. Harrison and M.M. Bryden (eds), *Whales, Dolphins and Porpoises,* pp. 142–159. Facts on File, New York.
Cumming, D.H.M., 1989.
   Review of Elephant Numbers and Trends: A Brief Summary Report of the Work of Iain Douglas Hamilton and Richard Barnes, In: Review of the Ivory Trade and the Future of the African Elephant, Volume 2, Ivory Trade Review Group, pp. 8–15. Prepared for the 2nd Meeting of CITES African Working Group, Gabarone, Botswana, July 4–8, 1989. IUCN, Gland, Switzerland.
Feeny, D., F. Berkes, B.J. McCay and J.M. Acheson, 1990.
   The tragedy of the commons: twenty-two years later. *Human Ecology* 18:1–19.
Freeman, M.M.R., 1990.
   A commentary on political issues with regard to contemporary whaling. *North Atlantic Studies* 2(1–2):106–116.
Freeman, M.M.R, 1993.
   The International Whaling Commission, Small-type whaling, and coming to terms with subsistence. *Human Organization* 53:243–251.

Freeman, M.M.R. and L.N. Carbyn (eds), 1988.
  Traditional Knowledge and Renewable Resources Management in Northern Regions. IUCN Commission on Ecology and Boreal Institute for Northern Studies, Edmonton.
Freeman, M.M.R., T. Matsuda and K. Ruddle (eds), 1991.
  Adaptive Marine Resource Management Systems in the Pacific. Harwood Academic, Chur, Reading, Paris, Philadelphia, Tokyo, Melbourne.
Freeman, M.M.R., E.E. Wein and D.E. Keith, 1992
  Recovering Rights: Inuvialuit Subsistence and Bowhead Whales in the Western Canadian Arctic. Canadian Circumpolar Institute, Edmonton.
Gaskin, D.E., 1982.
  The Ecology of Whales and Dolphins. Heineman, London.
Green, J. and J. Smith (eds), 1986.
  Native People and Renewable Resources Management. Alberta Society for Professional Biologists, Edmonton.
Hanks, J., 1979.
  The Struggle for Survival: The Elephant Problem. C. Struik Publishers, Cape Town, South Africa.
Hardin, G., 1968.
  The tragedy of the commons. *Science* 162:1243–1248.
Holling, C.S. (ed), 1978.
  Adaptive Environmental Assessment and Management. John Wiley, Chichester, U.K.
Hoydal, K., 1990.
  Management of marine resources. *North Atlantic Studies* 2(1–2):85–86.
IUCN, 1993
  Guideline for the Ecological Sustainability of Nonconsumptive and Consumptive Uses of Wild Species. IUCN/SSC Specialist Group on Sustainable Use of Wild Species and IUCN Sustainable Use of Wildlife Programme, May 1993.
IUCN/UNEP/WWF, 1992.
  Caring for the Earth. A Strategy for Sustainable Living. IUCN/UNEP/WWF, Gland, Switzerland.
Kalland, A., 1993.
  Management by totemization: whale symbolism and the anti-whaling campaign. *Arctic* 46:124–133.
Kalland, A. and B. Moeran, 1992.
  Japanese Whaling: End of an Era?. Curzon Press, London.
Kellert, S., 1986.
  Social and perceptual factors in preservation of animal species. In B.G. Norton (ed), *The Preservation of Species: The Value of Biological Diversity,* pp. 50–73. Princeton University Press, Princeton.
Klinowska, M., 1989.
  How brainy are cetaceans? *Oceanus* 32:19–20.
Klinowska, M., 1992.
  Brains, behaviour and intelligence in cetaceans (whales, dolphins and porpoises). In O.D. Jonsson (ed) *Whales and Ethics,* pp. 23–37. University Press, Reykjavik.
Lanfant, M-F. and N. Graburn, 1992.
  International tourism reconsidered: the principle of the alternative. In V.L. Smith and W.R. Eadington (eds), *Tourism Alternatives: Potentials and Problems in the Development of Tourism,* pp. 88–112. University of Pennsylvania Press, Philadelphia.

Lee, J.A., 1988.
Seals, wolves and words: loaded language in environmental controversy. *Alternatives* 15(4):21–29.

Ludwig, D., R. Hilborn and C. Walters, 1993.
Uncertainty, resource exploitation, and conservation: lessons for history. *Science* 260, 2 April: 17, 36.

Lynge, F., 1992.
Arctic Wars, Animal Rights, Endangered Peoples. University Press of New England, Hanover, N.H.

Manderson, L.M. and H. Hardacre, 1989.
Small-Type Whaling in Ayukawa: Draft Report of Research. Document IWC/41/SE 3. International Whaling Commission, Cambridge.

Manderson, L and H. Akatsu, 1993.
Whale Meat in the Diet of Ayukawa Villagers. *Ecology of Food and Nutrition* 30:207–220.

Martin, A.R., G.P. Donovan, S. Leatherwood, P.S. Hammond, G.J.B. Ross, J.G. Mead, R.R. Reeves, A.A. Hohn, C.H. Lockyer, T.A. Jefferson and M.A. Webber, 1990.
Whales and Dolphins. Salamander Books, London and New York.

Martin, R.B. and A.M.G. Conybeare (eds), 1992.
Elephant Management in Zimbabwe (2nd ed). Department of National Parks and Wildlife Management, Harare, Zimbabwe.

McCay, B.J. and J.M. Acheson (eds), 1987.
The Question of the Commons: The Culture and Ecology of Communal Resources. University of Arizona Press, Tucson.

Macphail, E.M., 1982.
Brain and Intelligence in Vertebrates. Clarendon, Oxford.

Makombe, K. (ed), 1994.
Sharing the Land: Wildlife, People and Development in Africa. IUCN/ROSA Environmental Issues Series No. 1. IUCN/ROSA, Harare, Zimbabwe, and IUCN/SUWP, Washington, D.C.

Misaki, S., 1993.
Japanese world-view on whales and whaling. In *Whaling Issues and Japan's Whale Research,* pp. 22–36. The Institute of Cetacean Research, Tokyo.

NRC, 1986.
Proceedings of the Conference on Common Property Resource Management. National Academy Press, Washington, D.C.

Ostrom, E., 1990.
Governing the Commons: The Evolution of Institutions for Collective Action. Cambridge University Press, Cambridge.

Owen-Smith, N., 1988.
Megaherbivores: The Influence of Very Large Body Size on Ecology. Cambridge University Press, Cambridge.

Parker, I.S.C and A.D. Graham, 1989a.
Elephant Decline: Downward Trends in African Elephant Distribution and Numbers (Part I). *International Journal of Environmental Studies* 34:287-305.

Parker, I.S.C and A.D. Graham, 1989b.
Elephant Decline: Downward Trends in African Elephant Distribution and Numbers (Part II). *International Journal of Environmental Studies* 35:13–26.

Payne, K., 1989.
  Elephant talk. *National Geographic* 176:264–277. ━

Peterson, J.H., 1993.
  Whales and Elephants as Cultural Symbols. *Arctic* 46:172–174.

Pinkerton, E. (ed), 1989.
  Co-Operative Management of Local Fisheries: New Directions for Improved Management and Community Development. University of British Columbia Press, Vancouver.

Ruddle, K. and T. Akimichi (eds), 1984.
  Maritime Institutions in the Western Pacific. National Museum of Ethnology, Osaka.

Ruddle, K. and R.E. Johannes (eds), 1990.
  The Traditional Knowledge and Management of Coastal Systems in Asia and the Pacific. UNESCO, Jakarta.

Schmidt, K., 1994.
  Scientists Count a Rising Tide of Whales. *Science* 263, 7 January, pp. 25–26.

Stuart-Hill, G.C., 1992.
  Effects of Elephants and Goats on the Kaffrarian succulent Thicket of the Eastern Cape, South Africa. *Journal of Applied Ecology* 29:699–710.

Western, D., 1993.
  The Balance of Nature. *Wildlife Conservation* March/April 1993.

Williams, E.H., L. Bunkeley-Williams, J.M.Grizzle, E.C. Peters, D.V. Lightner, J. Harshbarger, A. Rosenfeld and R. Reimschuessel, 1993.
  Epidemic misuse. *Nature* 364, August 19: 664.

Williams, N.M. and E.S. Hunn (eds), 1982.
  Resource Managers: North American and Australian Hunter-Gatherers. Westview Press, Boulder.

WWF, 1993.
  Sustainable Use of Natural Resources: Concepts, Issues, and Criteria. WWF International Position Paper, Gland, Switzerland.

Young, O.R. and G. Osherenko, 1993.
  International regime formation: findings, research priorities, and applications. In O.R. Young and G. Osherenko (eds), *Polar Politics: Creating International Environmental Regimes*, pp. 223–261. Cornell University Press, Ithaca.

2. ELEPHANTS AND WHALES AS RESOURCES FROM THE NOOSPHERE

IKE C. SUGG and URS P. KREUTER

"The issue is not sustainability. We remain philosophically opposed to resumption of commercial whaling." Tom Miller, Center for Marine Conservation, stating his organization's position on whaling (Walters 1993).

"I would rather see no elephants than elephants being culled . . . because I think it is morally unjustified to kill elephants. . ." Cynthia Moss, Amboseli Elephant Research Project, on: Can the Elephant be Saved? (NOVA, WGBH-TV, November 20, 1990).

INTRODUCTION

Since the beginning of recorded history, scientists and sophists alike have sought to categorize natural phenomena into useful constructs. Today, as a result, "the environment" is often reduced from complex, interrelated systems to convenient, individuated spheres. For instance, there is the geosphere (land), the hydrosphere (water), the atmosphere (air), and the biosphere (life). In addition to these physical entities there is also another, more important and more inclusive sphere: the *noosphere*.

The noosphere is, according to the construct's originator, "the thinking layer" (de Chardin 1959), and thus the realm in which natural resource use is decided. There can be no doubt that resources will be used; mankind, like all other animals, must use resources in order to live, let alone prosper. Moreover, such use invariably results in "pollution", defined by the second law of thermodynamics as the "excretion of entropy" (Lovelock 1988).

The issue, then, is not whether to use resources, but which resources to use; *how* to use them; and how best to deal with the consequences of such use. As every natural resource debate will attest, waste products are not the only effects of resource use. After all, no definition of "resource" could be meaningful without implying *value*. Thus, conflicts over resource use are inevitably conflicts over values. The debate over elephants and whales, for instance, is clearly a debate over values. But *whose* values are these? To whom are these resources valuable, and for whom are they to exist?

The noosphere is a useful construct with which to begin answering such questions. To posit a "thinking layer" is to place mankind in the natural world. That humans are part of nature, and therefore cannot be segregated from it, is a critical first premise. From this premise springs the chapter's general thesis: ideas have consequences, for both mankind and nature.

Just as every effect has a cause, and every action a reaction, ideas have consequences (Weaver 1948). For example, if a theory becomes generally accepted, it can impact the daily lives of millions of people and the resources they use. Voluntarily, individuals may choose which theories to accept; but through politics, groups can dictate which theories will govern individuals. Politics, therefore, can transform ideas into public policies that govern the individuals of an entire nation or, as in the case of preserving elephants and whales, ideas can govern the world at large. Once an idea is transformed into political belief, is institutionalized, and is then acted upon, its consequences can be profound.

Such has been the case with the idea that consumptive commercialization of elephants and whales will doom them to extinction. As a reputable American conservation biologist opined, "there will be few schemes of commercial conservation that can supply the world market without destroying the species and ecosystems they purport to save" (Ehrenfeld 1992). Such rhetoric, which is more anti-market than it is pro-conservation, symbolizes the present controversy over elephants and whales. Yet markets exist with or without legal commercialization. The question is whether, given suitable institutional arrangements, wild animals can be saved by commercializing them. Evidence suggests that they can (Goldstein 1991, Kreuter 1993, Simmons and Kreuter 1989, Sugg 1992).

Since the controversy over elephants and whales has arisen as much over conflicting views of economics as ethical concerns, this chapter examines resources and values in the context of economics, and then discusses them in ethical relation to elephants and whales.

ECONOMICS AND RESOURCES

Contemporary analysts have described the noosphere as "a philosophical term for human consciousness as it relates to the earth", suggesting that our belief systems, both figuratively and literally, "shape the world we inhabit" (Lyman 1993). Perhaps no where has this been more evident than in the former Soviet Union, where Marxist socialism has wreaked unparalleled ecological havoc (Feshbach and Friendly 1992, Goldman 1972, Peterson 1993, Smith 1982). Apparently, this political system affects ecosystems in much the same way as it affects economies, and for much the same reason: denying people the right to own, control and benefit from resources destroys the natural incentive to invest in and care for them (Yandle 1983). Stewardship, like entrepreneurship, generally goes wanting without such incentives.

While most environmentalists recognize that Marxist socialism has been ecologically disastrous, many still believe that socialism is more environmentally friendly than capitalism (Bookchin 1990, Commoner 1990) and that the economic productivity created by free markets is inimical to environmental quality (Funk 1992). Even though the arguments in support of this belief have been refuted in both theory and practice (Anderson and Leal 1991, Bernstam 1991, Simon 1981), the belief persists and has resulted in the adoption of market socialism by main stream environmentalists. Under market socialism the means of production are privately owned, but the ends of production are politically controlled. As two prominent American analysts observed, "the goal remains the same: to direct human behavior through state action" (Smith and Jeffreys 1993).

Market Failure versus Government Failure

As the controversies over elephants and whales chronicled in this volume illustrate, widespread distrust of markets has acted as a catalyst for government intervention. The intellectual rationale for such intervention has been the theory of market failure (Pigou 1912), which attempts to describe why markets may not guide resources to their socially optimal use (Rose 1986). However, every economic decision may negatively affect some agents in the economy (DeCanio 1993) and thus, based on this theory, no market outcome of consumer demand can achieve a socially optimal solution. Market socialists thus argue that every market transaction is a "failure" and that central planning authorities are required to control every economic decision. However, as the experience of the former Soviet Union has shown, creating and maintaining such a regulatory system is exceedingly costly and coercive, and is doomed to fail.

Yet the most recalcitrant problem intrinsic to intervention in markets is that values are by nature subjective, and thus dispersed throughout society in the minds of individuals (Hayek 1945). Since values represent the relative importance individuals place on any one thing among many, complete knowledge of total values cannot be obtained by any one mind or bureaucracy of minds. Preferences can only be revealed through individuals' expressions of value, as reflected by prices in a free market (Anderson and Leal 1992). Regulations, taxes and subsidies distort these values, by design.

When values do become politicized, regulatory bodies, such as the Convention for International Trade in Endangered Species (CITES) and the International Whaling Commission (IWC) are entrusted to determine them. However, because these bodies are susceptible to special-interest pressures, a fact made abundantly clear in this volume and others (Greve and Smith 1992), their evaluations can be skewed even beyond that which informational deficiencies dictate. Thus attempts by government to correct perceived market failures often reveal another phenomenon, namely government failure (Menell 1992, Rasker et al. 1992).

As governments have grown over time, many people have come to appreciate that command and control "solutions" create their own problems (e.g., Brodin 1992, Stein 1991). Yet, many remain wedded to the belief that values can be optimized by governments. "Scientific management", they say, can somehow produce a "bliss point" not achievable through the market place in a freer society (Anderson and Leal 1992: 299). An objective standard by which to accurately measure values for other people is, however, an illusion.

At bottom, people must choose between free markets, in which prices result from voluntary exchanges, and having governments alter prices for special interest groups that gain access through the political process. Nonetheless, while the appeal of economic central planning appears to be waning, that of ecological central planning is waxing because, to many, environmental problems provide the classic case of market failure (Cairncross 1991).

Tragedy of the Commons

It has also been argued, however, that it is not the existence of markets for wildlife products *per se*, but rather the lack of well defined and enforceable proprietary rights over wild animals that threatens them (Smith 1981). Indeed, no animal species that has been both economically valued and privately owned appears to have ever become extinct. For most of human history, elephants and whales have, however, been treated as open-access, common-pool resources.

According to Hardin's (1968) "tragedy of the commons" model, open access resources are doomed to over-exploitation and degradation. The model assumes humans to be self-interested beings, and is based on the proposition that the benefit of overusing an uncontrolled common-property resource is reaped by the individual, while the costs of individual overuse are borne by all users of the resource. In contrast to this negative third-party effect (externality), there is a positive externality; if an individual user decides to conserve the resource, then other users reap the benefits of the individual's conservation without cost. The net incentive for each individual with access to the resource is to take as much of the resource as possible before it is taken by another. Thus, Hardin concluded, "freedom in the commons brings ruin to all" (Hardin 1968: 1244).

After a decade of explaining his model, Hardin (1978) specified two solutions to the tragedy; privatization of the resource or socialism. However, in the case of elephants and whales, to pose such a choice presents a false dichotomy. Many users of elephants and whales have long-standing traditions of communal resource tenure, which do not necessarily result in resource exhaustion (Berkes et al. 1989) and can provide adequate resource protection (see, for example, Kisangani in this volume). However, the efficacy of communal systems has often been undermined by excessive human population pressure and political intervention, and their application to elephants and whales is frustrated by the transient nature of such species. Establishing either communal or private property rights in elephant and whale populations thus raises serious political and technological difficulties. These difficulties result from the nature of the resources and values involved, and must therefore be put into historical and political perspective.

PROPERTY RIGHTS TO ELEPHANTS AND WHALES

Wildlife and fisheries have generally occupied a special class of entity (along with air and water), the members of which are the property of no one; *res nullius* (Bean 1977). However, under Justinian law wild animals also fell under the broader rubric of things capable of being owned, *res commercio* (Pomerance 1929). Indeed, the Romans allowed game farming in several colonies (Deininger 1992). These findings conflict with the interpretation of Justinian law claiming that "things in a state of nature, *ferae naturae*, cannot be alienated as private property but must belong to all" (McCay 1990). With the rise of the modern state in western Europe, the crown's control over wildlife gradually merged with its sovereign power to own (Pomerance 1929:247). Hence the oft-quoted description of wildlife under traditional Western law as the "King's game" (Edwards 1992).

Today, scholars recognize that it is important to distinguish between the intrinsic nature of resources and the property rights regime under which they are held (Ostrum 1986). The property status of wildlife and fisheries varies among different nations (Smith 1981), though both are generally said to qualify as common-property resources (Feeny et al. 1990: 3). The two primary criteria used to classify a resource as common-property are: (1) controlling access to the resources may be costly or virtually impossible; and (2) individual use of such resources subtracts from the welfare of other users. While the first criterion is a physical attribute, alterable by economic incentives and technological advances, the latter is dependent on the legal property regime in which the resource occurs. In theory there are four progressively exclusive categories of property regimes: open access, state property, communal property, and private property (Feeny et al 1990: 4). The more exclusive the property regime, the less likely demand for the resource in question will lead to over-exploitation.

Demsetz (1967) observed that guiding incentives to achieve greater internalization of externalities is one of the primary functions of property rights (communal and private). Property rights evolve, he wrote, "when the gains of internalization become larger than the cost of internalization". The costs of developing property rights in a resource can be prohibitive in two ways: (1) for "natural" reasons, such as physical intractability of the resource; or (2) for political reasons, such as a law proscribing private appropriation of a resource or trade in it. Elephants and whales are generally regarded as satisfying both prohibitive criteria.

While African states are legally responsible for elephants occurring within their borders (Barbier et al. 1990), law enforcement is frequently ineffectual so that elephants, like other wildlife, have come to be regarded as an open-access resource in many areas. By contrast no country owns whales, even those that occur inside of the 200 mile coastal zones within which adjacent countries have exclusive fishing rights (Lyster 1985). The lack of clear and enforceable property rights to elephants and whales appears to have been a major factor leading to their over-exploitation.

Elephants

Africans and elephants co-existed for millennia under communal property regimes without causing each other notable harm. While human population pressures were low, such regimes did not appear to create a "tragedy" for elephants. However, colonization of Africa led to increased demand for ivory and in turn concerns about the future of elephants. Such concern was first raised by the Roman naturalist Pliny, who in A.D. 77 reported that large tusks were increasingly rare among African elephants (Harland 1990). It resurfaced in the 19th century after the European colonization of Africa, which was driven largely by the value of ivory to European consumers (Adams and McShane 1992). By the end of the 19th century, the British were importing some 450 tons of African ivory annually, and the Americans about half that much (Bonner 1993).

According to property rights theory (Demsetz 1967), such pressures should have resulted in the evolution of property rights to conserve the increasingly valuable elephants. However, sustainable exploitation of resources was often not an objective of colonialists. Moreover, colonial rule increasingly disrupted and often destroyed native land-tenure systems (Johnson and Anderson 1988), and usurped the rights to wildlife. For instance, in 1882, the British prohibited shooting elephants and hippos in the Cape Colony without a license (Bonner 1993: 46). Over time, Western wildlife laws were imposed throughout sub-Saharan Africa, and wild animals became "King's game." In many places, the subsequent creation of game parks further dispossessed African people of their ancestral lands (Isaccson 1991, Bonner 1993: 167). With such policies dictating the relationship between Africans and African wildlife, it is not surprising that more exclusive property rights arrangements did not evolve.

Despite the repressiveness of colonial wildlife regulations, most post-independence African governments have maintained the spirit and letter of these laws. Zimbabwe is however, a notable exception. It abandoned the "King's game" concept in 1975 when it reinstated the right of private landowners to control the wildlife on their property. Not long after gaining independence in 1980, Zimbabwe amended its wildlife law to enable District Councils to apply for the authority to manage wildlife on communal lands, a right that had been denied to rural Africans since 1898 (Zimbabwe Trust 1992). This devolution of wildlife usufruct rights to those people bearing the costs of living with wild animals has been the key to Zimbabwe's recent success in conserving wildlife (see Peterson, this volume).

The Zimbabwe example shows that with economic value, elephants are perceived as an asset to be protected; without such value, they are viewed as a competitor for resources and thus a liability. It furthermore shows that common property rights do not have to result in tragedy for wildlife. If given the rights and the incentives to exclude others from taking what is theirs, communal societies can decide for themselves which resources to use and how to manage them effectively.

Courtesy of Urs P. Kreuter

Plate 2.1 Elephant trophy bull with 36 kg tusks can provide substantial dividend, and thus an incentive to protect them in rural areas.

Unfortunately, however, international politics has frustrated such community-based wildlife conservation efforts by undermining positive conservation incentives. The Parties to CITES voted to ban trade in African elephant products in 1989 (see Kreuter and Simmons, this volume), thereby reducing the value of elephants to the people with whom they compete for resources.

Despite the lack of evidence that the trade ban has eliminated demand for ivory, or that it has terminated illegal hunting in areas without effective law enforcement, the ban has been maintained. Even though it was agreed at the 1989 CITES meeting that countries demonstrating successful elephant management strategies would be allowed to resume controlled trade in elephant products, the majority of delegates at the 1992 CITES meeting were unwilling to consider such a request from the four southern African countries with stable or increasing elephant populations (Bonner 1993: 281). Similar international interference in the rights of communities and countries to use mammalian resources occurring within their jurisdictional bounds has undermined sustainable whaling.

Whales

In the 17th century, Hugo Grotius argued that the ocean was incapable of being reduced to ownership (Jeffreys 1992). The apparent intractability of many marine habitats and the migratory nature of marine fisheries and mammals, seems to have resulted in a classic case of the tragedy of the commons. As human populations grew, the demand for fisheries products increased and technology improved to satisfy demand. Without exclusive property regimes, the rule of capture has prevailed in the marine commons, with over-harvesting of resources being a common result.

Gradually, however, modern coastal nations began to exert greater control over marine resources by claiming territorial rights further out to sea. In the late 1970's, the United Nations Law of the Sea Convention allowed countries to extend their exclusive economic zones to 200 miles. Today, more than 80 nations claim such zones, which together contain over 90 percent of the world's living marine resources (Jeffreys 1992:9).

While demarcating territorial boundaries may transfer authority over resources to specific nations, it is insufficient to ensure effective resource conservation if an unregulated rule of capture reigns within those areas. Yet, anthropologists report that, for years or even centuries, many traditional coastal communities have had property rights arrangements that have successfully averted the common's tragedy (Cordell 1990, McCay and Acheson 1990). Today, informal (e.g., Maine coastal lobster fisheries) and formal private property rights regimes (e.g., Japanese coastal fisheries) work well in this regard (Acheson 1990, Ruddle 1987).

The high seas appear more problematic. Hardin (1972:121) noted that "It is doubtful if we can create territories in the ocean by fencing. If not, we must . . . socialize the oceans (by forming) an international agency with teeth. Such an agency must issue not recommendations but directives; and enforce them." Thus the IWC, established in 1946, has evolved to effect such control.

Yet some believe that technological development and increasing consumer demand for finite stocks necessitate the privatization of marine resources

Courtesy of Mats Ris

Plate 2.2 Minke whales are examined during fisheries research, Northern Norway.

(Singer, 1986) and that electronic sensing devices offer the possibility of privatizing the marine commons (North and Miller 1973). Several authors have argued that technologies to track fugitive resources by satellite, sonar and radio already exist and could be employed to privatize whales (Anderson and Hill 1975, Jeffreys 1992, Loehnis 1990, Majewski 1987). For example, in "Whales '93", an undersea surveillance system is being applied to identify and track whales with significant success (Amato 1993, Broad 1993). While such technologies could provide a mechanism for defining and enforcing property rights to whales in the marine commons, their application is hampered because, as with African elephants, international politics reigns supreme over whales.

In 1982, the IWC imposed a moratorium on all commercial whaling, based on arguments that such harvests were placing whales at undue risk because data on whale stocks and dynamics were deficient, thus casting doubts on the effectiveness of trade controls (Butterworth 1992). The moratorium took effect in 1986 and was to last until 1990, by which time the IWC's Revised Management Procedure (RMP) was scheduled to be complete. Designed to ensure ecologically sustainable harvests, it was anticipated that the RMP would set scientifically defensible quotas, leaving whaling opponents no objective basis for continued support of the moratorium (Cherfas 1992, Mundell and Swinbanks 1992).

The moratorium was, however, extended in 1991 and 1992, because the computer model underpinning the RMP was incomplete, and it remains in effect despite the unanimous recommendation by the IWC Scientific Committee in 1993 that the completed RMP be adopted (Hamond 1993, IWC 1992). Yet despite the IWC's mandate to base its decisions on the best available scientific data, the IWC eschewed the findings of its Scientific Committee and maintained the worldwide ban on all whaling. This decision was made even in the face of the fact that, out of a total of 79 species of cetaceans, only the blue whale and the rights whales (together with three species of river dolphin) are, in fact, endangered (Klinowska 1991).

NOOSPHERIC VALUES

CITES and the IWC were both created to regulate the use of resources that are not adequately conserved by the users themselves. By banning whaling and the sale of ivory and hides from African elephants, the members of the IWC and CITES have exerted what amounts to proprietary control over elephants and whales. They, together with non-governmental organizations, have used the symbolic and emotive power of elephants and whales to inculcate the ideological precepts of an anti-consumptive-use philosophy into international environmental policies. In the absence of legal markets based upon legally defined, defended and enforced property rights, preservationist values are thus being imposed upon people who often do not share them.

Yet as Mostafa Tolba, the former executive director of UNEP, stated at the 1992 CITES convention in Kyoto, CITES "does not provide a legal basis for turning the world into a zoo or into a museum. The philosophy that underlies it is one of conservation and utilization, rather than outright preservation" (Tolba 1992). Similarly, the IWC Treaty to regulate whaling was created to "provide for the conservation, development and optimum utilization of whale resources", and directed the signatories to respect "the interest of the consumers of whale products and the whaling industry" (IWC Treaty 1946). Thus, despite preservationists' declarations to the contrary, utilization is firmly grounded in the environmental establishment. The World Conservation Union's treatise, "Caring for the Earth" (IUCN 1991), clearly states that people have "the right . . . to the resources needed for a decent standard of living, and hence a right to derive economic and other benefits from wild species." Such recognition has led to the principle of "sustainable utilization".

Sustainable Utilization

That which occurs "without compromising the ability of future generations to meet their own needs" (World Commission on Environment and Development 1987:8), is generally regarded as sustainable. "Caring for the Earth" set a target for encouraging all countries to adopt guidelines for sustainable use of wild species by the year 2000. In setting this target it acknowledged the principles of national sovereignty and people's rights to use and develop their own resources. Yet, the IWC and CITES have denigrated these principles, because they conflict with the values of powerful members and ideological interest groups, especially animal rights groups (Bonner 1993, Butterworth 1992).

For instance, the Species Survival Network, a coalition of animal welfare groups, was recently formed to resist inclusion of the sustainable-use principle in international wildlife policy. In its report on the proposed CITES listing criteria (NGO Working Group 1993), the coalition argues that utilization should be allowed only if sustainability is proven beyond doubt. However, abiding by such a criterion would effectively preclude use since conclusive, *a priori* evidence for the effects of use on resource stocks is unobtainable. Moreover, if only those resources "needed for a decent standard of living" are to be used, then many natural resources currently used by people might be subject to prohibition. As long as resource use can be politically controlled, even sustainable uses may be precluded. Thus, the importance of who decides which resources are to be used, and who determines if the uses are sustainable, cannot be overstated.

The stipulation that resources be used without compromising the ability of future generations to meet their own needs is itself problematic. Sustainability, so defined, is conceptually analogous to the Lockean Proviso of 1640, by which the original appropriation of resources is just only insofar as "there is enough, and as good left in common for others" (Laslett 1963:329). While the Proviso is a principle of justice relating to the initial acquisition of resources, sustainability is a principle of justice relating to their use or development. The common thread binding them is usage, for both ultimately require that future generations have sufficient opportunity to *use* resources.

However, under conditions of scarcity "every acquisition worsens the lot of others" (Bogart 1985), for "any appropriation diminishes to some tiny degree the amount of a resource that is available to others" (Rose 1987:429). Indeed, this interpretation suggests that establishment of any control over resources violates the Proviso, whether it be private (Sartorius 1984, Thomson 1976), communal or state appropriation (Schmidtz 1990).

Given that mankind must use resources to survive, the choice is between leaving them in an open-access commons, or appropriating them in some way. If the first alternative leads inexorably to "tragedy", then rather than ruling out appropriation, the Lockean Proviso actually requires it (Schmidtz 1990). This paradoxical but inescapable conclusion has obvious implications for the conservation of elephants and whales, both of which are generally considered to be common property resources. Rational conservationists who recognize that the tragedy of the commons arises due to the lack of property rights to natural resources, also understand that some form of local appropriation is necessary if future generations are to have such opportunities.

Conservationism versus Preservationism

Conservationists understand that, by definition, resources are to be used, but used sustainably so as not to preclude the opportunity of future generations to use them also. Conservationism allows both consumptive and non-consumptive use of resources, including elephants and whales, since it finds no overriding ethical content in the form a given use takes; the ethical issue is whether the resource can sustain usage over time. The extent to which usage can be sustained depends largely on the incentives for conservation, which are a function of the institutional arrangements under which the resource is managed, and the value placed on it.

Preservationists, by contrast, are generally opposed to consumptive use of natural resources, and thus to the very concept of "resources". They usually attribute natural entities with "option" and "bequest" values, which imply future use, or "existence" value, which implies non-consumptive use. Krutilla (1967) was among the first to propose the concept of existence value in environmental entities and amenities, noting that many people obtain satisfaction from the mere knowledge that wilderness and wild animals exist. Thus people repeat preservationist slogans such as "only elephants should wear ivory" and "save the whales." Such mantras are tantamount to the exhortation that species should be preserved for the sake of preservation.

Preservationism can thus be regarded as the antithesis of conservationism. Nonetheless, it is alluring because it provides a mechanism for controlling resources without incurring the costs and responsibilities associated with ownership. If preservationism is combined with the notion that some resources are "the common heritage of all mankind" and thus part of a "global commons," then there is no need to pay for what "we" already have. For example, as a senior official at World Wildlife Fund has stated, "There's no overriding human need for these whales.... They belong to the world, not to Japan or Norway" (Reid 1993). Bonner (1993) found a similar regard for elephants, noting the oft-quoted view of them as "a world heritage."

That is not to say that preservationists do not use natural resources; they employ them symbolically and free of charge for fund-raising, photography and tourism, and to foster taboos that curtail the consumptive use of resources. Yet, the preservationist argument, that for the sake of posterity humans should refrain from consuming natural resources, is disingenuous because it also prevents future generations from consuming resources since each successive generation would fall under the same prohibition.

A further problem with preservationism is that methods for accurately measuring option, bequest and existence values have been elusive because the actual amount people will pay for the existence of environmental amenities is unknown and unknowable without markets for people to express their preferences. Preservationists nevertheless "seek to employ formal economic methods to resolve matters of cultural symbolism and social ideology" (Rosenthal and Nelson 1992) because, if nonuse values can be shown to outweigh use values, they can assert that preservation is economically superior to utilization. Naturally, preservationists tend to select objects to preserve on the basis of perceived "naturalness" and rarely compare their artificially quantified existence values with the existence value of individuals, communities, cultures, or other similarly unquantifiable anthropocentric entities.

From Demand Values to Rights

The practical and philosophical problems facing preservationists stem from their desire for an objective basis on which to predicate their values. Since demand values, such as option and bequest values as well as market values, are anthropocentric, they are subjective and thus subject to change. This creates a problem for people wishing to preserve resources in perpetuity. "If the goals sought by environmentalists are supported only by their subjective, personal or culture-bound tastes, environmentalists will be no better than their opponents" (Norton 1992). Demand values are thus an unreliable basis for imposing preservationism.

Unless they are willing to admit that their principles are simply value preferences that they feel powerful enough to foist on others, preservationists must justify the coercion necessary to achieve their ends. In a fruitless attempt to avoid the conundrum of subjectivity, preservationists have thus sought to construct a theory of rights for nature (Ehrenfeld 1978, Nash 1989). "Unfortunately, no one could come up with a theory to support such rights attributions," laments one preservationist (Hardgrove 1992). Yet, efforts to do so continue, for, as Tom Regan the animal rights philosopher notes, "the development of what can properly be called an environmental ethic requires that we postulate inherent value in nature" (Regan 1981).

The concept of intrinsic value is the most common denominator underlying theories designed to trump demand values, especially market values. Strictly construed as values external to the human realm, intrinsic values are philosophical explications of non-subjective existence values. According to O'Neill (1992), to hold an environmental ethic is to take the position that non-human beings have non-instrumental value; they are not means to human ends, but have value in themselves. However, since people cannot view anything other than from a human perspective, even intrinsic values are products of human consciousness (Callicott 1989), and even non-anthropocentric value theories are, at bottom, anthropocentric (Hardgrove 1992) and therefore inherently subjective (Taylor 1984). This conclusion undermines the preservationist rationale for postulating intrinsic value because as Norton (1992) argues, "the main motivation for defenders of intrinsic value is to avoid cultural relativity by insisting that natural value is independent of human consciousness."

Callicott (1985) thus states that, "the central and most recalcitrant problem for environmental ethics is the problem of constructing an adequate theory of intrinsic value for non-human entities and for nature as a whole." Even preservationists acknowledge that this is problematical: "nonliving objects can only be defended on the grounds that they are instrumental to living centers of purpose that use them for their own intrinsically valuable ends" (Hardgrove 1992). Thus inanimate nature cannot have intrinsic value, but only instrumental value. Similarly, to argue that rare or endangered species (as elephants and whales purportedly are) or diverse ecosystems are intrinsically more valuable than common species or mono-cultural agrosystems, would also commit a relational fallacy. For something to be intrinsically valuable it must be "quite apart from everything else" and valuable "in absolute isolation" (Moore 1903). Given these conceptual constraints, no theory of intrinsic value can justify an environmental ethic that is simultaneously non-subjective and coherent.

The attempt to establish intrinsic value in nature grew out of frustration in attempting to establish inviolable rights in nature. After all, rights are protection against coercion, and thus the strongest moral currency. In Ronald Dworkin's terms, rights are moral "trumps" (Dworkin, 1977), overriding other concerns and values, and thus negating arguments for instrumental value. In other words, if wildlife or wild places have rights, "coercing nature" (manipulating wild entities) is on a par with coercing people, which contradicts human rights. Thus, the issue of utilization is negated, since consumptive use would be tantamount to consuming a rights-holder. However, "attributions of rights to collectives, while more likely to be adequate to preservationists' concerns, deviate from the logic of the traditional concept of rights so extensively as to undermine any significant analogies to human rights" (Norton 1987).

Peter Singer, the famous animal rights philosopher, does not actually attribute animals with rights; indeed, he denies their existence (Singer 1975).

Singer is a utilitarian philosopher with a keen concern for animal welfare. According to him, pain and suffering arise from sentience, and it is this experiential capacity that is deserving of respect and protection. Yet, life in the wild is impossible without pain, suffering and death. Moreover, Rolston (1992) found that "suffering in some sense seems co-present with neural structures; there are endorphins in earthworms, which indicates both that they suffer and that they are provided with pain buffers." Where are the lines to be drawn? As Regan (1984) concedes, full-fledged rights for animals would prohibit treating them as means to ends, consumptive or not, and would preclude management of wildlife entirely.

CONCLUSION

In thinking about whales and elephants, it is clear that, as biospheric resources, their noospheric value emanates from both the economic and ethical realms of the "thinking layer." Like the noosphere itself, values and rights are human constructs. Just as rights require rights-holders to respect rights, values require valuers to hold values. This leaves Mankind uniquely alone in the biosphere as sole possessor of moral agency. From this, several salient conclusions can be drawn about how human beliefs shape the world future generations inherit.

Mankind can confer rights upon "nature" and attribute intrinsic value to it, though no coherent theory of rights or values can support this. To credit natural resources with intrinsic value is to dispense with the concept of resources, and thus to put them on the "ends" side of the ethic ledger. Yet, such attribution would invariably negate the rights of people who do not share those beliefs. People can think that other animals have rights, and thereby deem them inviolate; but they will renounce mankind's role in nature, narrow the sphere of human rights to a sliver, and imperil their own species in the process. In other words, if environmentalists continue down the path of intellectual deceit, they will increasingly segregate values from valuers, rights from man, and man from nature.

Such behavior, however, creates "disvalues" in nature for people deprived of the rights to use natural resources. It would be far more sensible to recognize the folly of anthropomorphizing nature, of longing for lost innocence and the Garden of Eden. The primordial Garden was, after all, a garden. To live, mankind must harvest produce from the Garden; to prosper, people must take more than they need to merely exist. This elicits fundamental questions about the economics and ethics of limiting human rights to "subsistence" resource usage, a concept that "has become a powerful weapon in the service of imperialism" (Kalland 1993).

The International Covenant of Economic, Social and Cultural Rights states that: "In no case may a people be deprived of its means of subsistence"

(D'Amato and Chopra 1991:46). However, as D'Amato and Chopra argue, in most cases political intervention by preservationists threatens traditional life-styles, not lives. Either way, indigenous peoples are accorded rights not unlike those granted to elephants and whales — qualified rights to continued exist-ence. Such "rights" prohibit prosperity through commercial use of resources (Kalland 1993:41), and preserve people in aboriginal conditions of poverty just as surely as they seek to preserve "nature" in an arbitrarily chosen state of Edenic bliss.

Since they have the capacity to chose, people ultimately deserve the freedom to decide for themselves what values they will hold, what kind of world they want to live in. To argue otherwise is to deny the right to self-determination, to endorse subjugation. Freedom of choice necessitates economic freedom, wherein market values are the voluntary actualization of subjective individual values. Yet ethically, this can be acheived only through clear property rights, without which impositions of value deprive others of the means to achieve their own ends.

Thus, to treat the biosphere with respect is to first treat the noosphere with respect. It is to address the central issue of self-determination, of property rights. Those who own the Garden and the resources it produces should an-swer the question: *resources for whom?* While difficulties in establishing more exclusive property rights arrangements to elephants and whales exist, they pale in comparison to the obstacles created by politics. One day, perhaps, re-source use will be depoliticized, thus creating incentives to conserve the in-strumental value of natural resources, and to assume the responsibilities and costs of managing them in a sustainable manner, or letting them revert to a "wilder" state. Until that day when preservationists are forced to limit their control to what they own or lease, elephants and whales will suffer the vagar-ies of politics that alienate cost from benefit, resources from users, and man-kind from nature.

REFERENCES

Acheson, J.M., 1990.
    The Lobster Fiefs Revisited: Economic and Ecological Effects of Territoriality in Maine Lobster Fishing. In: B.J. McCay and J.M. Acheson (eds.), *The Question of the Commons*, pp. 37–65. University of Arizona Press, Tucson, Arizona.
Adams, J.S. and T.O. McShane, 1992.
    The Myth of Wild Africa: Conservation without Illusion. W.W. Norton and Co., New York and London.
Amato, I., 1993.
    A Sub Surveillance Network Becomes a Window on Whales. *Science* 262:549–50.
Anderson, T.L. and P.J. Hill, 1975.
    The Evolution of Property Rights: A Study of the American West. *Journal of Law and Economics* 18:163–179.

Anderson, T.L. and D.R. Leal, 1991.
   Free Market Environmentalism. Pacific Research Institute for Public Policy, San Fran-
   cisco, California.
Anderson, T.L. and D.R. Leal, 1992.
   Free Market Versus Political Environmentalism. *Harvard Journal of Law and Public
   Policy* 15:297–310.
Barbier, E.B., J.C. Burgess, T.M. Swanson and D.W. Pierce, 1990.
   Elephants, Economics and Ivory. Earthscan, London.
Bean, M.J., 1977.
   The Evolution of National Wildlife Law. Council on Environmental Quality, Office of
   the President, U.S. Government, Washington, D.C.
Berkes, F., D. Feeny, B.J. McCay and J.M. Acheson, 1989.
   The Benefits of the Commons. *Nature* 340:91–93
Bernstam, M.S., 1991.
   The Wealth of Nations and the Environment. Institute of Economic Affairs, London.
Bogart, J.H., 1985.
   Lockean Proviso and State of Nature Theories. *Ethics* 95:828–836.
Bonner, R., 1993.
   At the Hand of Man: Peril and Hope for Africa's Wildlife. Alfred A. Knopf, New York.
Bookchin, M., 1990.
   The Philosophy of Social Ecology: Essays on Dialectical Naturalism. Black Rose Books,
   Montreal.
Broad, W., 1993.
   Navy Listening System Opening World of Whales. The New York Times, August 23,
   1993:A–12.
Brodin, E., 1992.
   Collapse of the Swedish Myth. *Economic Affairs* February: 14–22.
Butterworth, D.S., 1992.
   Science and Sentimentality. *Nature* 357:532–534.
Cadieux, C.L., 1991.
   Wildlife Extinction. Stonewall Press, Washington, D.C.
Cairncross, F., 1991.
   Costing the Earth: The Challenge of Governments; The Opportunities for Business.
   Harvard Business School Press, Boston, Massachusetts
Callicott, J.B., 1985.
   Intrinsic Value, Quantum Theory, and Environmental Ethics. *Environmental Ethics*
   7:257–275.
Callicott, J.B., 1989.
   In Defense of the Land Ethic. SUNY Press, Albany, New Jersey.
Cherfas, J., 1992.
   Whalers Win the Numbers Game. *New Scientist* 134:12–13.
Commoner, B., 1990.
   Making Peace With The Planet. Pantheon Books, New York.
Cordell, J. (ed.), 1990.
   A Sea of Small Boats. Cultural Survival Inc., Cambridge, Massachusetts.
D'Amato, A. and S.K. Chopra, 1991.
   Whales: Their Emerging Right to Life. *American Journal of International Law* 85:21–51.
DeCanio, S.J., 1993.
   Carbon Rights and Economic Development. *Critical Review* 6(2–3):389–410.

de Chardin, P.T., 1959.
The Phenomenon of Man Harper and Row, London.

Deininger, F., 1992.
The Past, Present and Future of Game Farms in Central Europe. Paper presented at 3rd International Wildlife Ranching Symposium (27–30 October 1992). CSIRO, Pretoria, South Africa.

Demsetz, H., 1967.
Toward a Theory of Property Rights. *American Economic Review* 57:347–359.

Dworkin, R., 1977.
Taking Rights Seriously. Harvard University Press, Cambridge, Massachusetts.

Edwards, S.R., 1992.
Preface: Sustainable Conservation by and for the People. In: R. Littell, *Endangered and Other Protected Species: Federal Law and Regulation*, pp. vii–xv. Bureau of National Affairs: Washington, D.C.

Ehrenfeld, D.W., 1978.
The Arrogance of Humanism. Oxford University Press, New York.

Ehrenfeld, D., 1992.
The Business of Conservation. *Conservation Biology* 6:1–3.

Feeny, D., F. Berkes, B. McCay and J.M. Acheson, 1990.
The Tragedy of the Commons: Twenty-Two Years Later. *Human Ecology* 18:1–19.

Feshbach, M. and A. Friendly, 1992.
Ecocide in the USSR: Health and Nature Under Siege. Basic Books, New York.

Funk, W., 1992.
Free Market Environmentalism: Wonder Drug or Snake Oil? *Harvard Journal of Law and Public Policy* 15:511–516.

Goldman, M., 1972.
The Spoils of Progress. MIT Press, Cambridge, Massachusetts.

Goldstein, J.H., 1991.
Economic Incentives for Environmental Protection: The Prospects for Using Market Incentives to Conserve Biological Diversity. *Environmental Law* 21:985–1014.

Greve, M.S. and F.L. Smith, 1992.
Environmental Politics: Public Costs, Private Rewards. Praeger, New York.

Hamond, P., 1993.
Letter to R. Gambell, Secretary of the IWC, May 26, 1993.

Hardgrove, E.C., 1992.
Weak Anthropocentric Intrinsic Value. *The Monist* 75(2):181–207.

Hardin, G., 1968.
The Tragedy of the Commons. *Science* 162:1243–48.

Hardin, G., 1972.
Exploring New Ethics for Survival: The Voyage of the Spaceship Beagle. The Viking Press, New York.

Hardin, G., 1978.
Political Requirements for Preserving our Common Heritage. In: H.P. Brokaw (ed.), *Wildlife and America*, pp. 310–317. Council on Environmental Quality, Washington, D.C.

Harland, D., 1990.
Jumping on the "Ban" Wagon: Efforts to Save the African Elephant. *The Fletcher Forum* Summer 1990:284–300.

Hayek, H.A., 1945.
The Use of Knowledge in Society. *The American Economic Review* 35:519–530.

Isaccson, R., 1991.
Big Game Parks: What Future? *Green Magazine* June 1991:40–44.

IUCN, 1991.
Caring for the Earth: A Strategy for Sustainable Living. IUCN / UNEP / WWF, Gland, Switzerland.

IWC, 1992.
Chairman's Report of the 44th Annual Meeting. International Whaling Commission, Cambridge.

IWC Treaty, 1946.
The 1946 International Convention for the Regulation of Whaling, Article IX. 62 Stat. 1716, T.I.A.S. No 1849.

Jeffreys, K., 1992.
Who Should Own the Ocean? The Competitive Enterprise Institute, Washington, D.C.

Johnson, D. and D. Anderson, (eds.), 1988.
The Ecology of Survival: Case Studies from Northeast African History. Crook, London.

Kalland, A., 1993.
Aboriginal Subsistence Whaling: A Concept in the Service of Imperialism? In: G. Blichfeldt (ed.) *11 Essays on Whales and Man*, 39–41. High North Alliance, Lofoten, Norway.

Kreuter, U.P., 1993.
Politics and African Elephant Conservation. *NWI Resource* Spring 1993:22–25.

Klinowska, M., 1992.
Dolphins, Porpoises, and Whales of the World. The IUCN Red Data Book. IUCN, Gland, Switzerland.

Krutilla, J.A., 1967.
Conservation Reconsidered. *American Economic Review* 57(4):777–786.

Laslett, P. (ed.), 1963.
Two Treatises of Government by J. Locke, 1690. Cambridge Press, New York.

Loehnis, D., 1990.
Flotation to Save the Whale. The Sunday Telegraph, December 23, 1990:5.

Lovelock, J., 1988.
The Ages of Gaia: A Biography of our Living Earth. W.W. Norton and Co., New York and London.

Lyman, F., 1993.
What's New in the Noosphere? The Key to Healing the Planet Lies Within. *The Amicus Journal* 14(4):20–21.

Lyster, S., 1985.
International Wildlife Law. Grotius publications, Cambridge.

Majewski, J., 1987.
Own Your Own Whale. *Economic Affairs* October / November: 46–47.

Marquardt, K., 1993.
Norway Resumes Whaling. *The People's Agenda* July / August: 10.

McCay, B.J., 1990.
The Culture of the Commons: Historical Observations on Old and New World Fisheries. In: B.J. McCay and J.M. Acheson (eds.), *The Question of the Commons*, pp. 195–216. University of Arizona Press, Tucson, Arizona.

McCay, B.J. and J.M. Acheson (eds.), 1990.
  The Question of the Commons. University of Arizona Press, Tucson, Arizona.
Menell, P.S., 1992.
  Institutional Fantasylands: From Scientific Management to Free Market Environmen-
  talism. *Harvard Journal of Law and Public Policy* 15:489–510.
Moore, G.E., 1903.
  Principia Ethica. Cambridge University Press, Cambridge.
Mundell, I. and D. Swinbanks, 1992.
  Science Panel's Report Could End Whaling Ban. *Nature* 357:35.
Nash, R.F., 1989.
  The Rights of Nature: A History of Environmental Ethics. University of Wisconsin Press,
  Madison, Wisconsin.
NGO Working Group, 1993.
  CITES and the Revision of the Berne Criteria: An Unpublished Report to the Stand-
  ing Committee of CITES. International Wildlife Coalition, March 1, 1993.
North, D.C. and R.L. Miller, 1973.
  The Economics of Public Issues (2nd ed.). Harper and Row, New York.
Norton, B.G., 1987.
  Why Preserve Natural Variety? Princeton University Press, Princeton, New Jersey.
Norton, B.G., 1992.
  Epistemology and Environmental Values. *The Monist* 75(2):208–226.
O'Neill, J, 1992.
  The Varieties of Intrinsic Value. *The Monist* 75 (2):119–127.
Ostrom, E., 1986.
  Issues of Definition and Theory: Some Conclusions and Hypotheses. In *Proceedings
  of the Conference on Common Property Resource Management*, pp. 599–615. National Re-
  search Council, National Academy Press, Washington, D.C.
Peterson, D.J., 1993.
  Troubled Lands: The Legacy of Soviet Environmental Destruction. Westview Press, Boul-
  der, Colorado.
Pigou, A.C., 1912.
  Wealth and Welfare. MacMIllan, London.
Pomerance, R., 1929.
  Wild Animals:y Nature of State's Interest. *Cornell Law Quarterly* 14:245–250.
Rasker, R., M.V. Martin and R.L. Johnson, 1992.
  Economics: Theory versus Practice in Wildlife Management. *Conservation Biology*
  6(3):338–349.
Regan, T., 1981.
  The Nature and Possibility of an Environmental Ethic. *Environmental Ethics* 3:19–34.
Regan, T., 1984.
  The Case for Animal Rights. University of California Press, Berkeley, California.
Reid, T.R., 1993.
  World Whaling Body Riven by Dispute. The Washington Post, May 15,
  1993:A–17,A–26.
Rolston, H. III, 1992.
  Disvalues in Nature. *The Monist* 75(2):250–278.
Rose, C., 1986.
  The Comedy of the Commons: Customs, Commerce and Inherently Public Property.
  University of Chicago Law Review 53:711–781.

Rose, C.M., 1987.
  Enough and as good of What? *Northwestern University Law Review* 81:417–442.
Rosenthal, D.H. and R.H. Nelson, 1992.
  Why Existence Value Should Not be Used in Cost-Benefit Analysis. *Journal of Policy Analysis and Management* 11:116–122.
Ruddle, K., 1987.
  Administration and Conflict Management in Japanese Coastal Fisheries. FAO Technical Paper No. 273, Rome.
Sartorius, R., 1984.
  Persons and Property. In: R.G. Frey (ed.), *Utility and Rights.* University of Minnesota Press, Minneapolis, Minnesota.
Schmidtz, D., 1990.
  When is Original Appropriation Required? *The Monist* 73(4):504–518.
Simmons, R.T. and U.P. Kreuter, 1989.
  Save an Elephant Buy Ivory. The Washington Post, October 1, 1989:D3.
Singer, P., 1975.
  Animal Liberation. Random House, New York.
Singer, S.F., 1986.
  Fisheries Management: Another Option. In: J.G. Sutinen and L.C. Hanson (eds.), *Rethinking Fisheries Management*, pp. 199–203. Proceeding from the Tenth Annual Conference, Center for Ocean Management Studies, University of Rhode Island, June 14, 1986.
Smith, F.L. and K. Jeffreys, 1993.
  A Free-Market Environmental Vision. In: D. Boaz and E.H. Cran (eds.), *Market Liberalism: A Paradigm for the 21st Century*, pp. 389–402. Cato Institute, Washington, D.C.
Smith, R.J., 1981.
  Resolving the Tragedy of the Commons by Creating Property Rights in Wildlife. *Cato Journal* 1(2):439–468.
Smith, R. J., 1982.
  Privatizing the Environment. *Policy Review* Spring 1982:11–50.
Stein, G., 1991.
  Saying Farewell to Welfare: The End of Sweden's 'Third Way'. *Policy* Autumn 1991:2–5.
Sugg, I.C., 1992.
  To Save an Endangered Species, Own One. The Wall Street Journal, August 31, 1992:A–10.
Taylor, P.W., 1984.
  Are Humans Superior to Animals and Plants? *Environmental Ethics* 6:149–160.
Thomson, J.J., 1976.
  Property Acquisition. *Journal of Philosophy* 73:664–666.
Tolba, M.K., 1992.
  Counting the Cost. Statement by the Executive Director of UNEP to the 8th Meeting of the Parties of CITES, Kyoto, March 1992.
Walters, M.J., 1993.
  Whales: Have They Been Saved? *Animals* September/October, 1993:22–26.
Weaver, R.M., 1948.
  Ideas Have Consequences. The University of Chicago Press, Chicago.

World Commission on Environment and Development, 1987.
  Our Common Future. Oxford University Press, Oxford.
Yandle, B., 1983.
  Resource Economics: A Property Rights Perspective. *Journal of Energy Law and Policy*
  5(1):1–19.
Zimbabwe Trust, 1992.
  Wildlife: Relic of the Past, or Resource of the Future? Zimbabwe Trust, Harare, Zimba-
  bwe.

3. ECONOMICS, POLITICS AND CONTROVERSY OVER AFRICAN ELEPHANT CONSERVATION[1]

URS P. KREUTER and RANDY T. SIMMONS

INTRODUCTION

In emergent Africa you either use wildlife or lose it. If it pays its own way some of it will survive. (Myers 1981)

Cows grow trees, elephants grow grass. (Maasai expression — Western 1993)

This chapter investigates some of the economic and political factors affecting the future survival of the African elephant (*Loxodonta africana*). It concentrates on human responses to elephants as an open-access resource, the elephant population crisis reported in the late 1980's, and the resulting initiation, implementation and effect of the international ivory trade ban.

The African elephant is perhaps the ultimate common property resource in Africa because of its very large home range, which can vary from 20 km² to 3,000 km² depending on water and forage availability (Owen-Smith 1988). The legal title to elephants, as with most other wildlife, is generally vested with a designated government agency of the country in which they occur (Barbier et al. 1990). Given their fugitive nature, large home ranges, and widespread weak law enforcement, elephants have, however, effectively become an open-access resource in much of Africa (Harland 1990).

The African elephant is important for at least three reasons. First, due to its very large body size (3,000–5,000 kg for mature adults) and high food intake (around 1.0% of body mass) (Owen-Smith 1988), the elephant has a major impact on the savanna and forest ecosystems that it inhabits (Harland 1990). Ecologically, it is a *key species* because the high biodiversity characterizing African savannas "reflects a shifting, unstable interplay between, elephants, environment, habitat, and human activity on a grand scale" (Western 1993). The disappearance of elephants may therefore result in the extinction of species dependent on this interplay (Western 1989), thus compromising the maintenance-of-biodiversity objective of the World Conservation Strategy (IUCN 1980 1991).

Second, the African elephant is an important conservation *symbol*, having been claimed to be an important "conservation flagship" because "it evokes strong sympathy and can, given public support, protect the integrity and diversity of African ecosystems" (Western 1989). Third, the elephant's aesthetic and emotional appeal and its potentially high economic value have put it at the center of international conservation *politics*.

These three attributes have all contributed to the current controversy over elephant conservation — a controversy that is less a rational discourse than a conflict between widely differing socio-economic and symbol systems (Peterson 1993). It has evolved due to two diametrically opposed perceptions of the relationship between mankind and nature. These perceptions are embodied in the preservationist philosophy, where mankind is excluded from nature, and the utilitarian philosophy, in which mankind has dominion over nature. Utilitarianism is the basis for human survival through the conscious use and manipulation of the environment. Preservationism grew out of romantic urbanite desire for untamed nature in response to rapidly changing landscapes and the growth of cities in Europe and North America (Adams and McShane 1992, Tober 1981).

Preservationists generally imagine Africa as the Garden of Eden in which the development of humans and livestock are seen as unnatural and ecologically unsound (Adams and McShane 1992, Graham 1973). The resulting perception that "Africa is on the brink of ecological collapse" seemed to give license for intervention in rural development in Africa (Bell 1987) and for promoting of wildlife preservation in pristine habitats.

Utilitarianism, by contrast, promotes the sustained use of wildlife for human benefit as the best hope for conserving wild animals in Africa (Simmons and Kreuter 1989a, 1989b). This philosophy recognizes that poor rural people living with wild animals have no incentive to conserve them without having clear user rights and receiving tangible benefits from them. This is critical for elephant conservation since humans and elephants compete directly for resources through their preference for more fertile, wetter land (Parker and Graham 1989a). As human populations expand, human-elephant conflicts will increase and elephants will be excluded if they provide no tangible benefits for rural people.

HUMAN RESPONSES TO OPEN-ACCESS ELEPHANTS

The African Response

Since Africans have been deprived of hunting rights in much of Africa for over 50 years, first by colonial administrators and later by post-independence governments which retained western preservation models, Africans generally place little value on wildlife. Large animals, such as elephants and hippos, are frequently despised by African peasants who fear being trampled by them or who watch helplessly as their harvests are destroyed (Western 1993). Africans, who compete with wild animals for land, food and water, thus exploit wildlife whenever possible using all available means (Adams and McShane 1992). Since attempts to enforce wildlife laws (which are often rooted in colonial regulations) has led to violent conflict between people who share a common culture, conservation is regarded as less than an honorable profession by many Africans.

Courtesy of Richard E. Hoare

Plate 3.1 Subsistence farmers and district council representative examining a maize crop damaged by elephants in a communal farming area, Zimbabwe.

Courtesy of Richard E. Hoare

Plate 3.2 Elephant damage to a grain storage bin in a communal farming area, Zimbabwe.

The combination of such anti-conservation sentiment, antipathy for personally costly wildlife, and the income-earning potential of poaching led to an escalation in illegal hunting in post-independence Africa. While government officials have been involved out of greed, lowly-paid rangers and subsistence farmers have poached out of need. For example, in Kenya corruption occurred from the President, through ministers down to wardens and rangers (Bonner 1992, Lamb 1982), and despite Kenya's 1977 hunting ban, elephant and rhino poaching continued. Widespread poaching facilitated by corrupt officials consequently emerged as the most common factor associated with the illegal trade in ivory (Cumming and du Toit 1989). This compromised conservation actions and led to the rapid decline in elephant numbers during the 1980's (details in Appendix Table), a decline which prompted the call for an ivory-trade ban by western preservationists.

The Western Response

The root of western intervention in African conservation can be found in American culture. The United States has a strong historical bias in favor of unrestricted access to wildlife, and there is a perception that urban demand for game meat was the primary cause of declining wildlife in the country. These two aspects have produced legislation restricting the commercial use of wildlife (Tober 1981). Ownership to wildlife was formally assigned to the state in which it occurred, while landowners were accorded no property rights to wild animals.

With growing urbanization, increasingly more people became enamored with romantic images of nature. Such sentiment resulted in the rapid growth of the animal rights movement, especially after the 1975 publication of Peter Singer's "Animal Liberation", in which he argues that, because animals share with humans the desire to avoid pain, they deserve "equal consideration" with humans (Bonner 1993, see also Sugg and Kreuter in this volume).

The mix of legislated antipathy towards the commercial use of wildlife, together with America's increasing involvement in international policy formulation since 1945, created a national psyche which feels justified in imposing its non-consumptive values for wildlife on other sovereign countries. American preservationists who, in 1989, initiated the call for an ivory-trade ban (supposedly due to concerns over the rapid decline in elephant numbers) were merely acting out this psyche by exploiting Africa's open-access elephants to promote their own values. It was the reported rapid decline in elephant numbers during the 1980's that was used to justify the intervention.

STATUS OF ELEPHANT POPULATIONS

Recent Population Estimates

The African Wildlife Foundation was the first conservation organization to raise the alarm about declining elephant numbers (see Bonner in this volume).

However, the concerns over the future of elephants suddenly escalated in June 1989 when the Ivory Trade Review Group (ITRG), which was funded by Wildlife Conservation International and the World Wildlife Fund, released its preliminary report on the status of the African elephant. It stated that elephants had decreased from 1,343,340 in 1979 to 631,930 in 1989 (ITRG 1989). Regionally, the reported declines during this ten year period were 6% in West Africa, 44% in Central Africa, 77% in East Africa and 25% in Southern Africa.

These estimates are widely accepted as being factual, because the ITRG included several well known elephant specialists. The reliability of these estimates has, however, been questioned by other professional biologists (Cumming 1989a) since nearly 75% of the estimates were no more than informed guesses (Adams and McShane 1992, Douglas-Hamilton 1988). Moreover, the quality of population data has generally declined since 1981 (Cumming et al. 1990) though in some cases, such as the central African forests (Barnes 1989), surveys have allowed revision of elephant population estimates. Yet well documented increases in elephant numbers in Zimbabwe, Botswana and Kenya's Amboseli National Park were not clearly acknowledged by the ITRG which instead accepted differences between population estimates in 1987 and 1989 as established trends. Thus the ITRG's public statement that elephants were in rapid decline claimed a degree of precision in population estimates which did not exist, "perhaps to fit the crisis theme" (DNPWLM 1989). Indeed, Cumming (1989a), a former chairman of the African Elephant and Rhino Specialist Group and a reviewer for the ITRG, concluded that the elephant numbers game had reached questionable proportions.

The Role of the Ivory Trade on Declining Elephant Numbers

The decline of elephants during the 1980's has been directly linked to the growth in ivory exports from Africa (Barbier et al. 1990, Milner-Gulland and Mace 1991), 80% of which was illegal (AECCG 1989). For example, Milner-Gulland and Mace (1991) showed that an annual 12–13% harvest of tusked elephants, in conjunction with hunter preference for larger tusks, produces an ivory volume and tusk size distribution compatible with the international ivory-trade statistics of the late 1980's. Prior to 1990, the main consumers of worked ivory were Japan (40%), the Western Europe (20%) and the USA (15%) (Harland 1990), and at its peak trade is thought to have exceeded 1000 tons per annum (Bonner 1993). The increased demand has frequently been associated with the growth of expendable income in some East Asian countries.

Yet, despite the high demand for ivory, Parker and Graham (1989b) found no clear correlation between ivory prices and export volumes between 1925 and 1987 and reported similar population declines for other large African mammals that produce no ivory. They therefore attributed the increase in elephant killings and ivory exports directly to competitive exclusion by humans. Nevertheless, the reported five-fold increase in ivory prices during the 1980's, from an average of $60/kg in 1979, does mirror increased demand (Adams and McShane 1992). The ITRG (1989) report thus concluded that the "ivory trade poses a threat of extinction to the African elephant as a whole ... and the weak management and enforcement capacity of most of the range states leaves no grounds for optimism that conditions will improve."

CITES Criteria for an Ivory Trade Ban

The ITRG report led to calls by major western conservation groups to prohibit the international trade in ivory by listing the African elephant under Appendix I of the Convention for International Trade of Endangered Species (CITES). To qualify for such a listing "a species must be currently threatened with extinction" (Lyster 1985). The African Elephant and Rhino Specialist Group of the International Union for the Conservation of Nature (IUCN) estimated that the minimum viable population size for elephants is about 2,000 animals. Since there were at least 10 populations exceeding this number in 1989, the species as a whole was not facing extinction (Cumming 1989b), and thus did not qualify for Appendix I listing. One sports writer noted that declaring the African elephant as endangered is analogous to classifying the American elk as endangered simply because it has disappeared from some of its prior home range (Gresham 1991). Yet, despite the fact that there was no continent-wide biological emergency for elephants, the pressure for a ban on international ivory-trade mounted.

IMPLEMENTING THE IVORY-TRADE BAN

Western Engineering of the Ban

The Humane Society of the United Sates (HSUS) was the first to call for a trade ban in ivory when it petitioned the United States Fish and Wildlife Service (USFWS), at the beginning of 1989, to formally recommend Appendix I listing for elephants at the 1989 CITES meeting (Thomson 1992). But USFWS officials were reluctant to comply without the proposal emanating from an African country. In order to fulfil this condition the co-founders of the London-based Environmental Investigation Agency (EIA) (with financial support from the Washington-based Animal Welfare Institute) persuaded Tanzania's Wildlife Conservation Society to sponsor a letter to the Tanzanian President requesting him to propose Appendix I listing for elephants (Thomson 1992, Thornton and Curry 1991). The proposal itself was drafted by a senior program officer of the World Wildlife Fund in the United States (WWF-US) who had collaborated with the EIA (Thomson 1992).

Once the Tanzanian proposal was released, Kenya quickly followed in calling for a ban. As a result of the outrage fostered in conservation donors by the subsequent media horror show at the beginning of 1989, Western conservation organizations quickly rallied to the call (see Bonner, this volume). The momentum of the ivory ban movement increased further with President Bush's announcement, on June 5, 1993, that ivory imports into the United States were to be banned forthwith. His action was politically expedient since it placated environmentalists without incurring significant costs to the federal government or American business (Bonner 1993).

Although it was pressured to promote a trade ban, the CITES Secretariat circulated several documents prior to the 1989 meeting of CITES members, recommending improvement of the quota system for ivory trade because "transfer to Appendix I would not contribute to the conservation of the African elephant, and may in fact be counter-productive" (Harland 1990). The Secretariat, furthermore, emphasized that the criteria for Appendix I listing were not met by the African elephant at the species level. The final draft of the ITRG (1989) report nevertheless stated that:

> "The Group *believes* that ... the ivory trade poses a threat of extinction to the African elephant as a whole.... We suggest therefore that the danger to the elephant is real, continental and therefore justifies the transfer of the elephant to Appendix I."

On the basis of such *beliefs*, proponents claimed that the trade-ban would suppress demand for ivory because illicit ivory traders would not advance funds to poachers without secure markets to sell the ivory (Bohlen 1989). Opponents countered that it would only serve to drive the ivory market underground and thus further endanger the species, pointing out that poaching was most severe in East African countries where elephant hunting had already been banned (Simmons and Kreuter 1989b). Even officials of the United Nations Environment Program argued in mid-1989 that a complete ivory-trade ban was likely to fail in saving elephants (Harland 1990).

Voting for a Trade Ban for Ivory

Under intense pressure from Western conservation groups (Harland 1990), the majority of Parties at the October 1989 CITES meeting voted to list the African elephant on Appendix I, thereby initiating an ivory-trade ban. Politically, the voting patterns of the 91 member nations attending the meeting are revealing (see Table 3.1). Of the 76 nations endorsing a trade ban only 20% (19) had elephants while 63% (57) were non-African countries. By contrast 73% (8) of the 11 countries opposing the ban had elephants. Based on the ITRG's population estimates, elephants decreased by 74% between 1979 and 1989 in the predominantly East and West African countries voting for the ban, but they increased by 9% during the same period in the mainly southern African countries voting against the ban. In the latter group elephants comprised 52% of the total estimate for 1989.

These statistics emphasize that countries with no elephants (mainly non-African countries) or with rapidly decreasing elephant populations voted to im-

Table 3.1 Voting Patterns of Countries at the 1989 CITES Meeting (actual numbers and percent of total votes).

Country Category	For		Against		Abstained		Total	
	No	% tot	No	% tot	No	% tot	No	% tot
Elephant range (ER)	15	16.5%	8	8.8%			23	25.3%
African non-range	4	4.4%	1	1.1%			5	5.5%
Non-african	57	62.6%	2	2.2%	4	4.4%	63	69.2%
Total	76	83.5%	11	12.1%	4	4.4%	91	
% of ER countries		19.7%		72.7%				25.3%

pose an anti-trade policy on countries with conservation programs under which elephant populations were stable or increasing. The adoption of a world-wide trade ban for ivory, which negatively impacted conservation programs supported by the sale of ivory harvested from overabundant elephants, is particularly ironic in view of the total failure of a similar trade ban on rhino horn in reducing the decimation of the black rhino.

THE EFFECT OF THE IVORY-TRADE BAN

Effect on CITES and National Conservation Programs

The 1989 vote in favor of a trade ban has not only affected the elephants conservation status but has subverted the CITES treaty. The voting Parties were aware that several elephant populations were not endangered but the majority nevertheless voted for Appendix I listing because it was perceived that transfer of only the declining populations might compromise the effectiveness of an ivory-trade ban in reducing poaching. They willfully violated the CITES guidelines, thereby denying southern African states (with stable or increasing elephant populations) the protection of the legal instrument they had undertaken to uphold (Harland 1990).

In addition to undermining the principles of CITES, the ivory trade ban has also directly and indirectly compromised some national conservation programs in southern Africa. For example, in Zimbabwe and South Africa it has eliminated, without recompense, revenue derived from the international sale of ivory from elephants culled to prevent overpopulation in national parks. As a result southern Africa now has large stockpiles of ivory which cannot be converted into financial resources to protect wildlife. In addition, while the ban applies only to international trade in elephant body parts, it has also suppressed producer-country ivory markets which are aimed mainly at foreign tourists who may no longer return home with ivory. Furthermore, successful imposition of the ivory-trade ban has resulted in organizations, such as the HSUS, attempting to end the importation of sport-hunted ivory into the USA

(USFWS 1993). While, so far, this bid has failed, it is naive to assume that the ivory-trade ban cannot affect consumption-based conservation programs that depend on non-ivory revenue from elephants.

Effect on Ivory Demand and Prices

Evidence that the ivory-ban has caused demand to fall is mixed. Although legal American, European and Japanese ivory consumption was effectively terminated by the trade ban, internationally the demand for ivory has not been eliminated. For example, a 1990 WWF-US report claimed that the United States moratorium on ivory imports was having a profound impact on world commerce in ivory (O'Connell and Sutton 1990) but that both China and South Korea continued to import ivory. A more recent report from WWF International stated that raw ivory was being bought in Africa by dealers from North and South Korea, Taiwan, the United Arab Emirates, and African countries such as Ghana, Nigeria, Senegal, Somalia, and Zaire, though the volume was below pre-ban sales (Dublin and Jachmann 1992, Sugg 1992). Even in 1993 ivory shipments have been confiscated; Belgian authorities seized over 100 kilograms of ivory from Gabon destined for South Korea (EIS 1993), Chinese authorities seized a $5 million ivory shipment headed from Hong Kong to Japan (Ferrai 1993) and 153 tusks were confiscated at the Bangkok airport (Reuter Library 1993).

Evidence of the effect of the trade ban on ivory prices is limited because most ivory is now traded on the black market. The 1990 WWF-US report claimed that prices had dropped from a high of U$200–300 per kilogram in 1989 to as little $2 per kilogram in Central Africa (O'Connell and Sutton 1990). The drop in price has, however, enabled new consumers, who had previously been excluded from the ivory market by high hard-currency prices, to buy ivory (Barbier et al. 1990). For example, demand for ivory by tribal leaders in Cameroon has increased because they can once again afford to use it for ceremonies (Adams and McShane 1992). New entries into the market are likely to result in higher ivory prices especially if ivory is in short supply. Partial support for this contention is provided in Table 3.2 which shows that prices paid for ivory to hunters increased between 1989 and 1991 in Zambia, Tanzania and parts of the Cameroon, although they decreased in other parts of the Cameroon, Ivory Coast and Zaire (Dublin and Jachmann 1992). Officials from nine other African countries also reported decreases in ivory prices. To date, there is therefore no universal trend in ivory prices in response to the trade ban.

Effect on Illegal Hunting

Since the signatories to CITES gave its secretariat no enforcement powers, nations have to independently enforce its provisions but most have lacked the

Table 3.2 Pre-Ban (1989) and Post-Ban (1991) US$/kg Ivory Prices (Source: Dublin and Jachmann 1992).

Country		Pre-ban		Post-ban
Cameroon[1]	1)	4.2		16.8
	2)	9.5– 39.1		33.6– 50.4
	3)	86.2		34.4– 62.1
Ivory Coast[2]	1)	46.2– 56.1		36.3– 39.6
	2)	231.0–264.0		99.0–132.0
	3)			198.0–238.0
Tanzania[3]		6.0– 12.0		10.0– 26.0
Zaire[4]		16.3		5.9
Zambia[5]	1978:	15.0– 25.0	1990:	10.0– 12.0
	1983:	19.0– 32.0	1991:	14.0– 26.0
	1986:	18.0– 24.0		
	1989:	4.0– 10.0		
Malawi	Illegal ivory is still being traded at relatively high prices in the field.			

[1] Cameroon   1) paid to hunters in Savanna zone
2) paid on the open market in Savanna zone
3) paid to hunters in forest zone
[2] Ivory Coast   1) paid to hunters
2) to first middleman
3) ivory carvings
[3] Tanzania   paid for small to medium tusks to hunters.
[4] Zaire   paid for small to medium tusks to hunters.
[5] Zambia   paid for small to medium tusks to hunters.

ability or will to do so (Bonner 1993). Moreover, since ivory trade has always consisted of an illegal component and trade bans provide no incentives for increasing management expenditures, it was predicted that banning ivory trade would provide incentives for expanding its illegal component and that there would be little shift in supply (Barbier et al 1990). This hypothesis was corroborated by O'Connell and Sutton (1990) who stated that, due to the lack of vigor of government anti-poaching efforts, some illegal elephant killing continued in Africa, and by Adams and McShane (1992) who conclude that hunting has continued without interruption in Africa's tropical forests. Furthermore, in Zimbabwe's Zambezi Valley elephant poaching has reportedly escalated since 1989 because of increased collusion between poachers and some officials (EIA undated), due to the local abundance of elephants, the growth of the black market for ivory and reduced law enforcement capability (due, in part, to the loss of national revenues from ivory sales).

In contrast to these increases in illegal elephant hunting, Kenya (Africa's most vociferous ban proponent) has attributed the reported 32% increase in its elephants between 1989 (19,000) and 1991 (25,000) to reduced poaching as a result of the trade ban. Given a maximum population growth rate of around 4% for elephants (Hanks 1979) this claim is, however, spurious; the increase represents either immigration or inaccurate population estimates. Dublin and Jachmann (1992) found that noticeable decreases in poaching levels in Zambia, Tanzania, Cameroon, Ivory Coast, and Zaire were directly proportional to the increased level of law enforcement, which was facilitated by increased donor aid. Properly channelled law enforcement resources, not reduced demand nor reduced ivory prices, appear to have reduced poaching in some areas. It is thus important to recognize that the apparent success of the ivory-trade ban in closing most international markets and effective elephant conservation are not synonymous (Adams and McShane 1993).

FREE RIDING AND EXTERNALIZING COSTS

Capitalising on Western Mawkishness

Due to the fugitive nature of elephants and the consequent difficulty in limiting access to them, trade ban proponents have been able to impose their values on elephant-range countries by harnessing Western sentimentality over elephants. In so doing, they have usurped the sovereign rights of countries with elephants to manage their national resources for their own advantage (Kreuter and Simmons 1994).

Tanzania and Kenya initiated the proposal to list the African elephant on Appendix I (at the behest of Western preservationists) due to concerns about the impact of declining elephant populations on their tourist industries. Since the East Africans had suffered the greatest losses in elephants during the 1980's, they were effectively seeking a solution to a national problem through the closure of ivory trading everywhere. Zimbabwe, however, believed that such a listing would only divert attention from the real underlying cause for the decline of the African elephant, the lack of effective law enforcement. In response to coercive pressure to adopt the listing, Zimbabwe stated that it would not abandon its successful conservation programs "either to save face for other countries that have failed, or to provide more employment for London and Washington-based conservation ideologues" (Armstrong and Bridgland 1989).

Dependent on public donations, Western conservation groups rallied to the call for a ban. By merging the diverse characteristics of the well known African elephant, they created a mediagenic conservation totem through which *Elephant extermination imminent* is reported to have become the most lucrative slogan in the environmental movement's history (Bonner 1993, Gresham 1991, see also Bonner in this volume). Yet, while capitalizing on public sentimentality over elephants, Western conservation groups and governments have provided few resources to improve elephant protection (Harland 1990).

Paying for Elephant Protection

The costs of living with elephants are borne by rural Africans, who compete with elephants for resources, and by African governments, which reserve land for elephants. Since rural Africans can usually derive direct tangible benefits from elephants only through illegal hunting, there is little incentive in many parts of Africa to maintain elephant populations. Due to the well established black market for ivory, and the lack of local political will or financial resources for effective law enforcement, international regulations are unlikely to significantly affect illegal elephant hunting activities (Barbier et al. 1990). In addition, since the trade ban has resulted in a substantial decrease in the economic value of elephants, it may have effectively reduced incentives to protect elephants as a valuable resource. Yet, these deficiencies were almost entirely ignored by the proponents of the ivory ban.

To appease affected African countries and persuade them to support the trade ban, rather than trust market incentives for protecting elephants, the Western donor community recommended substantial funding for elephant range countries. In 1989, for example, the Senior Vice President of WWF-US wrote that "the fate of the African elephant cannot be left to the vagaries of the market place. The more effective response is the international effort ... to provide aid to wildlife departments..." (Bohlen 1989). However, a recent report concluded that, contrary to these promises:

> "It is a damning indictment of the donor community that, in general, they have not provided the critically-needed funding promised to these range states at the time of the Appendix I listing of their elephant populations" (Dublin and Jachmann 1992).

Although the African Elephant Action Plan states that its purpose is to assist range states in procuring the means to maintain viable elephant populations, it continues that "the bulk of the financial, political and practical burden rests with African governments" (AECCG 1990).

These comments imply that while Western preservationists, animal rights proponents, and governments (trying to appease these constituents) are eager to capture the political and economic benefits stemming from the implementation of the ban, they are unwilling or unable to pay for the direct costs of protecting elephants and the external costs resulting from policies that promote their own objectives. In economic terms they are free-riding on the existence of elephants at the expense of Africans living with elephants.

Even if the ban's proponents were willing to donate substantial funds to protect elephants, it is unlikely that they could muster sufficient resources to do so. Given an estimated investment requirement of US\$215/km^2/yr for zero elephant decline in Zambia (Leader-Williams and Albon 1988), some US\$100 million are needed every year for effective elephant protection in Africa's 500,000 km^2 of state parks and reserves (Adams and McShane 1992). Moreover, about 50% of Africa's elephants occur outside of state protected areas. These animals cannot be effectively protected through coercive law enforcement measures, regardless of the level of funding, because governments are unable to control human activity in these areas.

PROVIDING POSITIVE INCENTIVES FOR ELEPHANT CONSERVATION

The ivory-trade ban has failed to provide Africans with incentives to conserve elephants, or sufficient resources to control illegal hunting. Furthermore, sustainable conservation programs require more than a law enforcement response (Lewis et al. 1990). Rural community cooperation is indispensable since, to be effective, conservation policies must not only be ecologically sustainable and economically efficient, they must also be locally acceptable (Murphree and Cumming 1991). Bromley and Cernea (1989) emphasized that the success of project formulation and implementation depends on the inherent system of incentives and sanctions influencing the behavior of individuals who live in the local area and who depend on the natural resource in question. Key factors for successful wildlife conservation programs thus include the support of local leaders for the legal use of wildlife, and local participation in actual management efforts (Kiss 1990, Lewis et al. 1990).

While lucrative ivory prices combined with poor law enforcement has tended to intensify illegal elephant hunting, the main reasons for the decline of elephants appear to have been that the people competing with elephants for resources have generally not been able to legally derive revenues from elephants (Harland 1990); nor have they been included in management decisions affecting wildlife on their land (Murphree 1993). This hypothesis is based on observations that poaching rates seem to be a function of the proportion of local residents receiving benefits from wildlife management. For example, studies in Zambia (Lewis et al. 1990) suggested that, when wildlife benefits were limited to a few people, those residents not receiving benefits conspired to frustrate the programs success. Once enough community members received benefits, there was, however, rapid acceptance of the need to cooperate with the legal users of wildlife and poaching rates drop dramatically.

Community-based wildlife conservation programs, using the elephant as a key species, have been initiated both in Zambia and Zimbabwe. In Zambia, the Administrative Management Design Program involves communities near game reserves in the management of wildlife, and provides them with direct economic benefits through employment and community development (Lewis 1989). In the Luangwa valley a similar project generates income from the sale of "wildlife killing rights" to local people (Armstrong and Bridgland 1989). Since local poachers are highly skilled men who are admired for beating what is viewed as a repressive law enforcement system, "poaching" is one of 14 categories of killing rights. By giving wildlife management a positive image, local poachers have felt free to apply for local hunting licenses.

Courtesy of Urs Kreuter

Plate 3.3 Project CAMPFIRE is a wildlife-based community development initiative in Zimbabwe.

In Zimbabwe, the Communal Area Management Program for Indigenous Resources (CAMPFIRE), was initiated in 1982 by the Department of National Parks and Wildlife Management to devolve management authority over common property resources to communities who bear the costs of existing with them (Martin 1984, 1986). The objective was to encourage the establishment of co-operatives with territorial rights over well-defined communal resource areas. The program originally focused on wildlife management, but it has increasingly been applied to communal grazing and forest resources (Murombedzi 1991). It has generated substantial revenue, primarily from elephant safari hunting, in several communal areas. CAMPFIRE's success in encouraging rural Africans to regard wildlife as a valuable asset, and thus refrain from poaching, has reportedly varied depending on the level of involvement of local villagers in making decisions about wildlife management (Peterson 1991). Where decisions were made at the district government level, local people saw little relationship between conservation and wildlife-related benefits. However, where villagers were included in making decisions about the distribution of revenue from wildlife, there has been great support for wildlife conservation. Further details about CAMPFIRE are provided in the chapters by Peterson, Taylor and Thomas in this volume.

CONCLUSION

Under colonial rule traditional institutional arrangements in Africa were disrupted and sustainable common property regimes regressed to open access disorder. Facilitated by corrupt government officials, poachers have been able to exploit open access elephant populations at unsustainable rates. The resulting rapid decline in elephant numbers was subsequently used to justify the implementation of an international trade ban for ivory. The ban was championed mainly by Western preservationists and animal rights activists to promote their own agendas, but they provided few resources to protect elephants from poachers. There is no substantive evidence that the ivory-trade ban has been successful in eliminating demand for ivory or substantially reducing ivory prices. Elephant poaching appears to be continuing almost unabated except where the international aid for elephant conservation has been effectively channelled into law enforcement.

Repressive law enforcement, however, provides no positive incentives to local communities to protect the wildlife existing on their land. It is precisely such repressive measures under colonial and post-independence rule that has resulted in widespread African antipathy towards wildlife. The evidence strongly suggests that the best sustainable conservation strategy for African elephants is to promote them as a valuable resource which provides direct personal benefits to the people who face the cost of co-existing with them. To achieve this the full economic potential of elephants must be realizable, including revenue from ivory and other products. Moreover, local institutions, that are respected by local leaders and villagers alike and which include local residents in management and decision making processes, must be promoted in order to facilitate the future survival of Africa's elephants.

NOTE

1. Funding for the research leading to this chapter was provided by the Earhart Foundation, Ann Arbor, Michigan.

REFERENCES

Adams, J.S. and T.O. McShane, 1992.
  The Myth of Wild Africa: Conservation Without Illusion. W.W. Norton and Co., New York.
AECCG, 1989.
  African Elephant Action Plan. African Elephant Conservation Coordinating Group, WWF International, Gland Switzerland.

AECCG, 1990.
  African Elephant Action Plan (edition 4). African Elephant Conservation Coordinating Group, International Development Centre, Oxford, Britain.
Armstrong, S. and F. Bridgland, 1989.
  Elephants and the Ivory Tower. *New Scientist* 1679:37–41.
Barbier, E.B., J.C. Burgess, T.M. Swanson and D.W. Pierce, 1990.
  Elephants, Economics and Ivory. Earthscan, London.
Barnes, R., 1989.
  The Status of Elephants in the Forests of Central Africa: A Reconnaissance Survey. In: The Ivory Trade and the Future of the African Elephant. Report of the Ivory Trade Review Group to the 7th CITES Conference of the Parties, Lausanne, October 9–20, 1989. IUCN, Gland, Switzerland.
Bell, R.H.V., 1987.
  Conservation with a Human Face: Conflict Reconciliation in African Land Use Planning. In: D. Anderson and R. Grove (eds.), *Conservation in Africa, People, Policies and Practice*, pp. 79–101. Cambridge University Press, Cambridge.
Bohlen, C., 1989.
  How to Save an Elephant. Washington Post, Saturday 14 October, 1989:A22.
Bonner, R. 1993.
  At the Hand of Man: Peril and Hope for Africa's Wildlife. Alfred A. Knopf Publisher, New York.
Bromley, D.W. and M.M. Cernea, 1989.
  The Management of Common Property Natural Resources: Some Conceptual and Operational Fallacies. World Bank discussion papers: 57. The World Bank, Washington, D.C.
Cumming, D.H.M., 1989a.
  Review of Elephant Numbers and Trends: A Brief Summary Report of the Work of Iain Douglas Hamilton and Richard Barnes, In: Review of the Ivory Trade and the Future of the African Elephant, Volume 2, Ivory Trade Review Group, pp. 8–15. Prepared for the 2nd Meeting of CITES African Working Group, Gabarone, Botswana, July 4–8, 1989. IUCN, Gland, Switzerland.
Cumming, D.H.M., 1989b.
  Review of Proposals to Transfer the African Elephant (Loxodonta africana, Blumenbach) from Appendix II to Appendix I. In: Review of the Ivory Trade and the Future of the African Elephant, Volume 2, Ivory Trade Review Group. pp. 1–7. Prepared for the 2nd Meeting of CITES African Working Group, Gabarone, Botswana, July 4–8, 1989. IUCN, Gland, Switzerland.
Cumming, D.H.M. and R.F. du Toit, 1989.
  The African Elephant and Rhino Group Nyeri Meeting. *Pachyderm* 11:4–6.
Cumming, D.H.M., R.F. du Toit and S.N. Stuart, 1990.
  African Elephants and Rhinos: Status Survey and Conservation Action Plan. IUCN, Gland, Switzerland.
Douglas-Hamilton, I., 1988.
  African Elephant Population Study. GRID African Elephant Database Project — Phase 2. December, 1988. Unpublished typescript.
DNPWLM, 1989.
  International Moves to Ban the Legal Trade in Raw Ivory: The Zimbabwe Position. Unpublished Report, Department of National Parks and Wildlife Management, Harare, Zimbabwe.

Dublin, H.T. and H. Jachmann, 1992.
   The Impact of the Ivory Ban on Illegal Hunting of Elephants in Six Range States in Africa.
   WWF International Research Report, Wold Wide Fund for Nature Report, Gland,
   Switzerland.
EIA, undated.
   Under Fire: Elephants in the Front Line. Environmental Investigation Agency, Lon-
   don.
EIS, 1993.
   Traffic Denounces Trade in Protected Species. European Information Service, Europe
   Environment. March 2.
Ferrai, J., 1993.
   Customs seize $5M ivory shipment. South China Post, March 23.
Graham, A.D, 1973.
   The Gardeners of Eden. George Allen and Unwin, London.
Gresham, G., 1991.
   Elephants and Ivory. *Sports Afield* October 1991:142–143,152–155.
Hanks, J., 1979.
   The Struggle for Survival: The Elephant Problem. C. Struik Publishers, Cape Town,
   South Africa.
Harland, D., 1990.
   Jumping on the "Ban" Wagon: Efforts to Save the African Elephant. *The Fletcher Fo-
   rum* Summer 1990:284–300.
ITRG, 1989.
   *The Ivory Trade and the Future of the African Elephant. Volume 1: Summary and Conclu-
   sions.* International Development Centre, Oxford, Britain.
IUCN, 1980.
   World Conservation Strategy. IUCN, Gland, Switzerland.
IUCN, 1991.
   Caring for the World: A Strategy for Sustainable Living. IUCN/UNEP/WWF, Gland,
   Switzerland.
Kiss, A. 1990.
   Living with Wildlife: Wildlife Resource Management with Local Participation in Africa.
   World Bank Technical Paper No. 130. African Technical Department Series. The
   World bank, Washington, D.C.
Kreuter, U.P. and Simmons, R.T., 1994.
   Who Owns the Elephants? The Political Economy of Saving the African Elephant.
   In: T.L. Anderson and P.J. Hill, *Wildlife in the Marketplace.* Rowman and Littlefield
   Publishers, Inc. Savage, Maryland.
Lamb, D. 1982.
   The Africans. Random House, New York.
Leader-Williams, N. and S.D. Albon., 1988.
   Allocation of Resources for Conservation. *Nature* 336:533–535.
Lewis, D., 1989.
   Zambia's pragmatic conservation program. *Pachyderm* 12:24–26.
Lewis, D.M., A. Mwenya and G.B. Kaweche, 1990.
   African Solutions to Wildlife Problems in Africa: Insights from a Community Based
   Project in Zambia. *Unasylva* 41:11–20.
Lyster, S., 1985.
   International Wildlife Law. Grotius Publications, Cambridge.

Martin, R.B., 1984.
Communal Area Management for Indigenous Resources (Project CAMPFIRE). In: R.H.V. Bell and E. McShane-Caluzi (eds), *Conservation and Wildlife Management in Africa*, Proceedings of a U.S. Peace Corps Workshop, Kasungu National Park, Malawi, October 1984, pp. 279–295. U.S. Peace Corps, Washington, D.C.

Martin, R.B., 1986.
Communal Areas Management for Indigenous Resources. CAMPFIRE Working Document No. 1/86., Department of National Parks and Wildlife Management, Harare, Zimbabwe.

Milner-Gulland, E.J. and R. Mace, 1991.
The Impact of the Ivory Trade on the African Elephant (*Loxodonta Africana*): Population as Assessed by Data from the Trade. *Biological Conservation* 55:215–229.

Murombedzi, J. 1991.
Decentralizing Common Property Resource Management: A Case Study of the Nyaminyami District Council of Zimbabwe's Wildlife Management Program. Paper No. 30., Drylands Networks Program, IIED, London.

Murphree, M.W., 1993.
Communities as Resource Management Institutions. International Institute for Environment and Development, Gatekeeper Series No 36, London.

Murphree, M.W. and D.H.M. Cumming, 1991.
Savanna Land-use: Policy and Practice in Zimbabwe. CASS/WWF Paper 1, Centre for Applied Social Studies, University of Zimbabwe, Harare, Zimbabwe.

Myers, N. 1981.
A farewell to Africa. *International Wildlife* 11:36–46.

O'Connell, M.A. and M. Sutton, 1990.
The Effects of Trade Moratoria on International Commerce in African Elephant Ivory: A Preliminary Report. World Wildlife Fund, Washington, D.C.

Owen-Smith, N., 1988.
Megaherbivores: The Influence of Very Large Body Size on Ecology. Cambridge University Press, Cambridge.

Parker, I.S.C and A.D. Graham., 1989a.
Elephant Decline: Downward Trends in African Elephant Distribution and Numbers (Part I). *International Journal of Environmental Studies* 34:287–305.

Parker, I.S.C and A.D. Graham., 1989b.
Elephant Decline: Downward Trends in African Elephant Distribution and Numbers (Part II). *International Journal of Environmental Studies* 35:13–26.

Peterson, J.H., 1991.
CAMPFIRE: A Zimbabwean Approach to Sustainable Development and Community Empowerment Through Wildlife Utilization. Center for Applied Social Sciences, University of Zimbabwe, Harare.

Peterson, J.H., 1993.
Whales and Elephants as Cultural Symbols. *Arctic* 46:172–174.

Reuter Library, 1993.
Thai airport customs seize African elephant tusks. Bankok (dateline). January 5.

Simmons, R.T. and U.P. Kreuter, 1989a.
Herd Mentality: Banning Ivory Sales is no Way to Save the Elephant. *Policy Review* 50:46–49.

Simmons, R.T. and U.P. Kreuter, 1989b.
  Wildlife Preservation: Save an Elephant — Buy Ivory. Washington Post, October 1, 1989. p. D3.
Singer, P., 1975.
  Animal Liberation. New York Review, New York.
Stewart, G., 1992.
  The Elephant Crisis. *Petersen's Hunting* April 1992:30–34.
Sugg, I.C., 1992.
  Pachyderms in the Policy Pits. The Washington Times, March 23, 1992:E4.
Thomson, R., 1992.
  The Wildlife Game. Nyala Wildlife Publication Trust, Westville, South Africa.
Thornton, A. and D. Curry, 1991.
  To Save the Elephant: The Undercover Investigation into the Illegal Ivory Trade. Doubleday, London.
Tober, James A., 1981.
  Who Owns the Wildlife? The Political Economy of Conservation in Nineteenth Century America. Greenwood Press, London.
USFWS, 1993.
  Proposed Guidelines on African Elephant Sport-hunted Trophy Permits. *U.S. Federal Register* 58(25):7813–7814.
Western, D., 1989.
  Chairman's Report: Ivory Trade under Scrutiny. *Pachyderm* 12:2–3.
Western, D., 1993.
  The Balance of Nature. *Wildlife Conservation* March/April: 52–55.

Appendix Estimated National Elephant Population Sizes (*Source:* Cumming, 1989a).

Region & Country	1979	1981	1984	1987	1989	% Change 1979–89	Years of Census
West Africa							
Benin	900	1250	2300	2100	1700	88.9%	1981
Burkina Fasa	1700	3500	3500	3900	3500	105.9%	1977,81,83
Ghana	3500	970	1000	1100	2200	−37.1%	1979
Guinea	300	800	800	300	560	86.7%	None
Guinea Bissau	0	0	0	20	40		None
Ivory Coast	4000	4800	4800	3300	2800	−30.0%	1982
Liberia	900	2000	800	650	1000	11.1%	None
Mali	1000	780	700	600	660	−34.0%	?
Mauritania	150	40	0	20	80	−46.7%	?
Niger	1500	800	800	800	440	−70.7%	None ?
Nigeria	2300	1820	1500	3100	2500	8.7%	1976
Senegal	450	200	100	50	110	−75.6%	1982
Sierra Leone	400	500	500	250	300	−25.0%	1982
Togo	80	150	100	100	290	262.5%	1984
Subtotal	17180	17610	16900	16290	16180	−5.8%	
Central Africa							
Cameroon	16200	5000	12400	21200	26000	60.5%	None
Cen. Afr. Rep.	71000	31000	19500	19000	28000	−60.6%	1985
Chad	15000	?	2500	3100	2300	−84.7%	1985
Congo	10800	10800	59900	61000	25000	131.5%	None
Equit. Guinea	1300	?	1800	500	1800	38.5%	None
Gabon	13400	13400	48000	76000	93000	594.0%	1986
Zaire	371700	376000	523000	195000	102000	−72.6%	1987
Subtotal	499400	436200	667100	375800	278100	−44.3%	
Eastern Africa							
Ethiopia	900	?	9000	6650	8000	788.9%	None
Kenya	65000	65056	28000	35000	16000	−75.4%	1981,83,87
Rwanda	150	150	100	70	50	−66.7%	1983
Somalia	24300	24323	8600	6000	2000	−91.8%	1984
Sudan	134000	133772	32300	40000	22000	−83.6%	1980,81
Tanzania	316300	203900	216000	100000	76000	−76.0%	1980,83,85,86,88
Uganda	6000	2320	2000	3000	1600	−73.3%	1982,87
Subtotal	546650	429521	296000	190720	125650	−77.0%	
Southern Africa							
Angola	12400	12400	12400	12400	18000	45.2%	None
Botswana	20000	20000	45300	51000	67000	235.0%	1985,86,87,88
Malawi	4500	4500	2400	2400	2800	−37.8%	1984
Mozambique	54800	54800	27400	18600	17000	−69.0%	None
Namibia	2700	2300	2000	5000	5000	85.2%	Annual census
South Africa	7800	8000	8300	8200	8200	5.1%	Annual census
Zambia	150000	160000	58000	41000	45000	−70.0%	1979,84,85
Zimbabwe	30000	47000	47000	43000	49000	63.3%	Annual census
Subtotal	282200	309000	202800	181600	212000	−24.9%	
Total	1343340	1192331	1182800	764410	631930	−53.0%	

4. WESTERN CONSERVATION GROUPS AND THE IVORY BAN WAGON[1]

RAYMOND BONNER

INTRODUCTION

The battle among conservationists over the utilization of resources is almost as old as the organized conservation movement. Though 'conservation' has become an all-embracing generic term meaning all efforts to save the environment and resources, initially it meant the wise and planned use of resources — wise and planned, but use nonetheless. In the United States this was the foundation for the conservation movement and for the first conservation laws and came to be known as the 'utilitarian' approach to conservation. In an opposing camp individuals, such as John Muir, argued that nature should be left undisturbed by man for aesthetic or ethical reasons. They were called the 'preservationists.'

Thirty years ago, Julian Huxley, in his seminal articles in *The Observer* about wildlife conservation in Africa, encouraged "proper utilization" of wildlife for the benefit of local people (Huxley 1960). He described how the Waliangulu of Kenya, had been allowed to hunt elephants moving out of Tsavo National Park, and how most of the proceeds from the sale of elephant meat and ivory in the park were given to the tribal council. Huxley suggested that similar practices needed to be introduced in other areas. He thought it would be possible to implement utilization programs by replacing the "negative notion of preservation with the positive concept of conservation." His ideas of allowing Africans to continue shooting wildlife were far too radical, however, and went nowhere. If local people had been given some benefits from the wildlife, poaching of elephants and rhino may never have reached the alarming levels that it did. However, it was not until twenty years later that utilitarianism became part of the conservation orthodoxy, and even then the battle was far from over.

In 1980, the 'sustainable use' concept was finally incorporated in to the World Conservation Strategy, a landmark conservation manifesto which bears the imprimatur of World Wildlife Fund (WWF), the International Union for Conservation of Nature (IUCN), and the United Nations Environment Programme (UNEP). The authors of the document concluded that one of the principal objectives of an effective conservation policy should be "to ensure the sustainable utilization of species and ecosystems." In adopting this principle, conservationists faced the reality of the Third World. They understood that poor people were going to use the resources around them in order to survive, notwithstanding philosophical and ethical appeals by wealthy Westerners.

Therefore, people in the Third World had to be encouraged to use their resources in a sustainable way.

The most valuable resource in most African countries is wild animals. One
way to 'utilize' them is through tourism, but this is not necessarily the most
profitable or efficient method. Other forms of utilization mean that animals
will be killed: crocodiles, leopards and zebra for their skins; impala, eland,
wildebeest for their meat; lions and buffalo for hunting trophies; and elephants
for ivory. Killing animals for commercial reasons is not a pleasant thought.
However, the reality of an impoverished continent — of children covered with
flies and dying from malnutrition; and of women stooped over in fields, strug-
gling to grow enough food for their families' survival — is not pleasant either.

TO KILL OR NOT TO KILL

The question is whether man should leave nature alone, or intervene. Many
conservationists believe that culling is sometimes necessary for ecological reasons — to save trees and forests, to preserve the habitat of other species, and
even to save elephants from their own destructiveness. Sometimes there are
simply too many elephants. They spend up to sixteen hours a day eating, and
can consume up to three hundred pounds of leaves, bark and grass and fifty
gallons of water every day. A herd of elephants therefore goes through an area
like a slow tornado, snapping off branches and uprooting trees, leaving devas-
tation behind. However, the general public, understanding little about ecosystem dynamics, recoils at the idea of killing elephants.

Laissez-faire Preservation

The most intense debate about culling occurred in the late 1960s in Kenya's
Tsavo National Park. The elephant population in and around the park had
reached about 40,000, and virtually every scientist agreed that the range could
not support that number. Some conservationists and wildlife officials wanted
to cull 3,000 elephants. Others argued that nature should be allowed to take its
course. It was another skirmish in the long war between those who believe in
managing wildlife and laissez-faire preservationists. The latter prevailed. Then
the area was hit by a severe drought. A science writer who visited the park in
the early seventies described the resulting horror (Rensberger 1977). "Much of
Tsavo looks as if a war had just ended there. It is a ravaged land. The elephants shuffling about, scrounging bits of grass under fallen trees, are both
the war's orphans, and its combatants" (see also Parker and Amin 1983).

At least 9,000 elephants and several hundred rhino died of starvation, the
elephants having destroyed not only their own food supply but the rhino's
also. Though the drought was severe, many scientists claim that if elephants
had been culled, far fewer animals would have died, because competition for
the reduced forage supply would have been less. In addition, ivory and hides
from the culled elephants could have been sold and the proceeds could have

Courtesy of Stephen J. Thomas.

Plate 4.1 Progressive ring-barking of *Colophospermum mopane* woodland by elephants in Mana Pools National Park, Zimbabwe.

Courtesy of Stephen J. Thomas.

Plate 4.2 Victim of Zimbabwe's 1992 drought, elephant carcass in the Gona re Zhou National Park, Zimbabwe.

been used for conservation or for the benefit of the subsistence farmers living near the park, who were also suffering from the drought. Instead the elephant carcasses rotted in the hot sun, feasted on by hyenas and vultures. This fuelled the resentment of the local people toward parks, wildlife and conservation laws.

Human Intervention

In recent years, Zimbabwe has been the focus of arguments about killing elephants and of attacks from ivory-ban advocates. In the past two decades, its wildlife department has culled at least 44,000 elephants. Zimbabwean wildlife officials say that without this measure the elephant population would now exceed 134,000, roughly one elephant for every square mile, or the equivalent of 60,000 elephants in England and Wales. This number far exceeds the estimated carrying capacity of about 34,000 elephants in Zimbabwe's parks and protected areas; an estimate based on the requirements of other fauna and flora that the country wants to preserve in its parks.

The idea of killing elephants is anathema to animal rights proponents, who focus on saving each and every animal rather than on the larger question of saving species or habitats, which may require sacrificing individual animals. In one of his attacks on Zimbabwe, the director of the London-based Environmental Investigation Agency (EIA), one of the most active animal rights groups, wrote: "There is no scientific information that has ever been independently scrutinised which states that the 'cull' of elephants in Zimbabwe is necessary" (Thornton 1989). In fact, many scientists and reputable conservationists thought that elephant population reductions in Zimbabwe were ecologically justified. The biggest problem with the EIA's accusation, however, was the word "necessary." Necessary for what?

What if Zimbabwe decided to kill elephants not for scientific reasons but simply because people were complaining that there were too many elephants? The issue is not just numbers, but sovereignty. Does the international community have the right to tell a country what to do about its wildlife? Would New Yorkers tolerate Africans telling them how many bison there ought to be in Central Park? Conservation inherently involves value judgments. Setting aside land for a national park requires a judgment that the land will not be used for farming, livestock grazing or housing. Such judgements are increasingly being foisted on Africans by Western policy makers.

CONFLICTING OBJECTIVES IN WWF

Searching for a Policy

As one of the largest conservation organizations, WWF has played a central role in shaping international policy for conserving elephants. It is committed to the principle of sustainable utilization, and it funds programs which encourage

African communities to benefit through hunting and the sale of wild animal products. However, the organization fears that most people donate money for conservation because they want to see animals preserved, not utilized. The chairman of the board of WWF-US acknowledged the dilemma in 1991 by saying: "We're trying to bring our members along on utilization, but our development people, the fund-raisers, are very nervous because there is no question that the great majority of our membership are animal lovers and have difficulty in making the evolution to a more sophisticated understanding of conservation" (Russell Train, personal communications, May 1991). The ensuing debate over the desirability of a ban on ivory trading exposed and intensified the conflict between good conservation and fund-raising. Conservationists at WWF who knew anything about Africa opposed a ban, arguing that it would be bad conservation and bad for Africans, and they asked how WWF could endorse a ban and at the same time support the doctrine of sustainable utilization.

Six months after WWF had convened a meeting of the heads of major African wildlife departments and its own people in Lusaka in April 1988 to formulate a policy on ivory trade and elephant conservation generally, it had still not done so. The organization was trying to balance its commitment to sustainable utilization and its knowledge that, in some areas, reducing elephant numbers was ecologically necessary with its scramble for members and donations. It could have tried to explain the importance of sustainable use and culling to its members, but it was difficult to explain such complex issues with catchy slogans. So instead WWF's creative energy went into crafting a policy that would not offend the public.

Pressure for an Ivory Ban

By the fall of 1988, with the West becoming more emotional about Africa's elephants, WWF was under great pressure to provide its national organizations with a policy for responding to the expanding volume of letters from members and questions from journalists. WWF-US received thousands of letters, more than it had received on any other issue, and the overwhelming majority of those who wrote wanted an ivory ban. Finally, WWF adopted a policy which opposed killing elephants "except where absolutely necessary for the conservation of the species." Several national organizations found this policy acceptable but WWF-US balked.

In a letter to the officer at WWF-UK who was coordinating the effort to develop a policy, the president of WWF-US explained: "Based on our survey research in this country, there is a major problem with any policy that condones killing elephants — when the elephant population is dwindling — in order to save them. It has too much of the flavour of what the U.S. military used to tell the American public about why they destroyed Vietnamese villages — to protect them." It was a catchy analogy, but the reality was that it might be necessary to destroy elephants in order to save them, as the experience in Tsavo had shown. The letter revealed what was really bothering the organization: "I know that U.S. animal rights groups — which have already directed considerable criticism at us for our position on the legal ivory trade — will misrepresent our stance and cloud our efforts to advise WWF members about our position."

During the entire ivory-ban controversy, WWF found its members and the general public confronted with conflicting interests. Shrill language from animal rights groups wanted a ban on ivory trading, whereas science and good conservation dictated against one. An internal WWF-US memorandum written in May 1988 captured the predicament: "Most of the (450,000 WWF-US) members are traditionally oriented towards species 'preservation', and there is little understanding of the complexities of conservation in Africa in the 1980s, particularly where wildlife utilization programs are concerned." The same battles between money and science haunted leading British conservationists. The scientists and conservationists at WWF-UK supported sustainable utilization and culling, but there were fears that the organization might lose 20 percent of its members if they were aware that their money went for such activities.

Cynics consequently claimed that conservation organizations promoted the ivory-ban issue in order to perpetuate their existence. There are, however, legitimate reasons for an organization putting a premium on recruiting members and raising money. After all, it has to raise money to stay in business, to carry out its good works even beyond the immediate cause that brings in money. Surely, however, given the brainpower and imagination of direct mail fundraisers, the means could be found to sell conservation principles like sustainable utilization to the public. It would, however, take moral courage to come up with, for example, a slogan like "Save an Elephant — Buy ivory" (Simmons and Kreuter 1989) which is what sustainable utilization is all about. In the end, the public might not understand the complexities of conservation; conservation organizations might lose members by supporting culling and sustainable utilization. However, if explaining these concepts meant losing members, at least the organizations would have remained true to conservation principles. Instead they sacrificed their principles, and abandoned most Africans as well.

AFRICA BESIEGED

Africans Against the Ban

In November 1988, wildlife officials from a dozen African countries, those with nearly all of the elephants, met in Nairobi in an effort to find a solution to elephant poaching. For three days, African wildlife officials and conservationists listened to ominous reports about the future of the African elephant. One animal population biologist, using computer modelling, predicted the "commercial" end of the elephant in East Africa within seven years, and in all of Africa by 2010 (Caughley 1988, see also Caughley et al 1990). Yet Iain Douglas-Hamilton, often referred to as "Mr. Elephant" and who was attending the meeting as a European Economic Community and WWF consultant, stated that "an all-embracing moratorium would ignore the fact that sustainable yield is a reality in parts of Southern Africa, on which a whole policy of conservation and development is based". He was referring to Botswana, South Africa and Zimbabwe, where elephant populations were "well managed and well protected," and the countries made money from the sale of ivory. Therefore, he said, it might be argued that even a moratorium of five to ten years, far from being a solution to the problem, might "endanger elephants." Once again the sentiment of African governments and most of the Western conservationists present, was against a ban on ivory trading.

Even by the beginning of 1989, not one African country was in favour of a ban nor had a single Western country or any of the major conservation groups called for one. At IUCN there was virtually no debate; opposition to a ban was unanimous. At WWF-US, the conservationists and scientists were still prevailing over the fund-raisers. The African Wildlife Foundation's (AWF) position was that people should voluntarily refrain from buying ivory, but the organization was not calling for a formal ban.

Fuelling the Call for an Ivory Ban

Despite the lack of support for an ivory-trade ban among scientists and Africans alike, the pressure from members of conservation groups grew. The shift of positions came fast, amidst heavy lobbying and behind-the-scenes manoeuvring. In the West, the outcome might be viewed as the result of democracy at work, with governments responding to public pressure. Many Africans had never seen anything quite like the way in which the pro-ban advocates got their news into the press and lobbied, and they felt powerless to counter it. They did not have the money or political experience to engage in public relations campaigns.

AWF was the first major conservation organization to actively campaign against buying ivory. On February 12, 1989, it placed a full-page advertisement in the *Sunday New York Times*. Appearing two days before Valentine's day and hoping to deter men from buying ivory jewellery for their wives and girlfriends, it declared that "Today, in America, someone will slaughter an elephant for a bracelet." The organization claimed that in the previous ten years Africa's elephant population had been "sliced" from 1,300,000 to fewer than 750,000 and that 70,000 elephants were killed each year to meet the demand for ivory. At that rate, there would be no elephants left in Africa by 1999 AWF asserted. Thus "in ten short years we could have to explain to children why there are no more elephants." The advertisement had been prepared free of charge by an international advertising conglomerate, and a shorter version appeared frequently in the *International Herald Tribune*, which donated space. AWF also placed appeals in consumer magazines and the organization received a phenomenal amount of free publicity on television.

Four days after the appearance of the AWF advertisement, a coalition of animal rights groups, including the US Humane Society, Friends of Animals, and the Animal Welfare Institute, held a press conference in Washington, D.C.. They announced that they were filing a petition with the United States Interior Department to have the elephant declared an endangered species under United States law. Elephants had been on the "threatened" list since 1977, which meant that ivory could not be imported without a permit; raising it to "endangered" status would have stopped all ivory imports.

Meanwhile, the members of Defenders of Wildlife were reading a surprising article in the March issue of *Defenders*. It was surprising because the organization was founded in 1947 to protect species and habitats in the United States, but in 1989 it took up the cause of the African elephant. The article characterized the poaching of African elephants as "genocide," not only overstating what was happening, but cheapening the word.

In April, the International Wildlife Coalition, whose purpose is to "protect wild animals from cruelty and needless killing", entered what seemed to be degenerating into a contest for the most excessive language. "African Chainsaw Massacre" was the headline of its half-page advertisement in the *New York Times*, which juxtaposed a picture of a happy elephant family with one of elephant carcasses. The organization claimed the elephant would be extinct by 1997, and it went on: "The last elephant to die will likely be a baby.... The last thing this baby will see before it dies, will be its mother being killed and mutilated with a chainsaw."

Three days later, it was time for Friends of Animals. Their immediate target was Sotheby's, which was planning to auction two large pairs of elephant tusks, worth at least $16,000 a pair. "Why auction elephant tusks in the midst of an elephant holocaust?" the organization demanded in the *New York Times*. "Hyperbole? Judge for yourself. Since 1979 alone, nearly one million elephants — three quarters of Africa's wild elephant population — have been wiped out."

ELEPHANT MONEY

Fund Raising

The appeals on behalf of Africa's elephants, were however more than just loud hyperbole; they were very successful in raising funds. There was more money to be made from elephants than from whales, dolphins, gorillas, or any other species. Elephants had become the environmental fad.

AWF was a small organization with 22,000 members when it launched its elephant campaign in 1988. In its advertisement, AWF asked readers to help stop "the slaughter" of African elephants by sending tax-deductible donations. Twelve hundred people responded, and the organization received $42,526. Within a year, its membership had nearly doubled, and AWF had become a major player in the African conservation game. Although the Defenders of Wildlife had fewer than 80,000 members, the appeal to the readers of its magazine for donations to help the elephant campaign brought in $40,000, making it the most successful fund-raising campaign in its history. The International Wildlife Coalition and its "chainsaw massacre" article further emphasized the value of the mediagenic elephant for raising money. The organization had previously concentrated on saving whales, spending more than $1.3 million. It spent only $52,000 on elephants in 1989, yet its advertisement raised more money than it cost, which is highly unusual for a public interest advertisement.

In Britain the fund-raising campaigns were even more successful. WWF-UK sent out a four-page letter noting the alarming decline of elephants in Kenya, Tanzania and Zambia. The mailing went to 50,000 people and raised £300,000. After sending out a second elephant appeal in July 1989, WWF-UK received an additional £200,000. The total was twice as much as people had sent in response to pleas to save sea turtles and gorillas.

Climbing on the Ban Wagon

With the newspaper articles, press conferences, and appeals by celebrities, the rumblings that began in early 1988 were becoming a stampede. WWF feared it would be buried. The besieged director of the WWF office in Kenya noted, "You cannot believe all the film crews that are here right now doing elephants." Meanwhile, in Washington, D.C., the chief fund-raiser for WWF-US warned that with the media coverage given to AWF and other groups, WWF was in danger of losing its leading position on elephants. Her fears were not that elephants were threatened by this hyperactivity, but that WWF was. To raise money, an organization has to show itself to be better than the others, to be doing more.

June 1, 1989 was victory day for the fund-raisers at WWF-US. At a press conference in Washington, D.C., the organization announced that it "strongly endorsed" the proposals to put the African elephant on Appendix I. WWF International went along and sent a memorandum to all WWF national organizations instructing them to "follow the line we are taking as closely as possible in order to avoid any further stories of splits in the WWF Family over the ivory issue." Split WWF certainly was, with the majority of national chapters being against a ban. Senior officers in the organization said the about-face had been forced by public pressure and media interests. It hardly speaks well for the leadership of WWF and IUCN, which was coerced into supporting the call for a ban, that they did not have the courage to stick with what they considered good conservation, but gave way to public emotion.

After the 1989 CITES meeting in Switzerland, AWF declared the vote to classify elephants as endangered as "a victory for elephants." However, many conservationists feared that the ensuing ivory-trade ban would doom elephants to extinction because hungry Africans could not afford to tolerate the damage that elephants cause without receiving some economic benefit from them.

Paying for the Ivory Ban

Despite all the attention paid to the African elephant, the most critical question was not answered: Who would pay for elephant conservation? When southern African governments stated that they needed ivory revenue for anti-poaching efforts, the United States and European Community promised to substitute the lost income. At the 1989 CITES meeting, for instance, the United States delegation reported that Congress had authorized $2 million for African elephant conservation, continent-wide. A Zimbabwe delegate retorted that this was not even sufficient to support Zimbabwe's wildlife department for two years.

Six months later, seventeen Western governments, all of which had voted in favour of the ban, met in Paris to discuss African elephant conservation. Tanzania's director of wildlife, reported to them that East African countries required $84 million for their anti-poaching programs alone. Altogether, the conference was told, African countries with elephants needed an infusion of at least $500 million to protect their elephants.

The donors issued a declaration "recognizing the importance of the African elephant both as a key species in African ecosystems and the symbol of the maintenance of the continent's biological diversity, . . . (and) the global importance of this species as part of the world heritage." However, when it came to providing money to match this lofty rhetoric, to preserve this world heritage, the response was pathetic. The United States offered $2.5 million, France about $1 million, and Britain said it had already budgeted £3 million for conservation in Africa. If the African countries wanted more for conservation, the donors said, it would have to be in lieu of money for other development projects. In other words, the burden to sponsor conservation was placed on Africans.

Perez Olindo, a former director of Kenya's wildlife department, was bitter about the West's response, noting that African countries had set aside large areas for parks, land that would be economically more valuable as ranches and farms. He said that African governments protected wild animals, "which raid people's farms, kill their cattle, kill their next of kin. That is a very high cost being paid by the Africans. If the developed nations don't come to the aid of these developing countries, let them not moan afterwards that they didn't know at what price the African elephant could have been saved from extinction."

THE AFRICAN REALITY

Most Western proponents of the ivory ban operated as if elephants existed in some kind of human-free environment, ignoring the Africans living with them. Yet it is African people, not wealthy Westerners, who will bear the burden for preserving the continent's wildlife heritage. Without being allowed to benefit from the wild animals that destroy their crops and homes, Africans are encouraged to eliminate them. Poachers have, accordingly, taken their toll on elephants, rhino and other game. However, even if illegal hunting is brought under control, wildlife will continue to vanish because Africa is changing. Human populations are growing and Africa's desire to modernize cannot be halted; lions and elephants cannot be allowed to roam about in shopping malls.

Keeping people on the land is no solution for conservation either because smallholders may lose their entire harvest to wild animals overnight (Dublin and Rottcher 1988). "Elephants and man cannot coexist where agriculture is practised because elephants will eat and trample crops" (Moss 1988) and "Hungry African farmers . . . would gladly see all elephants eradicated" (Western and Cobb 1985). This is a reality few Western champions of Africa's wildlife contemplate, and one which starkly contrasts with the view that no elephants should be killed for any reason. The real conundrum will arise if the ivory ban actually works and elephant populations recover and grow. Simply stated, the choice facing African countries will be whether to feed their people, or to keep thousands of elephants because they are a "world heritage" which Westerners want to see.

Africans will conserve wild animals only if they are permitted to profit from the sale of wildlife products, but this concept confounds Westerners who, for decades, have been preaching tourism as Africa's salvation. For example, in his *Observer* articles in 1960, Julian Huxley wrote: "I would prophesy that the revenue to be derived from tourism in East Africa (which already runs to well over £10 million) could be certainly increased five-fold, and probably ten-fold, in the next ten years." He was right that there was big money to be made in tourism, but there are questions about the implication for parks, for conservation and for Africans. Many of Kenya's parks have become over-crowded with people, vehicles and lodges, leading to their ruination in the eyes of many visitors (Douglas-Hamilton et al. 1988). However, more troubling from a conservation perspective is that the benefits from tourism does not trickle down to local communities.

Richard Leakey, Director of the Kenya Wildlife Service, has furthermore realized that "a fundamental obstacle to conservation is the very small fraction of tourism revenue which goes to local people." Consequently, in 1990 he announced that 25 percent of park entrance fees would be shared with people living around the country's parks. He noted that these parks are "ecologically viable only because (the people living on the borders) tolerate the wildlife on their land". Underlying this statement is the fact that many species need space to roam. For example, an elephant clan's home range may exceed 1,500 square kilometres and they cannot be easily contained within a park's boundaries.

Moreover, contrary to the general perception in the West, 80 percent of Kenyan wildlife lives outside parks. Therefore, if wildlife preserves are to remain "wild" the people living around them must tolerate the wild animals moving onto their land. The future for African wildlife is in the hands of Africans, and depends on their willingness to tolerate it.

CONCLUSION

Huxley had the right ideas, but they were never implemented because too few Western conservationists cared about Africa's people. Conservation organizations, such as WWF and AWF, paid little attention to human needs until the mid-1980s when they finally began to realize that wildlife would not be saved unless they did something for the people. While both organizations have adopted the concept of sustainable utilization, they do not have the courage of their convictions. They fear attacks from animal rights activists, who champion the rights of animals but neglect the rights of people.

Common sense and sound conservation principles have not changed. What has changed is the popular and emotional appeal of elephants and wildlife, and the reluctance of conservation organizations to stand up to public emotion by providing honest explanations about the realities of living with wild animals. If Africa's wildlife is to be saved, it will not be with celebrity appeals, more firearms for anti-poaching units, or ivory bans. It will require radical policies and changes in attitudes. Westerners who contribute to conservation organizations will have to understand and accept sustainable utilization. Conservation organizations will have to stand up for their conservation principles and not be intimidated by the fund-raisers.

NOTE

1. This chapter is an extract from various parts of the book *At the Hand of Man: Peril and Hope for Africa's Wildlife* by Raymond Bonner and published by Alfred A. Knopf, New York, in 1993. The chapter has been reproduced with the kind permission of the publishers.

REFERENCES

Caughley, G., 1988.
    A Projection of Ivory Production and its Implications for the Conservation of Africa's Elephants. CSIRO Consultancy Report to CITES., Gland, Switzerland.

Caughley, G., H. Dublin and I. Parker, 1990.
  Projected Decline of the African Elephant. *Biological Conservation* 54:157–164.
Douglas-Hamilton, I., H.T. Dublin, D. Rottcher, M.A. Jama and P.V. Byrne, 1988.
  Identification Study for the Conservation and Sustainable Use of the Natural Re-
  sources in the Kenya Portion of the Mara-Serengetti Ecosystem. Unpublished report
  to the EEC.
Dublin, H. and D. Rottcher, 1988.
  Tourism and Wildlife in Kenya. Background Report for USAID/Kenya Natural Re-
  sources Support Project. USAID, Nairobi, Kenya.
Huxley, J., 1960.
  The Observer, November 13, 20, and 27, 1960.
Moss, C., 1988
  Elephant Memoirs: Thirteen Years in the Life of an Elephant Family. Elm Tree Books,
  London.
Parker, I.S.C. and M. Amin, 1983.
  Ivory Crisis. Chatto and Windus, London.
Rensberger, B., 1977.
  The Cult of the Wild. Anchor Press/Doubleday, Garden City, N.Y.
Simmons, R.T. and U.P. Kreuter, 1989.
  Wildlife preservation: Save an elephant — Buy ivory. Washington Post, October 1,
  1989. p.D3.
Thornton, A., 1989.
  The Guardian, June 26, 1989.
Western, D. and S. Cobb, 1985.
  A Review of the Ivory Trade and Policy Options. *Pachyderm* 11:11–12.1

5. SOCIAL DIVISIONS AND THE MASSACRE OF ELEPHANTS IN ZAIRE[1]

EMIZET KISANGANI

INTRODUCTION

Hardin's (1986) "tragedy of the commons" (or social dilemma) is a concept which is highly applicable to the elephant hunting grounds of Zaire. Elephants in Zaire are a non-exclusive resource characterized by open access due to unrestricted entry to the hunting grounds. This has encouraged unsustainable ivory hunting because the personal benefits of hunting exceed the personal costs.

Zaire covers 2,344,885 km² and has over 30 million people. The country is tropical, with more than 1,500 mm of rainfall per annum and broadleaf evergreen forest characterizing the Congo Basin, and seasonal rainfall and savanna with mainly broad-leaved deciduous trees typical of the surrounding uplands. Elephants are actively hunted in more than half the country, particularly in areas where the relatively poor distribution of water supplies results in seasonal migration and congregation.

The nature of the forest and the size of the elephant, has made hunters operate in groups. Group hunting arose due to the strong belief in traditional Zairian communities that the hunting ground must benefit the whole community. This belief was institutionalized through mythology and ritual as a result of the need to respect nature as the source of life. However, the introduction of new hunting technology, such as motorized vehicles and guns, and an expanding human population have provided individuals both the capability and motivation to exploit the hunting grounds for personal profit. For centuries, the BaMbuti pygmies have been effective owners of the hunting ground because of their familiarity with the forest and other tribes' fears of getting lost. The right to use the commons was determined by the ability to exclude potential competitors. With the arrival of Europeans and their technology, traditional selective exclusion became more difficult.

Another attribute of the resource is divisibility. Although the hunting ground may be divided among users, user rights to specific elephants cannot be easily defined because they migrate across regional boundaries. Nevertheless, the fact that elephants tend to live in herds near rivers, may create temporary ownership of a given herd by a hunting group living close to the area. This possibility has been increased with the introduction of guns, vehicles, and airplanes with which a given community may be able to contain a number of herds in specific locations. Traditionally, the hunting tribes had to pursue the elephants wherever they migrated. However, the elephant hunting grounds became more accessible to the technologically superior Europeans, who were able to exclude the natives.

This chapter presents the manner in which open access to Zaire's hunting grounds has emerged, and it describes the regimes that have controlled elephant hunting, especially in the northeastern regions. The first part examines hunting activities in the traditional setting. Subsequent sections investigate the processes that have resulted in the sequential shift in ownership of the hunting grounds from traditional indigenous groups to colonial settlers, and then from settlers to an ambiguous post-colonial public trust.

TRADITIONAL, PRE-COLONIAL ERA

Until the European colonization of the Congo in the 18th century, the natives of Zaire hunted elephants primarily for food, with one elephant furnishing sufficient meat for an entire village. Ivory had only limited use and exchange value. It was used by the BaMbuti pygmies as a hammer, although they traded it for metal and agricultural crops. Since the 12th century, they had also used it to buy protection from their powerful Bantu and Sudanic agricultural and pastoral neighbors (Turnbull 1965). This relationship resulted in the adoption of the "patron" Bantu language by the BaMbuti, but did not influence other aspects of their lives or livelihoods.

Decision-Making Structures

As hunters, the BaMbuti may be divided into net-hunters and archers. This technological distinction has shaped their social, economic, and political organization with respect to the management of the hunting grounds. The net-hunting bands were larger, consisting of 15 to 30 families, since the use of nets to capture elephants required more cooperation, whereas the archer hunting bands ranged from 7 to 15 families. In the net-hunting groups, women and children were usually associated with hunting activities, whereas in the archer bands they specialized in gathering operations. The lack of specialization in net-hunting technique was associated with a social system that was more egalitarian than that of the archers (Turnbull 1965). Mutual trust and cooperation, enhanced by the small size of the hunting bands, sustained intra-group cohesion and harmony.

Obligation to the community was balanced against self-interest, and the elders were respected as the source of wisdom and hence authority. However, since individual skill in the archer bands was a sign of maturity, a skillful young archer could gain a voice in the council of elders in vetoing a decision or in conflict resolution. Another mechanism for resolving conflicts was a system of "flux" that occurred at either end of an arbitrarily determined period called the "honey season." This involved the constant changeover of people between local groups and the frequent shifts of camps (Turnbull 1968). The honey season was the period of plenty and a time to dissociate antagonistic members for the net-hunters, but for the archers it was a period of poor hunting, during which they hunted together and after which they split into small groups. Fluctuating social relations served, not only to resolve conflict, but also de-emphasized stability in personal relations and weakened the concepts of unilinear descent and affiliation.

In the traditional society of the BaMbuti, informal rules, arising from the technological capabilities of the hunters and from traditional beliefs, regulated exit and entry into the hunting grounds. For example, the net-hunting technique produced the surest and largest supply of meat, but it frightened the animals, while the archer's technique usually did the opposite. Therefore, the former technique necessitated frequent changes of site resulting in a limited system of inheritance and little personal ownership. The archers, by contrast, were more attached to their hunting grounds. In both cases, however, every band had access to several hundred square kilometers of forest territory, which were usually separated by natural obstacles or by the proximity of agricultural tribes (Turnbull 1965). The net-hunters and archers believed that while they held an ancestral, and therefore inalienable, right to the hunting grounds, an elephant became property once it had been killed.

The fact that the surrounding tribes feared the forest made the BaMbuti less open to external control in their exploitation of the hunting grounds for elephants. One major way the Bantu tried to exert influence was to arrange intra-pygmy marriages. Through these relationships they claimed hereditary rights over pygmy offspring through patrilineal descent (Turnbull 1965). But the BaMbuti developed devices to evade the system by changing bands during flux periods or by calling for Bantu specialists to perform magic, in which the pygmies hardly believed, to improve hunting. Calling for the magicians was a political maneuver to avoid over-commitment to trade with the Bantu (in which the pygmies provided elephant meat and ivory in exchange for metal, food crops, and protection), because they could persuade their patrons that the Bantu magic had failed and, consequently, they had very little to trade, even though their hunting efforts were in reality highly successful.

Patterns of Human Interactions

During the pre-colonial era, identifiable patterns of interaction occurred within each hunting band, between hunting bands, and between hunting tribes and their patrons.

Intraband patterns of interaction were characterized by reciprocity, which made the free-riding behavior, that is central to the "tragedy of the commons", a costly choice. Moreover, mutual cooperation resulting from consanguineous ties provided hunters with a sense of community membership, enhancing the perception of the right of others to extract a livelihood from the environment. Interband interactions similarly lacked serious conflicts. The constant flux permitted bands to exchange members, thereby resolving animosity. If joint hunting was rare, interband cooperation was encouraged by attending other bands' festivities, which were regulated by the lineage elders (Schebesta 1941).

By contrast to the close intra- and interband kinship ties, under which reciprocity was long-term and often intangible, trade relations between the Bantu patrons and the pygmies were distant and tangible. Cooperation lasted only as long as protection was provided by the patrons. In other words, the consanguineous bonds were based on affiliative obligations which superseded individual choice, whereas non-kin bonds were of varying duration and based on self-interest.

Changing Relationships

In the traditional setting, the sustainable yield of the elephants was never threatened. The hunting tribes' belief systems, primitive hunting techniques, and the system of flux, kept their consumption patterns at levels corresponding with their daily needs. The pygmy system of reciprocity was noteworthy insofar as sick or weak individuals were never deprived of their benefits as members of the community. For example, after an elephant was killed, sick and weak hunters were given their shares before active hunters received theirs (Turnbull 1965).

The trade patterns with the Bantu enabled the traditional, symbiotic system to operate efficiently through specialization. The hunting ground was left to the pygmies, who had a comparative advantage in hunting, while the arable land was used by the Bantu for agricultural purposes. Trade allowed the Bantu, unfamiliar with the forest, to avoid the risk of getting lost or being killed during a hunt, and it allowed the pygmies to avoid the high cost of settling down to a sedentary agricultural life.

The traditional patterns of exchange began to disintegrate around the second half of the 18th century when ivory began to be traded intensively. It is reported that only ivory and slaves were traded for *turkedi*, the largest handmade textiles in the 18th century (Jeannin 1947). Four attractive *turkedi* were worth one slave or four 30-kilogram tusks. The massive influx of European settlers during the 1800's resulted in two impacts on the hunting grounds and the tribes exploiting them. First, it unbalanced the composition of human population, consumption patterns, and the technology of the traditional regime. Second, an increasing demand for ivory produced a supply of elephant meat which exceeded the demand for it.

LEOPOLDIAN ERA IN ZAIRE[2]

Decision-Making Structures

Contact between Europeans and natives in the Congo basin in the late 19th century brought increasing animosity. The agricultural tribes resisted the invaders, and the hunting tribes that could locate the elephants in the forest refused to cooperate with the alien hunters. Confronted with this behavior, the

Belgians' first reaction was to break up traditional family ties and the BaMbuti-Bantu associations by introducing a coercive labor system for hunting and gathering operations, and per capita taxation. Ivory became the principal means with which the citizens could fulfill their tax obligation. This meant that hunting and gathering activities were no longer controlled by the hunting tribes. With the increase in the number of alien hunters and the disruption of social ties and obligations, native people began acting out of personal interest in order to pay their taxes and to survive.

The second European reaction in the face of occupational difficulties was Leopold II's decree of July 1, 1885, expropriating land from the natives, without compensation. This enabled the Leopoldian representatives to begin selling land at four Belgian francs per hectare. The new individual or charter-company[3] owner, received preemption rights in surrounding areas, including exclusive hunting and gathering rights (Merlier 1962). The new system was totally alien to the natives, for whom land was not a marketable commodity. To gain the natives' compliance, the Leopoldian bureaucrats appointed chiefs who were willing to cooperate with the settlers, and executed those who refused to accept the new system. Through this an increasingly powerful centralized bureaucracy emerged which implemented laws drawn up in Belgium to regulate the hunting grounds in Zaire (James 1943).

Patterns of Human Interactions

Relationships between the bureaucrats and the charter companies were cooperative while the interests of the charters were safeguarded. However, Leopold II's 1891 decree that all ivory belonged to the crown resulted in the opposition by the companies to those who tried to implement it. Meanwhile, subjugation to forced labor and coercive taxation caused natives to try to maximize individual acquisition of ivory, whereas in the past the individual had acted as part of a corporate group. As individuals become more isolated, rural discontent rose which erupted into revolts against the Leopoldian system in 1895. The revolts subsided only in 1911 after an estimated 15 million native people (more than half of the total population) had been massacred (James 1943: 305).

Consequences for Elephants

The introduction of gunpowder, sophisticated means of transportation, and a monetary economy combined with coercive taxation destroyed the equilibrium that had existed in the tribal economies. The increasing size of the European population reduced the arable land available to the natives and resulted in human encroachment into the hunting grounds. Moreover, since the settlers spoke only French, intergroup communications broke down. The introduction of a monetary economy fragmented the tribal groups and gave a more limited meaning to the concept of "family". This was because compulsory taxation and unpaid labor forced the men to work in new enterprises, leaving the subsistence agriculture to women and children.

Levi and North (1982) argue that the formation of new property rights (such as the new exploitation rights for the hunting grounds) are associated with scarcity of resources. However, elephants were plentiful and roamed almost everywhere and the European settlers have represented less than four percent of Zaire's total human population. Contrary to the claim by Levi and North (1982), the Zairian elephant example shows that scarcity may result from a change in existing property rights. In fact, the laws changing the property rights were enacted in an attempt to eliminate constraints to exploitation, such as unknown customs or uncooperative behavior, and to enrich a small group of people at the expense of the majority.

The Leopoldian regime heavily favored the minority Belgian settlers at the expense of the natives in its laws regarding the use of the hunting grounds. The new demand for ivory resulted in a massive increase in the number of elephants shot. In only 23 years (1885–1908), over 200,000 elephants were killed for the ivory trade (Fallon 1944). Leopold's II's abuse of authority and selective enforcement of the new rules by his bureaucrats created inequity through the arbitrary exclusion of natives from their land and subsequently precipitated revolts. Although the revolts were crushed, awakening metropolitan concern over the blatant inequity hastened the demise of the Congo Free State, and the annexation of the Congo by Belgium in November 1908.

POST-LEOPOLDIAN COLONIAL ERA

The annexation of the Congo by Belgium was the beginning of a new form of exploitation. Since the colony was forced to finance itself, the sale of ivory became one of the main means of supporting the economy. Although ivory was no longer used to pay taxes to Belgium, forced labor to collect ivory continued.

Decision-Making Structures

The first colonial move to manage the elephant hunting areas was the declaration of the Institut des Parcs Nationaux du Congo Belge in November 1934, 26 years after the colony was formed. This decree established the boundaries of the first national parks, in which hunting, fishing, and tree felling were prohibited. By 1940, the parks covered an area of 45,026 km^2, whereas the elephant reserves spread over 15,544 km^2.

Large fines and two-month prison sentences were imposed for killing elephants in the parks, but the punitive measures increased when hunters had no hunting licenses. Such licenses were regulated by an April 1937 decree, which stipulated the payment of $2.00 for a one-year, state-issued license, and a \$5.00 tax for every mature elephant killed outside the parks. In the 1930's the Belgians began domesticating elephants near Gangala na Bodio to facilitate farming activities. This program was used to justify the protection of elephants and to increase the per elephant tax to \$20.00 in 1956. Since the tax level exceeded the annual income of most native hunters it excluded the majority of them from hunting.

To enforce these policies, a corps of Belgian officers and native employees was created in 1937. These officers had to prevent people from killing elephants either in protected areas or during the rainy season, the calving period for elephants in the savanna zone.

Patterns of Human Interactions

The colonial regime was a coercive system ruled by three interconnected entities: the administration, which enforced the law; the capitalist economy, which commercialized ivory; and the Catholic church, which was responsible for reporting all illegal activities. The administration was a centralized bureaucracy with Belgians occupying all positions. The villagers were no longer autonomous entities, but instead were treated as second-class citizens without legal rights to the hunting ground. The coercive nature of the system and cooperation between the charter companies and the church elicited native cooperation and sharply reduced smuggling incentives.

Consequences for Elephants

National parks and hunting regulations were justified as being ecologically necessary to maintain elephant populations at carrying capacity. Culling was introduced to prevent destruction of woodlands and forests by locally overabundant elephants. In this context and in order to supply meat and ivory to national and international markets, respectively, the Belgian authorities decided to crop some of the mature elephants each year. The annual off-take was to be less than five percent of the population, an amount large enough to relieve pressure on the ecosystem but small enough not to detrimentally affect herd dynamics (Hanks 1979). From 1937 to 1959, at least 200,000 elephants (mainly animals exceeding 55 years in age and bearing at least 15 kilograms of ivory) were killed and over 3,000 tons of ivory were exported (Institut National de la Statistique 1950-1959, Jeannin 1947, Offermann 1951). Fewer elephants of reproductive age were killed since permits for the export of ivory from such animals were seldom provided by the authorities.

Although the exploitation of elephants under colonial rule was controlled and efficient, the creation of parks resulted in the exclusion of native people from their traditional lands without compensation. This forced many native men into mining and plantation agriculture, which resulted in families being split over long periods of time. Prohibition of hunting, fishing, and tree felling in parks applied to an area of nearly 4,100 km² and was ruthlessly enforced. Merlier (1962) reported that, for the sake of preserving the flora and the fauna in Albert National Park (now Virunga Park) in the Kivu region, many villages were eradicated after residents had hunted, fished or set fires in the park.

Colonial law transformed the structure of property rights in order to allow the politically dominant settlers to exploit common property resources at the expense of the native people who had previously depended on them. Native Zairians had limited access to the reserves and parks, and no access at all to the institutions that created and maintained these areas. The colonial conservation policy was designed to serve the economic and recreational interests of European settlers. However, with the declaration of independence on June 30, 1960, the Zairians regained a measure of access and political control over the parks, reserves, and hunting grounds.

POST-COLONIAL NATIONAL BUREAU FOR IVORY

The extent of elephant killings in Zaire between 1960 and 1972 is almost unknown, due to the lack of data or an elephant policy during that period. It is clear, however, that the political and economic instability following Zairian independence encouraged erratic and uncontrolled exploitation of resources. The first post-colonial elephant policy emerged in January 1973 and resulted in the establishment of the monopolistic National Bureau for Ivory (NBI), but it survived for only three years.

Decision-Making Structures

The law creating the NBI explicitly recognized the common property character of the hunting grounds. It specified that the exclusive goal of the agency was to buy ivory from native hunters and sell it on international markets. Since the law was mute about the content of colonial decrees, the elephants living in the national parks were still protected by the state. However, the hunting grounds, which were erratically exploited after independence, could not be controlled by merely creating the NBI. The Zairian government, fearing local initiatives and decision making, imposed centralized controls on local authorities. Thus, like the former colonial statutes, the law did not differentiate licensed hunters from hunting tribes. Anyone wanting to hunt elephants had to purchase a hunting license, without which possessing ivory was punishable by imprisonment and fines. This, together with the fact that the law forced hunters to sell ivory at a government regulated price to the NBI, prevented the hunting tribes from resuming their traditional lifestyle. Yet fear of the forest made licensed hunters rely on the hunting tribes for information in order to locate elephants.

Local NBI bureaucrats were supposed to act as communication channels between different communities and central government, but instead they operated out of self-interest. As Crowe (1969) pointed out: when national regulatory agencies are used as the only administrative apparatus for managing common property resources, small, highly organized groups tend to corrupt the regulatory process to their own advantage. The NBI bureaucrats thus became involved in smuggling activities and the agency became the instrument of those in power.

Patterns of Human Interaction

The fixed ivory price paid to hunters by the NBI was below the open market price and, therefore, encouraged illegal hunting and smuggling of ivory. Furthermore, because the cost of bribing customs officials or police officers was considerably less than the rewards for poaching and smuggling ivory, smugglers easily crossed the Zairian borders to trade ivory for new trucks and other goods in the neighboring Sudan.

The increase in smuggling that followed the formation of the NBI in 1973, clearly indicated that colonial decrees no longer effectively controlled hunting. The post-colonial state lacked the tripartite unity of the bureaucracy, a capitalist economy, and the Catholic Church, which had facilitated law enforcement during the colonial era. Moreover, the NBI was designed to solve regulating ivory sales rather than to manage the elephant hunting grounds. Lastly, the use of this state apparatus for personal gains led even high-ranking officials to hunting, which resulted in the politicization of the NBI.

Consequences for Elephants

The bribery and corruption associated with the NBI resulted in a massive increase in elephant hunting. Police collusion with illegal ivory traders was encouraged by the fact that their real wages fell below subsistence levels. Moreover, since hunting regulations were discriminatory and favored the higher echelons of the populace, complying with "illegal" elephant hunting became analogous to rendering coercive services during the colonial era.

State controls over ivory activities therefore became ineffective throughout the country. By 1976, more than 90% of Zairian ivory sold was reportedly smuggled out of the country (Banque du Zaire 1976) and the NBI had become an ivory marketing agency without ivory. Accordingly, the Zairian authorities opted to liberalize ivory activities by abolishing the NBI. The prey, the hunters, had become so big that they ate the predator, the NBI (d'Arge and Wilen 1974).

CONTEMPORARY AFTERMATH

Decision-Making Structures

The law abolishing the NBI was meant to induce Zairians to create small co-operative hunting groups which would have the incentive to conserve elephants as their source of income. Unfortunately, it did not clearly specify how this objective was to be achieved. Rather than fostering sustainable use, the law permitted open access to elephants living outside parks which has led to wholesale slaughter. Furthermore, since the NBI had formerly also been responsible for protecting the parks, elephants were also freely killed in the national parks.

In the absence of any agency regulating the use of the hunting ground, local authorities were left to act according to their discretionary power. In the Upper Zaire and Kivu regions, for example, this led the declaration of an arbitrary "official reproduction period" during which elephants could not be hunted. People caught with ivory during this period were jailed if bribes were not sufficiently high. Such activities led to an alarming report in 1978 regarding the status of elephants (Departement de l'Economie 1978). This resulted in the total prohibition of ivory-related activities in April 1979. However, consuming elephant meat was not outlawed, and the acquisition of meat has become the ruse by which hunters circumvent the ivory law.

Patterns of Human Interaction

Wolpe (1980) stated that, when social structures evolve through interactions between capitalist and non-capitalist modes of production in which capitalism dominates, the conservation of non-capitalist modes is justified in terms of monetary gains. This has had two consequences for Zairian elephants since the NBI was abolished. First, illegally procured ivory can be sold through official markets for large profits. Second, the coexistence of illegal and legal means of ivory production has created differential access to ivory-related revenue. People with high-ranking connections or with sufficient money to bribe officials are not severely punished when caught with ivory. By contrast, less well-connected or poorer people who break the law are subjected to punishment. In other words, while poaching and smuggling ivory are prohibited by the state, the state and the illegal activities are connected because holding government office provides access to the hunting ground.

Paradoxically, the present exploitation of elephants is well-organized because the presidential clique, highly placed government officials, and people connected to these groups dominate elephant hunting. What can be expected from this free-riding behavior of the ruling group? The tendency of special interests free-riding on national resources always results in the decline of the entire economic and political system (Olson 1982).

Consequences for Elephants

The dramatic increase in elephant hunting associated with the ivory rush after 1976 could result in the extinction of elephants in Zaire. At independence in 1960, Zaire contained approximately 150,000 elephants scattered throughout the country, but by 1976 they had disappeared in several districts (including South Kivu, Lualaba, and North Shaba) and in some parks (Epulu, Djungu, and Faradje). In addition to human predation, drought and increased demand for arable land have, in some areas, significantly reduced the former elephant range areas. Yet, herds of four to eight elephants are still found in some parts of northeastern Zaire. This capacity for survival may in part be due to the belief by some Zairian tribes that wildlife parts should not enrich people or be worn as adornment. However, the recent increased use of poison to kill elephants is threatening their reproductive capacity, since poisoning water and fruit kills not only mature elephants, but also female elephants and calves.

Finally, the decrees written since 1976 have had only one purpose: to eliminate competition for the well-organized, well-capitalized, well-connected hunters. Moreover, they have failed to specify areas within which tribes may exercise traditional hunting rights. Thus politically and economically powerful groups have used the state apparatus for excluding the less fortunate Zairians from the elephant hunting ground.

CONCLUSION

This chapter has attempted to provide a perspective on the ways in which different regimes in Zaire have tried to profit from elephants. In general, the regulations aimed at managing common property resources have discouraged cooperative behavior but have promoted individual exploitation. The politically powerful have increasingly displayed their unbridled self-interest without consequences. The outcome has been the creation of a bureaucracy as a tool, not for the execution of national or community objectives, but for the fulfillment of special interest (ethnic) loyalties. Externally imposed laws without internal support are bound to produce political and economic dysfunctionality. For Zaire's elephants this has had disastrous consequences. The only way that common property resources, such as elephants, can benefit communities, and therefore be used sustainably, is to place the unit with which the individual is most familiar — the community — at the center of the decision-making institutions which determine the use of natural resources.

NOTES

1. This chapter is an edited version of the following paper: Kisangani, E., 1986. A Social Dilemma in a Less Developed Country: The Massacre of the African Elephant in Zaire, In: Proceedings of the Conference on Common Property Resource Management, National Research Council, National Academy Press, Washington, D.C. It is reproduced with the kind permission of the National Academy Press.
2. By the General Act of Berlin signed at the conclusion of the Berlin conference held in late 1984, the European powers agreed that activities in the Congo Basin should be governed by freedom of trade and navigation and by principles of neutrality in the event of war, suppression of slave trade traffic, and improvement of the condition of indigenous people. The conference recognized Leopold II of Belgium as the sole sovereign of the new Congo Free State. He later became involved in one of the bloodiest operations ever known (James 1943).

3. For more information concerning the charter companies, see Merlier 1962: chapters 1 through 6.

REFERENCES

Banque du Zaire, 1976.
    Rapport Annuel. Kinshasa, Zaire.
Crowe, H., 1969.
    The Tragedy of the Commons Revisited. *Science* 166:1103–1107.
d'Arge, R.C. and J.E. Wilen, 1974.
    Governmental Control of Externalities, or the Predator. *Journal of Economic Issues* VII(2):353–272.
Département de l'Economie, 1978.
    Conjoncture Economique. Kinshasa, Zaire.
Fallon, F., 1944.
    L'éléphant Africain. *Memoires de l'Institut Royal Colonial Belge* 13(2):1–51.
Hanks, J., 1979.
    The Struggle for Survival: The Elephant Problem. Mayflower, New York.
Hardin, G., 1968.
    The Tragedy of the Commons. *Science* 162:1243–1248.
Institut National de la Statistique, 1950–1959.
    Commerce Extétier du Congo Belge et du Ruanda-Burundi. Brussels, Belgium.
James, S., 1943.
    South of the Congo. New York: Random House.
Jeannin, A., 1947.
    L'Eléphant d'Afrique. Payot, Paris.
Levi, M. and D.C. North, 1982.
    Towards a Property Rights Theory of Exploitation. *Politics and Society* 11:315–320.
Merlier, M., 1962.
    Le Congo: De la Colonisation Belge á l'Indépendance. Maspero, Paris:.
Offermann, P., 1951.
    Les éléphants du Congo. *Service des Eaux et Forêts, Chasse et Pêche* 9:15–95.
Olson, M., Jr., 1982.
    The Rise and Decline of Nations: Economic Growth, Stagflation and Economic Rigidities. Yale University Press, New Haven, Connecticut.
Schebesta, P., 1941.
    Die Bambuti-Pygmaen von Ituri: Die Wirtschaft der Ituri-Bambuti. *Mémoire do l'Institut Royal Colonial Belge, Section Sciences Morales et Politiques* 2(2):1–284.
Turnbull, C.M., 1965.
    The Mbuti Pygmies: An Ethnographic Survey. *Anthropological Papers of The American Museum of Natural History* 50(3):145–282.
Turnbull, C.M., 1968.
    The Importance of Flux in Two Hunting Societies. In: Lee, R.B. and DeVore, I. (eds), *Man the Hunter.* Aldine Publishing Company, Chicago.
Wolpe, H., 1980.
    Introduction. In: H. Wolpe (ed), *The Articulation of Modes of Production: Essays from Economy and Society.* Routledge and Kegan Paul, London.

6. CULTURAL PERCEPTIONS AND CONFLICTING RIGHTS TO WILDLIFE IN THE ZAMBEZI VALLEY

RICHARD HASLER

INTRODUCTION

This chapter examines vested interests and rights in wildlife as manifested at the community level in a remote part of the Zambezi Valley. It deals with the question of "bundles of rights"[1] (Maine 1894) over wildlife by focusing on a case-study in Chapoto Ward[2] (Hasler 1991, 1993a, 1993b). The Department of National Parks and Wild Life Management (DNPWLM), District Council, safari operators, wildlife committees, chief, spirit mediums, and subsistence hunters have all claimed rights of control or use over wildlife. These ambiguous rights have led to conflicts between local cultural perceptions and the perceptions of the State and private sector.

Chapoto Ward is situated in the Dande Communal Land on the south bank of the Zambezi River and is bordered by Zambia in the north and Mozambique in the East. In the west and south Chapoto Ward borders the Chewore and Dande Safari areas, which are hunting areas administered by the DNPWLM. The relatively small human population (approximately 1500) competes directly with wild animals, especially buffalo and elephant, in order to obtain a harvest from their river-side gardens and fields. Some of these people formerly lived in the Chewore and Dande Safari areas before their land was expropriated to create the reserves.

There are two main ethnic identities within Chapoto Ward. The VaChikunda are the major ethnic group. They are historically agriculturalists, fishermen and hunters who are the descendants of the slave armies which protected the Portuguese plantations in Mozambique. The VaDema were migratory foragers in the Chewore hills, but they have been forced to settle within the last fifteen years because their gathering activities were curtailed by anti-poaching units attempting to prevent elephant and rhino poaching in their former foraging areas.

During the late nineteenth century, the VaChikunda warlord, Kanyemba, controlled large portions of the Zambezi River, which was a trade route for ivory, and other commodities. Together with his bothers, Chihumbe and Nyanderu, Kanyemba was the founding ancestor of the VaChikunda in Chapoto Ward. He is said to maintain a persistent dialogue between Zimbabwe's pre-colonial past and the present by communicating through his spirit medium. Thus, despite the State's legal jurisdiction over the land, the spirit of Kanyemba is locally considered the *de facto* landowner.

This chapter juxtaposes the rights associated with the royal ancestral spirits and those of the local safari operator, the District Council and the DNPWLM. It compares the degree of control that the spirit is said to exert over wildlife and illegal hunting with the control exerted by contemporary authorities. The multiple jurisdictions (Buck 1989) exhibited in two specific elephant hunts are used to illustrate these competing rights.

THE SPIRIT AND AUTHORITY FOR WILDLIFE

Authority, the Medium and the Spirit

Three weeks before this case-study began, the long standing spirit medium, Mr. Tauro, who was possessed by the spirit of Chihumbe (brother of Kanyemba), was gored by a buffalo whilst in his field on the western side of the local Mwanzamtanda river, where wildlife is abundant. Mr. Tauro had been warned by the spirit, which possessed him, not to move his house there and not to continue his illegal hunting activities. According to a reliable informant, it was his wife who wanted to live west of the river.

It was also said that the spirit wanted him to engage in agriculture and to obey the law, and that he had previously been gashed by a buffalo as a warning from the spirit not to hunt. The conflict of aspirations led the spirit of Chihumbe to warn Tauro of his impending doom. It was after this proclamation that he was gored, while trying to protect his crops from a buffalo, and subsequently died. Metaphorically, the spirit facilitated the death of the spirit medium through the vehicle of a buffalo.

The chairman of Chapoto Ward's wildlife committee is a paternal great grandson of Kanyemba (the incumbent spirit). He presides over ritual possessions in the *Dendemaro*, a shrine in which the herbs, walking sticks, and axes of the various spirit mediums are kept. In this shrine, the spirit of Kanyemba is consulted on a wide range of matters, including natural resource management, pest problems, problem animal control and hunting.

During one seance, the spirit was asked who controls crop-raiding animals. Before the medium herself could answer, the great grandson of Kanyemba interjected that the spirit could control certain animals, such as warthog and bush buck, but not larger animals, notably buffalo and elephant. However, controls emanating from the shrine appear to be situational and responsive to prevailing political realities. If the spirit overtly claims control over large mammals which are valued for game viewing and safari hunting (such as elephant, rhino and buffalo), it could find itself in conflict with the legal wildlife authorities because jurisdiction over wild animals is vested in the State[3]. Furthermore, by making sweeping claims about its power over animals the spirit could discredit itself.

Belief in the Power of the Spirit

There is significant evidence that the spirit is perceived to have control over a wide array of wild animals. For example, employees of big-game safari hunters reportedly consulted the spirit medium prior to successful hunts and gave meat to the spirit medium and to the chief in appreciation after the hunt. Local illegal hunters also consult the spirit medium for herbs to protect them and facilitate hunting. Knowing the outcome of hunts, the spirit denounces hunters who do not thank it for successes or do not supply the parts of the carcass designated for the spirit.

The spirit is also perceived to control pests and problem animals. For example at the end of 1989, a celebration was held to honor and thank the spirit of Kanyemba for having fulfilled promises to control pests (locusts, birds, and rats) and crop-raiding animals, especially baboons. The ritual for controlling pests and crop-raiding animals usually involves selecting magical herbs to make concoctions which are placed strategically where two roads or paths cross. However, many people accept that the ritual may fail to purge pests or problem animals if the community disobeys the spirit's laws or neglects its wishes.

The Chapoto community is, however, split over the professed belief in the offices of the spirit medium. The most significant impact is that of the local Faith Apostolic Mission Church, which fails to separate cultural and theological issues and accordingly condemns belief in the spirit and its medium. Despite this condemnation many of its members do honor the ancestral spirits in times of adversity. The Catholic Church does not condemn belief in such spirits and lets the individual decide whether attendance at the shrine is evil or not. As the local parish priest said, it is probably more evil to watch the television show "Dallas" than to attend the sessions at the shrine.

The institution of the spirit medium is a powerful cultural phenomenon (Garbett 1969, Lan 1985). As Schoffeleers (1979) pointed out, the spirits of the land are the guardians of the land and its living resources. Through Kanyemba, requests for vital rain are sent up the ladder of history to the original owners of the land. The spirit is the protector of the people, but may also punish people by withdrawing this protection. It is thought that if people do not attend to the spirit or obey his commands, all kinds of imbalances can occur. Lions may wage war on the people (which occurred in the 1940's in Chapoto Ward), the rains may not fall, honey may not be found in the forest, wild animals which are usually in abundance may disappear, or people may be plagued by crop-destroying animals. From many people's point of view it is therefore extremely important that correct attention be paid to the spirit. Many also perceive that the appropriate local authority for wildlife is the spirit whose voice may be heard through the spirit medium.

Cultural Symbolism and Control of Wild Animals

Problem Animal Control has historically referred to the activities of the DNPWLM in exterminating animals which are reported to be threatening life or property. However, as has been shown, there are cultural mechanisms for dealing with problem animals. When discussing problem animals, hunting control, and crop-raiding it is important to recognize the cultural distinction between charmed spirit animals and forest animals.

These categories are established by common sense and by referring to the possessed spirit medium. Spirit animals belong to the cultural order and often receive ritual treatment. Most wild animals are, however, forest animals and become spirit animals only when they exhibit unusual behavioral qualities which set them apart from the purely biological manifestations of nature. Informal groups of people often discuss the happenings involving "unusual" animal behavior, though those church-goers who do not attend the sessions at the ancestral shrine, claim that such happenings are perfectly normal.

Animals that behave in a manner that is uncharacteristic for their kind include: aggressive guinea-fowl, baboons that scratch themselves like dogs, and elephants that inveterately raid granaries and come close to human habitation. Explanations for such anomalous behavior are cast in terms of cultural entities and forces. For example, the guinea-fowl and baboons were reportedly sent by Kanyemba's spirit to castigate those people who had disobeyed his decree to observe Friday as *Chisi* (day of rest), while the elephant was inhabited by a troubled ancestral spirit. These explanations were based on observations that a baboon, scratching himself like a dog, sat in front of a church-goers home; guinea-fowls attacked church-goers who ignored the Friday decree; and the elephant no longer raided people's granaries after ritual porridge, made out of grain and herbs, was placed at the cross roads.

Certain animals have particular cultural significance. For example, lions that bear particular markings are said to be the physical manifestations of the royal ancestral spirits. The Lion of Kanyemba reportedly has spots like a leopard. Lions which are unusually aggressive or passive, or leave a kill uneaten near human habitation, or have other unlikely interactions with people are also reported to have spiritual significance. For example, upon being consulted about an incident in which a lion reportedly took the blanket off a sleeping man, the spirit medium ascertained that the lion had been sent by a royal ancestor from a neighboring area to find out if the use of witchcraft or herbs was a factor in the dispute between two individuals. Hyenas and crocodiles also have great spiritual significance, both being the agents of witchcraft and malevolence.

Perceptions of who controls animals are inexorably linked to existing beliefs and cultural practices. These beliefs change as the institution of spirit possession waxes and wanes in response to social currents. The issue of who controls animals is ambiguous, since the spirit has not claimed *de jure* control but has exhibited *de facto* control in the minds of many people. This represents a mechanism for controlling access to animals, a mechanism that has been ig-nored under centralized natural resource management. As the government ex-tends its policy of transferring responsibility for wildlife management to local communities, the possessed spirit medium may become more open in making decisions about wildlife resources and sanctioning access to animals.

BUNDLES OF HERBS AND THE RIGHT TO HUNT

The Role of Herbs in Hunting

Knowledge and use of "herbs" with magical qualities, are considered ex-tremely important to successful hunters. Herbs usually consist of pieces of root or foliage but they can also include animal products, such as lion dung, or other magical items. An important political and economic aspect of herbs is that not everyone is endowed with them. Renown hunters are said to be suc-cessful because they have access to herbs. From a local cultural perspective hunting and herbs are in fact synonymous since good hunters take herbs. Po-litical power is similarly ascribed to herbs, since Kanyemba himself is reported to have taken them. Herbs or knowledge of them can be acquired from the possessed spirit medium. They can also be obtained from a herbal specialist who may share knowledge with a close acquaintance or who will exchange in-formation for reward. Finally, they can be gained through an apprenticeship with a close patrilineal relative, or in a dream, in which a family spirit or *sahwira* (bond friendship) provides information about which herbs to use, where to find them and where to hunt or snare animals.

The main point is that the magical qualities of herbs facilitates successful hunting and gives hunters the courage and skill to hunt. From a cultural per-spective, the right to hunt is therefore normatively governed by broader indig-enous knowledge about the environment. However, although culturally appro-priate, such rights are illegal. The State has appointed the District Council as the "appropriate authority" for wildlife in the area. In consultation with the DNPWLM, and the safari operator, it has the legal right to decide what will be hunted and who will hunt.

The Rights of the Safari Operator and Client

The safari operator leases a hunting concession, which includes Chapoto Ward and the Dande Safari area, from the district council. The operator's use of wildlife is set by the bundle of rights associated with the concession. This in-cludes the rights of the DNPWLM, which has ultimate authority over wildlife and sets the hunting quotas; the District Council, which is the local govern-ment authority for land use and negotiates the conditions of the lease; and the client, who pays for the right to hunt and retains the trophy.

The client hunter obtains the right of ownership for the animal he has shot by paying a trophy fee, which can be as much as US$7,500 for an elephant. This money is paid by the safari operator to the District Council, which is supposed to use the money for community development projects or household dividends in the wards where it was generated. At the time of the research, however, the Council had not fulfilled all its obligations to include the ward in decision making and benefits.

As manager of the concession area, the safari operator's assistant has authority to control poaching activities, which has led to aggravating harassment and detention of people, mainly the VaDema. He is also responsible for supervising the skinning, removal of tusks, trunk, feet and ears, which are usually kept as trophies by the client. At the time of the research, the client decided what should be done with the meat. The meat of large animals such as elephant, hippo and buffalo, was often given to the local people since it was often impossible to remove the carcasses from the site of the kill before they rot. This practice has produced further conflicts between cultural and legal rights, as demonstrated in the following case study.

A PROBLEM OF OWNERSHIP AND USE

Introducing the Owners

On the afternoon of 4 April 1990, the trainee safari operator working in the concession area returned to Headman Mugonapanja's home after one of his two German clients had killed an elephant. The kill had occurred in the southwest of the ward, approximately nine kilometers away at the foot of the massif which forms the boundary with the Chewore Safari Area. Soon the clients arrived, one bearing an elephants tail like a fly swish.

They sat outside Mr. Mugonapanja's hut in deck chairs, discussing the problem of poaching. The main client, the one who had shot the elephant, agreed that it was a good idea that his money should be returned to the local people. However he did not believe that these people could manage natural resources, nor that a proposed electric fence would effectively separate agricultural areas from wildlife areas. Having shot 39 elephants in Zimbabwe over the previous eight years, he knew that elephant movements could not be controlled by an electric fence. In the middle of the conversation, Mr. Mugonapanja joined the group seated in the shade of his main hut, and greeted everyone by politely clapping his hands in traditional style and asking how they were. The client was introduced to Mr. Mugonapanja as the hut's owner but he did not respond or greet him. It was later discovered that the client owned an international construction business which employed 15,000 people in Germany alone.

Dismembering the Carcass

By mid-morning the next day local people, who had gathered to procure some scarce meat, together with the safari operator's skinners and trackers were cutting off the ears, trunk, feet skin and tusks. This exercise was conducted in a comparatively orderly manner since these parts belonged to the client. Once the elephant had been skinned on the uppermost side, the skinners started cutting up pieces of the tender and sought-after meat from the jowls and cheeks of the elephant.

While the safari operator's assistant was putting this meat in the Land Rover, one of the VaChikunda surreptitiously started cutting meat for himself and threw it back to his wife who quickly put it in a sack. At first only a few others started helping themselves, but they were told by the safari workers to refrain. However, when they saw that the best meat continued to be taken by the safari operator's personnel, about ten men started hacking at the meat on the head. Immediately, the entire carcass was covered with over a hundred men wielding knives and axes, tearing and cutting chunks of bloody flesh and throwing them back to women. Those who did not have access to the carcass jostled those who did, while those who were not cutting held their knives up so as not to stab someone, but some people did get cut. Blood and pieces of flesh were splattered on people's clothing, on their faces and in their hair. At times the carcass would wheeze like a punctured hot air balloon, but the noxious stench from its innards did not distract people from their tasks. When most of the meat had been cut from the flank, people climbed onto the carcass and men stood with their trousers drenched in the contents of the elephants stomach. Pieces of the innards which smelled particularly offensive seemed to be sought after.

At times the safari skinners brandished branches to control the mob and to recover choice meat, skin, and the bullets used to kill the elephant. Once most of the meat had been removed from one side, they placed chains around the carcass and turned it over with the aid of the Land Rover. Skinning was carried out, whilst safari personnel brandished branches to keep people back. The clients arrived soon after the innards had been reached. They were visibly amused, even enthralled at the sight of the orgy of flesh cutting, but they did want to take a picture of the carcass without any people on it. However, even shots fired in the air by the hunter apprentice failed to drive the people off sufficiently long and the client had to take pictures of masses of people wielding knives. After refreshments the clients and safari personnel stood and watched the butchering for a long time, and left the scene only after most of the meat had been taken and the bones were chopped open to extract the marrow.

Some people had walked up to sixteen kilometers in the hope of procuring meat. Though the kill was situated within five kilometers of VaDema households, there were fewer VaDema in the fray than VaChikunda. The VaDema did not seem to take the kill seriously and their headman sent only one of his young sons to get meat. Moreover, the VaDema seemed to be restricted to the fringes of the carcass and generally procured substantially less meat than the VaChikunda cutters. The strategy among successful cutters was family co-operation, with the most aggressive men maintaining positions of access to the carcass, and throwing chunks of meat back to other family members. Those who lacked this system, notably the VaDema, were unlikely to keep much of the meat they cut since meat lying around was quickly picked up by others. Old people and widows were also disadvantaged, and few were successful in obtaining much meat. The most successful cutters were also the more politically influential community members. This meant that most of the meat went to a small number of socially and politically significant VaChikunda households.

One of the meat-laden VaChikunda leaders, who is knowledgeable about local history, was asked whether meat had been distributed in the same manner by Kanyemba. The answer was a definite no. He stated that Kanyemba himself was less interested in the meat than in the tusks, though he usually also kept choice parts, such as the trunk, and valuable items, such as the skin. The flank of the elephant that fell on to the ground was reportedly given to the chief and the spirits, since they were the owners of the land, and was subsequently distributed to kin and to neighboring headmen. The rest of the meat was cut into portions by designated cutters and distributed to those people present, while some portions were retained for absent headmen. However, this account was disputed by other informants, who claimed that Kanyemba was not interested in the meat remaining after the flank had been given to the chief. The account may have been a glamorization of the distant past, based on more recent experiences. For example, in the 1970's, a white hunter used to distribute meat to both the chief and the spirit mediums and he designated cutters to distribute equal bundles of meat for all families present at the kill.

INVENTING A PROBLEM ANIMAL

The Disciplined Butchery

The events surrounding the killing of a problem elephant at Mr. Charuma's house in August 1990 contrast dramatically with the April elephant hunt described above, because the members of the community viewed the meat as belonging to them. One early morning gunfire thundered through the Mwanzamtanda valley and a wounded elephant lumbered towards inhabited areas, driven by a DNPWLM Problem Animal Control (PAC) team at the request of the villagers. Children ran towards the commotion, bullets ricocheted through the dust and the elephant finally crashed to the ground dangerously near Mr. Charuma's huts and granary. The prospect of obtaining meat from

Courtesy of Urs P. Kreuter

Plate 6.1 Elephants provide large amounts of meat, critical to protein deficient subsistence farmers.

the carcass outweighed concerns that the villagers may have had about the safety of their lives and property.

After the elephant collapsed, the secretary of the ward wildlife committee, who had escorted the PAC unit, nominated my research assistant, who had been recently elected to the wildlife committee, to supervise butchering of the animal. The DNPWLM official, who had killed the animal, went to report the incident to the police and later returned to collect the tusks for storage by the DNPWLM. The research assistant appointed those who had gathered to the various tasks, and other members of the committee, including the chairman, involved themselves in the activities. The meat was cut up into medium sized chunks and laid on leafy branches and portions of the meat were set aside for the chief and the spirit medium. Compared to the April kill, everything was conducted in a very orderly and organized manner.

While the carcass was being cut up the DNPWLM vehicle and the ward tractor and trailer arrived, the latter bearing a number of people from as far as twenty kilometers away. The DNPWLM vehicle was loaded up with some of the best meat, which was to be sold at Angwa Bridge and the nearby administrative center at Mashumbi Pools. A member of the wildlife committee accompanied it to ensure that the meat was sold in good faith. Next the ward tractor and trailer were loaded up with meat for sale at other centers in the ward. Workers who had helped in cutting up the carcass, were given portions of meat or intestines for their efforts. By the time the intestines had been reached, it became difficult to effectively cut up large portions of meat. What remained was handed over to those who had been waiting for an opportunity to cut meat for themselves. Although a free for all ensued, there was not quite the same commotion as had occurred in April because many people had accompanied the good meat loaded on the tractor. The next day nothing was left of the huge carcass except a few broken bones, flattened grass and soil drenched in the elephant's liquids.

Motive for the Killing

Surprisingly, the elephant was shot after the ward wildlife committee had lodged a formal complaint to the DNPWLM about crop-raiding buffaloes. The PAC unit at Mashumbi Pools arrived in Chapoto Ward in response to this complaint and had spent some time tracking buffaloes but were unable to get a clear shot. Near dawn they came across the spoor of a large elephant and tracked it down. Members of the wildlife committee encouraged the PAC team to kill it, not because it had damaged any crops or property, but because they realized that the sale of "problem animal" meat could earn the ward substantial money. Not long before, meat from a problem buffalo had been sold for over Z$400.

The elephant was a trophy bull and would have been worth Z$7,500 to the district council if it had been shot by one of the safari operator's clients. The research assistant tried to explain that the ward was loosing money by killing the elephant as a problem animal, but the other wildlife committee members disagreed since the District Council had not involved the community in decision-making nor direct monetary benefits derived from the trophy hunting of wildlife. They reckoned that the ward stood a greater chance of receiving some benefits by killing the elephant as a problem animal.

However, the meat was sold mainly in centers outside the ward for Z$2 per kilogram, a price which few people in the ward could afford, least of all those who needed meat the most. The distribution of meat within the ward was therefore sparse, and even the people who participated in butchering did not benefit significantly. The total revenue from the sale of meat was about Z$1,000 (US$250), which was retained by the District Council, allegedly to help finance community projects in the future. Although the community members had viewed the meat as belonging to them, they received little immediate reward for their efforts.

However, there is more at stake than the material benefits; there is also a symbolic value associated with the kill (Berkes 1989, McCay and Acheson 1987). A large portion of time in Chapoto Ward is spent trying to prevent wild animals from destroying the hard work in the fields. People are constantly aware of the wildlife menace but will be arrested for poaching if they deal with the problem. Thus, aside from the hope of obtaining meat or other benefits, there is a sense of catharsis in the death of crop-raiding animals. For example, when the elephant fell to the ground one man, who had been with the hunters, jumped onto the carcass with glee and stood there like a conqueror. As one informant put it, "it is a revenge for what they or their families have done to us".

DISCUSSION AND CONCLUSION

Families, relatives, neighbors and friends who share the proceeds of illegal hunting, do so secretly. Hunters from these groups, who may team up to provide meat to people loyal to them, compete for wildlife resources with people who are not loyal, who may report them to wildlife authorities or claim a portion of the meat for not doing so. Thus illegal hunting is conducted secretly, and wildlife becomes "common property" only when it is shared clandestinely among people belonging to a closed circle. Legally, the hunters are taking a resource which belongs to the State, without concern for its management, but in cultural terms, they are acting as successful hunters. They have acquired knowledge about hunting and herbs through dreams, or through experience as apprentices to their fathers or grandfathers. Provided they obey certain restrictions laid down by the spirit medium, and provided they keep the secret, they will be left alone to enjoy the fruits of their labors.

In the colonial past, attempts were made to control hunting. For example, the chief is reported to have provided guns to a few well known hunters who would then go on expeditions in the Chewore Safari Area. After a kill, meat was set aside for the owner of the gun, the hunter who killed the animal, and the people who attended the hunter (usually their young sons or close relatives). The rest of the meat was distributed among the remaining hunters. The system was, however, abused and the chief did not always get the choice or largest share to which he was culturally entitled as the living descendant of the royal ancestral spirits. Local illegal hunting subsequently emerged as an individual, competitive and secret activity because there was, and still is, ambiguity over control of the resource, as a result of conflicting bundles of user and ownership rights. This may explain the manner in which meat was distributed after the two legal elephant hunts.

In the April hunt, the client had legal right of usufruct over part of the wildlife resource, subject to the guidance of the safari operator and the authority of the District Council and the DNPWLM. However, according to customary understanding, the spirits of the land facilitate an abundance of wildlife in the area and allowed the kill to occur. The spirit has a significant but presently ambiguous control over large animals and illegal hunters in the area. Such control is ill-defined because, if the spirit openly claimed greater control it would conflict with the legal governmental control.

Since the safari operator's employees were responsible for retrieving the tusks, skin, and other wanted body parts for the client, they supervised the proceedings after the kill, but they had no independent authority over owner-ship. The local people, competing with each other as illegal hunters for a lim-ited supply of elephant meat, tried to obtain as much meat as possible for themselves at the expense of others. Technically, such behavior is consistent with the tragedy of the commons, in which there is said to be no effective con-trol over the right of access. But was the carcass really a common property re-source?

A complex hierarchy of agencies legally control the user rights over wildlife during the various stages of resource "management". When elephants destroy people's fields in Chapoto Ward, the controlling agents are the DNPWLM, the District Council, the safari operator and, less obviously, the illegal hunters. Once an elephant is being tracked, the agent in control is the safari operator. The client becomes the owner and therefore controlling agent once the el-ephant has been shot. Once the carcass has been given to the local people, they become the owner/agent, under the supervision of safari personnel, but only after the meat has been cut is ownership established. In all of these stages the spirits of the land are said to ultimately determine the effectiveness of the agents or owners by influencing the behavior of animals.

The coincidence of the activities of controlling agents can result in tension between them due to conflicting interests. For example, local people were gen-erally not happy with the way the elephant meat was distributed after the first kill. Their explanation for the inequitable distribution of meat was that neither they nor the wildlife committee had the authority to allocate the meat because the carcass belonged to the client. However, members of the wildlife commit-tee claimed that they would be able to control distribution if given the author-ity to do so, which is precisely why the organization of the second kill was conducted in a more orderly fashion.

The unsatisfactory distribution of meat at the first kill resulted from the am-biguous nature of user rights, ownership and control of wildlife in Zimbabwe. People competed for rights of access because no clearly defined internal au-thority was recognized and external state and private sector rights impinged on local claims. A more equitable distribution mechanism emerged after the second hunt because the local people felt that they owned the animal, enabling their representatives to allocate the meat.

Apart from underlining the various rights of access to elephants and other wildlife resources, the chapter has illustrated how the institutional develop-ment for community-level resource use is affected by legal and perceived con-trol mechanisms for the multijurisdictional resource.

NOTES

1. Bundles of rights refers to nested user rights. For example, in many African cultures wives often have rights to use land owned by their husbands.

2. The administrative and political system in rural areas of Zimbabwe consists of districts, which are comprised of wards, which in turn are made up of villages. Each village is led by a village development committee, each ward by a ward development committee and each district by a district council headed by a district administration and comprised of the ward committee chairmen.

3. The State, through the Department of National Parks and Wildlife Management, has been the legal custodian for wildlife since colonial times. However, the passage of the 1975 Wildlife Act and its subsequent modification in 1982, aimed at decentralizing control over wildlife, has allowed District Councils to obtain "appropriate authority" for managing wildlife in the communal lands they administer. Appropriate authority status is granted by the DNPWLM and empowers the Council to manage wildlife resources on behalf of the wards and villages which it represents.

REFERENCES

Buck, S.J., 1989.
Multi-Jurisdictional Resources: Testing a Typology for Problem Structuring. In: F. Berkes (ed.), *Common Property Resources: Ecology and Community based Sustainable Development*. Belhaven Press, London.

Berkes, F. (ed.), 1989.
Common Property Resources: Ecology and Community-Based Sustainable Development. Belhaven Press, London.

Garbett, K., 1969.
Spirit Mediums and Mediators in Valley Korekore Society. In: J. Beattie and J. Middleton (eds), *Spirit Mediumship and Society in Africa*. Routledge and Keegand Paul. pp 104–127.

Hasler, R., 1993a.
Political Ecologies of Scale and The Multi-tiered Co-Management of Zimbabwean Wildlife Resources under CAMPFIRE. Centre for Applied Social Science, University of Zimbabwe, Harare.

Hasler, R., 1993b.
The Cultural and Political Dynamics of Zimbabwean Wildlife Resource Use in the Zambezi Valley: A Case Study of Chapoto Ward. Phd dissertation, Michigan State University, East Lansing, Michigan.

Hasler, R., 1991.
The Political and Socio-Economic Dynamics of Natural Resource Management: CAMPFIRE in Chapoto ward (1989–1991). Centre for Applied Social Science, University of Zimbabwe, Harare.

Lan, D., 1985.
Guns and Rain Guerrillas and Spirit Mediums in Zimbabwe. University of California Press, Berkeley.

Maine, H., 1884.
Ancient Law (10th Edition). Henry Holt, New York.

McCay, B.J. and J.M. Acheson (eds), 1987.
The Question of the Commons — The Culture and Ecology of Communal Resources. University of Arizona Press, Tucson, Arizona.

Schoffeleers, J. M. (ed), 1979.
Guardians of the Land. Mambo Press, Gweru.

7. SUSTAINABLE WILDLIFE USE FOR COMMUNITY DEVELOPMENT IN ZIMBABWE

JOHN H. PETERSON, JR.
(Compiled by Urs P. Kreuter)

PREFACE

This chapter is an abbreviation of parts of an extensive paper written by John Peterson in 1991 while on sabbatical leave at the Center for Applied Social Sciences, University of Zimbabwe. The paper is entitled "*CAMPFIRE: A Zimbabwean Approach to Sustainable Development and Community Empowerment through Wildlife Utilization*" (Peterson 1991a) and is a thorough and comprehensive overview of Zimbabwe's Communal Areas Program for Indigenous Resources (CAMPFIRE). This chapter serves as an introduction to the two subsequent papers which address various aspects of CAMPFIRE in greater detail. For this reason it does not deal exclusively with elephants but examines the broader wildlife conservation issues on communal lands, which support much of Africa's wildlife, including elephants.

The origins and principles of CAMPFIRE are described and the applications of these principles in the Mahenye Ward and the Beitbridge District are detailed. These two areas were selected because they have played a central role in the evolution of the CAMPFIRE concept and because within each elephants have been the source of much of the wildlife revenue accruing to local communities. The Mahenye wildlife program preceded the formal inception of CAMPFIRE while the Beitbridge program has arguably applied the programs concepts more effectively than any other wildlife program in Zimbabwe's communal lands. The two areas are alike in that both have a limited wildlife resource base, and both have responded by emphasizing the development of the local communities producing the wildlife revenue. Figure 7.1 presents the location of 11 CAMPFIRE programs existing in Zimbabwe.

INTRODUCTION

The World Commission on Environment and Development (1987:31) in writing about sub-Saharan Africa noted: "No other region more tragically suffers the vicious cycle of poverty leading to environmental degradation, which leads in turn to even greater poverty." In the semi-arid lands of south and east Africa, a partial solution is beginning to emerge to the problems of under-development and environmental degradation due to cultivation of marginal lands and overstocking by cattle.

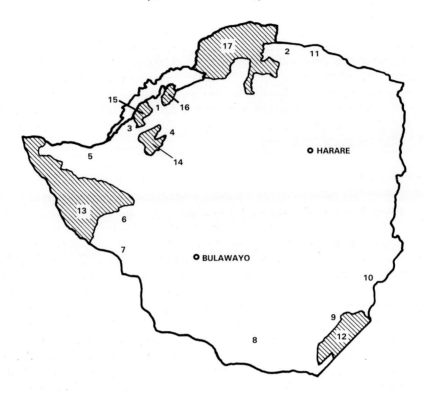

Districts with appropriate authority status for wildlife management

1 — Nyaminyami
2 — Guruve
3 — Binga
4 — Gokwe
5 — Hwange
6 — Tsholotsho
7 — Bulilima Mangwe
8 — Beitbridge
9 — Gaza Khomanani
10 — Gazaland
11 — Mzarabani

Major Zimbabwean protected areas

12 — Gonarezhou National Park
13 — Hwange National Park and neighboring safari and forest areas
14 — Chizarira National Park & Chirisa Safari Area
15 — Chete Safari Area
16 — Matusadona National Park
17 — Zambezi Valley wildlife complex

Figure 7.1 Location of major CAMPFIRE areas and protected areas in Zimbabwe.

This solution is based on the concept that wildlife or wildlife in conjunction with domestic species may be the most ecologically sound, sustainable, productive form of land use in semi-arid areas. Wildlife and multispecies production schemes are difficult to implement where serious overcrowding by people and livestock has already begun to destroy the environment. However, in drought prone areas where human and cattle population have remained relatively small and wildlife has not been eradicated, such schemes may be feasible. In the past seven years CAMPFIRE has evolved in Zimbabwe as a viable wildlife-based approach to sustainable development. The program is leading the way in the application of the principles of sustainable development, community empowerment and conservation through utilization. It was conceived within the Department of National Parks and Wild Life Management (DNPWLM) in response to the growing antipathy of rural Africans towards wildlife during the colonial era.

The Colonial Legacy

In pre-colonial times, the human population was much lower than today and wildlife was accessible to most Zimbabwe communities. However, under colonial rule, wildlife became state property. National Parks and safari hunting areas were proclaimed in some of the better wildlife areas in the Zambezi Valley in the north, in the Lowveld along the Mozambique border in the southeast and the Limpopo River in the south, and the area adjacent to the Botswana border in the west. Land was expropriated from Africans living in these areas to create the preserves. With a greatly diminished land base, and denied any legal access to wildlife, community-based traditions of land and wildlife management fell into disarray. Furthermore, people on communal lands adjacent to these parks and hunting areas were subjected to uncompensated crop and livestock damage by wild animals. Thus wildlife conservation took place at the expense of rural Africans, who came to view the damages they suffered as theft by the government's animals. Illegal hunting became the response to this "theft" in an attempt to retrieve the "stolen goods".

Meanwhile, the larger wild animals of interest to sport hunters, especially elephants and buffalo, flourished in the protected areas. The establishment of these areas under colonial rule also contributed greatly to the romantic Western image of Africa as a wilderness with a plethora of wild animals for the enjoyment of white hunters and tourists. The compression of the Africans into an increasingly restricted land base to provide land for white settlers and wildlife reserves for recreation was ignored.

THE ORIGINS OF CAMPFIRE

Zimbabwe was the first African country to recognize that landowners had to be given the right to benefit from wildlife if wild animals were to survive outside of the National reserves. It accordingly adopted a policy of "conservation by utilization" on private lands. In the last decade before Independence, the Parks and Wildlife Act of 1975 transferred effective control over wildlife on private land from the state to the landowners. While the Act was initially aimed at commercial farmers and ranchers, it was amended in 1982 to contain a provision which enabled the Minister of Environment and Tourism to designate the District Councils as the "appropriate authorities" for the management of wildlife on communal lands under their jurisdiction.

It was, however, not easy to develop policies permitting communal farmers to manage wildlife for profit because African farmers have no interest in protecting wild animals which endanger them. Reversing the trend of wildlife depletions on communal lands required changing both the pattern of wildlife use and governance in communal areas. Rural people had to become involved in managing their resources. This meant that the post-Independence structures of local government and the concept of conservation-by-utilization had to be linked.

The District Councils Act, as amended in 1980, established the District Council as the local authority of government in communal lands. Communal areas were consolidated into 55 districts, each divided into wards of approximately 6000 people who are represented by an elected councillor on the District Council. These Councils have equivalent land management authority to private landowners, and they are empowered to engage in business enterprises and to expand their revenues by limited taxation and by issuing business licenses. Thus the District Councils are the lowest level of government authority for communal lands.

To increase local participation in development planning, the Prime Minister's 1984 Directive on Development established the Ward Development Committee (WADCO) and Village Development Committee (VIDCO). Each WADCO is comprised of the ward representative in the District Council, who acts as the WADCO chairman, and the elected chairmen of the VIDCOs. In sparsely populated more traditional areas, wards and villages may also coincide with traditional social groupings such as chiefdoms.

Based on devolution of communal resource management authority to this point in time, one of the more remarkable sustained community wildlife programs developed in Mahenye Ward. Community-based wildlife management has operated in this area for ten years and many of the principles now central to CAMPFIRE were pioneered there without outside technical assistance, subsidies, or models.

THE MAHENYE WARD PROTOTYPE

Deprivation of User Rights

Mahenye Ward forms the south end of Gazaland District (commonly referred to as Chipinge District) on the east side of the Save River in southeast Zimbabwe. It is inhabited by the Shangaan people who migrated to Zimbabwe from South Africa settling around the confluence of the Save and Runde rivers. The

Shangaan originally inhabited the lands of both the Gonarezhou National Park in Zimbabwe and Kruger National Park in South Africa. Traditionally they practiced some agriculture, largely cultivated by women, but their primary subsistence was from hunting and large scale fish drives on the region's major rivers.

In 1966, much of the area inhabited by the Shangaan in Zimbabwe was incorporated into the Gonarezhou National Park and the people were evicted. Mahenye Ward thus consists of a mixed community of people some of whom were originally on the east bank of the Save River and others who were evicted from the park. The recency of the loss of their traditional homeland has had a continuing negative impact on the people's attitude towards the park and its animals.

The lack of interest by the post-independence government to return Gonarezhou National Park to them, created further resentment since land redistribution had been a central issue during the preceding civil war. Government justification for not returning the land was the national need for the foreign earnings from tourists who visited the park. This created additional incentives to poach wildlife in the park. The Shangaans felt that without wild animals there would be no tourists and therefore no need for a park. Poaching thus intensified.

The resulting conflict with the park officials came to a head at a critical meeting between the Mahenye community and the Warden of Gonarezhou National Park in February 1982. The community elders pointed out that they had been evicted from the park and were forced to live by agriculture in a drought prone area were wild animals, especially elephants, constantly raided their crops. One elder commented that "If they would control their animals, we could grow our crops. There would be no poaching." A local commercial rancher, speaking in Shangaan, suggested that wild animals crossing the Save River from the park into the communal land might be considered to belong to the community much as wildlife on a private land "belonged" to the landowner. He continued that if the community could occasionally sell some animals to safari hunters and the meat and some of the hunting fees were given to individual households, the people might be more willing to tolerate animals which raided their fields.

Elephant Hunting Permitted

Learning of the developments in Mahenye, the Director of National Parks and Wildlife Management approved a one year trial and a permit was issued for two elephant hunts during the 1982 hunting season. In August, two Americans successfully hunted the two elephants. Traditional customs were adhered to in distributing the meat to over five hundred villagers who congregated after the hunt. The elephants were presented to the chief and meat was given to each villager by the sub-chiefs. This ensured equitable distribution of the meat and involved the community leaders in the inception of the wildlife program.

The distribution of the meat had an immediate effect on the incidence of poaching. While waiting for the hunts to occur, an anti-poaching team had arrested 81 people for poaching everything from fish to elephants. By contrast in June 1983, a similar unannounced anti-poaching raid resulted in only eight arrests, all of which were for illegal fishing. This provided clear evidence of the value of the wildlife-utilization program for both the park administrators and the Mahenye community.

The further developments were even more remarkable. Ngwachumeni Island, which splits the Save River, lies within Mahenye Ward along the border with the Gonarezhou National Park. The greatest movement of elephants from the park into the ward was to Ngwachumeni Island, which is covered with riverine forests and which was the home of seven villages. The commercial safari operator who had solicited the two American hunters pointed out to the Mahenye people that more wildlife would move onto the island if the people would move off it. This was discussed among the Mahenye people and all seven villages agreed to move. This was a major step towards preserving wildlife as a community benefit and was a remarkable decision for people who had lost their traditional homeland to wildlife less than 25 years before. The reduction in the poaching and the vacating of Ngwachumeni Island resulted in an increase in wildlife in the area and the hunting quota was increased from two to four elephants in 1986.

District Council Interference

While the Mahenye community received meat, they did not immediately receive the money generated by the elephant hunts. By law hunting fees were transferred to the national treasury. To retrieve this money, the Mahenye people had to submit a proposal for its use that was acceptable to the District Council, the Ministry of Local Government, Rural and Urban Development (MLGRUD), and the DNPWLM.

In the first years after hunting began, the Chipinge District Council would not approve project proposals which benefitted only Mahenye Ward. The councillors of wards without wildlife did not understand the importance of returning wildlife revenue to the producing ward. However, Chief Musikavanhu, from the northern end of the district, did support the Mahenye position saying, "My children, what we have heard is the truth. We have no claim on this money. We did not sleep in the fields to protect the crops from elephants, as the people from Mahenye did. The elephants are theirs, not ours." This resulted in the District Council giving full support to Mahenye's wildlife program after 1986 by allowing Mahenye to fund projects from its wildlife proceeds. Chipinge District Council thus became the first in Zimbabwe to commit itself to the principle that proceeds from wildlife must go to the producing wards.

In February 1987, the Chipinge District Administrator presented the first cheque from wildlife revenue to the Mahenye community. The ward Councillor observed that "Wildlife was something which we had to use, something which was part of our way of life. But wildlife had no actual value. Now wildlife has value since wildlife can be exchanged for other things." Further impetus was given to the wildlife program in Mahenye by the early linkage of the wildlife program with the need for a school. The wildlife project and the school project jointly began leading the Mahenye people into a future based on self-development under their initiative supported by their District Council. Through 1988, the Mahenye story remained one of the greatest successes of wildlife-based community development (Environmental Consultants 1990:20). However, in 1990 many members of the Chipinge District Council were replaced in national elections. The new council again saw wildlife revenue produced in Mahenye as a resource to be allocated by the District Council with little regard for the people in the producing community.

The experience of Mahenye demonstrated the great potential for communal people to curtail poaching and actively promote community development through wildlife utilization. However, the cumbersome channels through which wildlife-based development proposals and wildlife revenues had to pass was compromising positive incentives. A more efficient mechanism for managing wildlife in communal areas had to be formalized. CAMPFIRE was conceived and has evolved in response to this need.

CAMPFIRE: A NEW PARADIGM

Objective and Methods

By 1986, the Government of Zimbabwe had created new statutes to encourage active local participation by rural people in community development. However, without well defined user rights, resources in communal lands are generally overexploited. The objective of the CAMPFIRE program was "to initiate . . . long term *development, management and sustainable utilization of natural resources in the Communal Areas* . . . involving forestry, grazing, water and wildlife" (Martin 1986:iv,17). This was to be achieved in the following manner:

1) Obtain the voluntary participation of the communities in a flexible program which incorporates long term solutions to resource problems.
2) Introduce a system of group ownership with defined rights of access to natural resources for the communities resident in the target areas.
3) Provide the appropriate institutions under which resources can be legitimately managed and exploited by the resident communities for their own direct benefit;
4) Provide technical and financial assistance to communities which join the program to enable them to realize these objectives.

The CAMPFIRE document envisioned the establishment of local natural resource cooperatives (based on "a cohesive entity with common objectives", such as a district, ward or village) and of a separate agency to implement the program. The DNPWLM, however, had neither the manpower nor the resources to initiate the process. Furthermore, without district level or even ward level support, CAMPFIRE would be doomed to remain an innovative program without local community involvement. A means had to be found to return wildlife benefits to communities which were incurring the negative costs of co-existing with wild animals. Without funds for implementing CAMPFIRE implementation, the DNPWLM turned to Non-Government Organizations (NGOs) to provide the necessary technical assistance and extension services. Three local organizations eventually collaborated with the DNPWLM: the Center for Applied Social Sciences (CASS) of the University of Zimbabwe, the Multispecies Animal Production Systems Project of the World Wide Fund for Nature (WWF), and the Zimbabwe Trust. By using essentially local NGOs, Zimbabwe avoided the problem of allowing external donors to dictate conservation policy, which is a problem that has frequently affected other African countries.

CASS's role was to undertake initial socio-economic research in potential CAMPFIRE areas and to discuss the program with District Councils and local communities. Zimbabwe Trust played an extension role by providing skill training and managerial, technical and financial support (CASS/WWF/ZimTrust 1989:243). The WWF program added a critical dimension of ecological and economic analysis and research to the socio-economic and community development work of CASS and Zimbabwe Trust. WWF could identify technical means of achieving wildlife management goals, and monitor the economic feasibility of such efforts.

Adaptive Management

The CAMPFIRE document emphasized that implementation of the program would have to be flexible in order to accommodate the diverse geographic, demographic and socio-economic factors in the communal areas of Zimbabwe (Martin 1986:32). This view conformed with the DNPWLM's policy of adaptive management, a concept which recognizes that wildlife managers have to continuously make decisions without full knowledge of relevant variables (Bell 1984:5). Under this concept, policy and management evolve in response to the results of decisions made during program implementation. CAMPFIRE has accordingly evolved in response to the experience of districts and wards attempting to manage their own wildlife resources.

The first two District Councils to be granted the authority to manage wildlife on the communal land under their jurisdiction were those in Nyaminyami and Guruve, both of which occur in the Zambezi Valley. The CAMPFIRE programs in both areas have generated significant revenues from wildlife but they have also received substantial technical and financial support while community involvement in decision-making has been limited. In the Gokwe District, delays in the implementation of the program, during a period of rapid human influx from other areas, has greatly compromised wildlife-based community development. The program which has perhaps most effectively demonstrated the possibility of developing a wildlife project without major outside assistance, and which most clearly adheres to the CAMPFIRE principle of producer community self-determination, is that found in the Beitbridge District (Child and Peterson 1991).

BEITBRIDGE: CAMPFIRE EPITOMIZED

Program Development

The Beitbridge District consists of the communal lands along the Limpopo River bordering South Africa. Drought (leading to ten years of crop failures), livestock pressure, and poaching have greatly reduced wildlife populations except in the wards at either end of the District. In the eastern corner of Chipise communal lands elephants and other wildlife move in from Kruger National Park in South Africa, and in the west there is a significant wildlife population in the Tuli Block.

   Interest in a community wildlife project began when the District Council Chairman and the Senior Executive Officer attended the first meetings of the newly formed CAMPFIRE Association in 1989. Shortly afterwards the Zimbabwe Trust held a CAMPFIRE workshop for the District Councils from southern Zimbabwe (including Beitbridge). Attendance at these meetings led to the formation of a wildlife committee by the Beitbridge District Council and subsequently the first sale of the DNPWLM-approved hunting quota (including three elephants, two buffalo, one lion and several less valuable species) to a commercial rancher in 1990. In February 1991, district representatives began discussing the distribution of the 1990 hunting proceeds, the 1991 hunting contract, employment for local people in the wildlife program, and the development of better water supplies.

Distribution of Wildlife Revenues

In a series of meetings held in March 1991, the Beitbridge District acted upon these discussions by accepting the recommendation of their Wildlife Committee that only 10 percent of the gross 1990 wildlife revenue be retained by the Council and that the remainder be returned to the producer wards. It was decided that, of the 12 wards in the district, only the three where wildlife revenue was earned (mainly from elephant hunting) would receive revenue, and that in one of them only the three VIDCO areas which generated revenue would benefit. This decision was challenged by some of the councillors from wards without wildlife. One chief responded saying "Communities that suffer damage from animals (must) get the revenue from wildlife. Those communities back from the river suffer no damage. Perhaps if they allowed the vegetation to be restored, they would have more animals."

The District Council approved the Wildlife Committee's recommendation to return Z$60,000 of the hunting proceeds to the Chikwarakwara VIDCO (in whose area 87% of the total wildlife revenue had been earned) and to give at least Z$5,000 to the four other wards and VIDCOs where wildlife had been shot, to encourage future wildlife production. It was also agreed that the people of Chikwarakwara should decide on the proportional allocation of their wildlife revenue to community projects and to household dividends. The next day, the Chikwarakwara VIDCO and Wildlife Committee met to discuss how the community should allocate the allotted funds. Some people wanted to distribute all the money to individual households but others disagreed. At one point, the District Council Chairman stated what he believed was the key issue (Peterson 1991b:65):

> "This money comes to you from your wildlife. You did not have to work for it. It is your money. The decision is yours. You must develop your own community. You cannot wait for government. It is your decision. You have the money. You can develop your own community according to how you decide."

The committee members accepted this challenge by voting to allocate Z$30,000 (50%) for a grinding mill and the construction of school facilities and to provide each household with a Z$200 dividend. At the community mass meeting the following day, the community accepted the committees' recommendation that community projects be given first priority of funding. What constituted the household was, however, discussed at length, and it was decided that every household that received wildlife dividends would have to immediately pay overdue school construction fees. The community members demonstrated their responsibility by carefully weighing factors such as definitions of household, needs of households and the needs of the community.

The wildlife revenue was finally distributed the following day at a prestigious event attended by the District Administrator and district level representatives of various ministries. The full $60,000 allocated to Chikwarakwara was placed in a wire basket and placed on a table so that it was visible to all the community members. The meeting started with numerous opening speeches about the benefits of wildlife, and much singing and dancing. Thereafter, each registered household representative was called to collect the $400 allocated per household, place their $200 contribution to the community projects in another basket, and pay the money owed to the school construction fund.

Since the message of wildlife conservation for sustainable development was accompanied by visible benefits and active community involvement in allocating them, the local people were remarkably quick to grasp the basic issues. They recognized that protecting existing wildlife was critical if wildlife revenue was to continue and increase.

DISCUSSION

The growth in interest in the CAMPFIRE concept exceeded the expectations of all cooperating agencies and organizations. By January 1990, less than four years after the publication of the CAMPFIRE document, 11 districts had been granted authority to manage their own wildlife and nine additional districts were preparing requests for such authority. Such interest and the common needs of the various communal wildlife programs resulted in the formation of a CAMPFIRE Association in August 1989.

The first meeting brought together 14 district representatives to consider the impact of the international ivory-trade ban on local communities. At this meeting it was decided that membership to the Association required full subscription to the "principle of the devolution of the custodianship of wildlife resources to producer communities in the communal lands." By 1991, the Association had 12 regular members (districts with authority to manage their own wildlife) and four associate members (districts with an interest in obtaining such authority). This represented over 30% of the 52 District Councils in Zimbabwe. The Association began seeking formal approval by the Ministry of Local Government and formal relationships with the Association of District Councils and the Wildlife Producers Association. With the recognition of this Association by government organizations, CAMPFIRE has grown beyond a DNPWLM program supported by NGOs. It is now an association with an established constituency representing communal farmers who seek sustainable development through wildlife utilization.

While sub-district units cannot legally represent wildlife producer communities, decision-making and revenues have in effect been transferred to such units by various CAMPFIRE Association members. The importance of the producer community, the ward or village, as the most significant social and resource production unit is one of the unique features of the CAMPFIRE program. The unity of such communities usually represents not just peoples' social unanimity, but also homogeneity of their subsistence economy and natural environment. For these reasons, CAMPFIRE identifies the producer community, not the district, as the key unit for natural resource management in communal areas.

Unless the people in the producer communities receive benefits from wildlife, they will make personal decisions about cattle, immigrants and poaching which will destroy wildlife on their land. Neither District Councils, the CAMPFIRE Association, nor ministries supporting CAMPFIRE can preserve wildlife on Zimbabwe's communal lands without the support of the local people. It was recognition of this fact by the DNPWLM that resulted in community empowerment (which gives people the right to decide how to use their own resources) being placed at the center of the CAMPFIRE concept. Community empowerment is the antithesis of centralized control for development, since it assumes that the people best understand their own needs and the natural resources at their disposal.

CONCLUSIONS

CAMPFIRE is a program for managing common property resources for the development of local rural people. It has developed in an African context to meet the needs of African people. It is specifically Zimbabwean in origin. Operating under a concept of adaptive management, CAMPFIRE has evolved as a set of principles, based on experience, rather than as a centrally designed program. The response by rural communities and District Councils to the program became a major criteria for directing the development of the program. Local Zimbabwean NGOs facilitated interaction between rural communities, District Councils and government. Finally, and of greatest significance, the districts with wildlife programs formed an Association to further promote the three CAMPFIRE principles of community empowerment, sustainable development and wildlife utilization. Without these three principles, local communities will not have the incentives to protect the wild animals in their areas.

Although CAMPFIRE is a Zimbabwean concept it is attracting increasing international attention, especially in neighboring African countries, because it presents an African perspective on wildlife which opposes the long standing Western perspective. The African perspective recognizes that people and wild animals continue to inhabit the same land over large areas in Africa while the Western perspective views wildlife as exhibits to be preserved for posterity in parks and zoos. African experience demonstrates that conservation is carried out best through utilization of wild animals as resources.

The promise of CAMPFIRE is the further transfer of responsibility for natural resource management to the producer communities. However, conflicts between district authorities and producer communities remains a problem for the sustained use of communal resources. The fundamental issue facing CAMPFIRE is whether government is based on decisions made BY the people or FOR the people. Put simply, can people be trusted to conserve their own resources?

REFERENCES

Bell, R.H.V., 1984
    Adaptive Management. In: R.H.V. Bell and E. Mcshane-Caluzi (eds), *Conservation and Wildlife Management in Africa*. Proceedings of a Workshop, Kasungu National Park, Malawi, October, 1984. U.S. Peace Corps, D.C. pp. 1–8.
CASS/WWF/ZimTrust, 1989.
    Wildlife Utilization in Zimbabwe's Communal Lands: Collaborative Program Activities. Center for Applied Social Sciences, WWF Multispecies Animal Production Systems Project, Zimbabwe Trust Coordinating Committee, Harare.
Child, B. and J. H. Peterson, Jr., 1991.
    Campfire in Rural Development: The Beitbridge Experience. Working Paper No. 1/91. Department of National Parks and Wildlife, and Center for Applied Social Sciences, University of Zimbabwe, Harare. pp. 7–50.

Environmental Consultants, 1990
People, Wildlife and Natural Resources: The CAMPFIRE Approach to Rural Development in Zimbabwe. Zimbabwe Trust, Harare.

Martin, R.B., 1986.
Communal Areas Management Program for Indigenous Resources (CAMPFIRE). Department of National Parks and Wildlife Management, Harare.

Peterson, J.H. Jr., 1991a.
CAMPFIRE: A Zimbabwean Approach to Sustainable Development and Community Empowerment through Wildlife Utilization. Center for Applied Social Sciences, University of Zimbabwe, Harare.

Peterson, J.H. Jr., 1991b.
The Week that Demonstrated Development: CAMPFIRE Deliberations at Council and Vidco Levels in Beitbridge District, March 25–30, 1991. In: B. Child and J. H. Peterson, Jr., *Campfire in Rural Development: The Beitbridge Experience*. Working Paper No. 1/91. Department of National Parks and Wildlife, and Center for Applied Social Sciences, University of Zimbabwe, Harare. pp. 51–86.

The World Commission on Environment and Development, 1987.
Our Common Future. Oxford University Press, Oxford.

8. ELEPHANT MANAGEMENT IN THE NYAMINYAMI DISTRICT, ZIMBABWE: TURNING A LIABILITY INTO AN ASSET[1]

RUSSELL D. TAYLOR

INTRODUCTION

A major source of conflict between wild animals and African people is the damage inflicted by wildlife upon crops and property, and injury or death caused to livestock and on occasion to human life. This is especially true of elephant, but can also apply to other large dangerous game. Consequently, rural people are intolerant of wildlife and farmers tend to inflate estimates of damage to crops and cultivated fields in anticipation of meat supplies from animals shot under crop protection programs (Taylor 1982).

The traditional and continuing response of wildlife management authorities to problem animals, especially dangerous game, is to harass and/or shoot the culprits. The success of such action has not been critically evaluated even though many thousands of animals have been killed, especially in colonial Africa, in the course of problem animal control (Bell 1985; Parker and Graham 1989). The nature of the problem needs careful assessment, especially where the economic value of problem animals potentially exceeds their nuisance value, and where their sustainable use is threatened by excessive control measures.

Under the Zimbabwe Government's CAMPFIRE program (Martin 1986; Anon. 1987) responsibility for wildlife was conferred on the Nyaminyami District Council of Kariba in northern Zimbabwe when it received "appropriate authority" status from the Department of National Parks and Wild Life Management (DNPWLM) in January 1989. The District Council is charged with administering and managing wildlife resources in the area for the benefit of the people of Nyaminyami. This paper outlines how the district is attempting to manage elephant in the Omay Communal Land so as to minimize conflict and increase the tolerance of local people, improve the livelihoods of such people through sustainable wildlife use, promote sustainable land use and enhance biological conservation.

Omay Communal Land in Nyaminyami District (which is a part of the Northern Sebungwe region on the southern shores of Lake Kariba), surrounds the inland boundaries of Matusadona National Park and covers 2,870 km² (Fig. 8.1). Omay has a population exceeding 20,000 people centered around four chieftainships, Mola, Negande, Nebiri, and Msampakaruma. Each chieftainship comprises two wards which are made up of a number of villages and

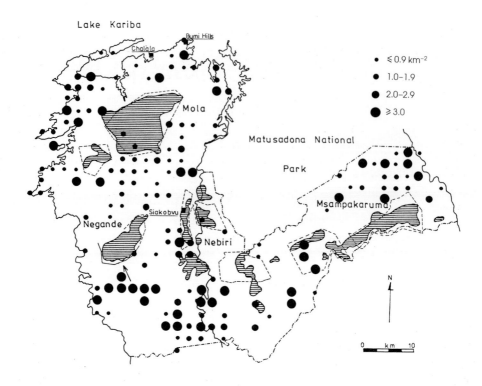

Figure 8.1 Omay Communal Land in Nyaminyami District, Zimbabwe, and dry season distribution of elephants. Hatched areas shows the major settlements within the Mola, Negande, Nebiri and Msampakaruma chieftainships and dots show the distribution and density of elephants. Completed and proposed electrified game fences are also indicated (– – –).

households. Economic growth, based on tourism and fishing, is focused on Bumi Hills and Chalala, and Siakobvu is the district's administrative center.

The environment is semi-arid with variable and seasonal rainfall amounting to 650 mm per annum, falling between November and March. The climate is hot with maximum temperatures exceeding 40°C and minimum temperatures rarely falling below 17°C. Agriculture is limited to subsistence cultivation and livestock are mainly goats since cattle have been precluded, until very recently, by the presence of tsetse fly (*Glossina spp.*). Large wild herbivore populations are typical of the Zambezi valley (Taylor 1988a) and include 2,000 elephant (*Loxodonta africana*), 6,000 buffalo (*Syncerus caffer*), 15,000 impala (*Aepyceros melampus*) and lesser numbers of 12 other species (Taylor 1991a; Taylor, Cumming and Mackie 1992).

Annual elephant censuses have been conducted in the Omay Communal Land for the past 13 years. The mean number estimated over ten counts is 2,098 ± 25% (95%C.L.) and the data indicate a long term upward trend. Notwithstanding the variability of individual estimates, this trend closely fits an expected 5% growth rate per annum. The mean crude density of elephants is $0.75 \pm 25\%/km^2$ but distribution is clumped and closely associated with uninhabited terrain so that localized densities may be as high as 3.0 elephant/km^2 (Fig. 8.1). Although overall, densities of elephant in the adjacent Matusadona National Park and Omay do not differ markedly between the two areas there are differences in distribution, ecological density, group size, home range size and movement (Taylor 1988b). This is largely a reflection of the management treatments to which elephants are subjected in the two areas. While elephants enjoy a high degree of protection from human disturbance in the National Park, they are subjected to hunting, harassment and human activities in the communal land.

MANAGEMENT OF ELEPHANTS

Safari Hunting and Problem Animal Control

Big game trophies in Africa are highly sought after by foreign clients, particularly Americans and Europeans. Wildlife in Omay has been used successfully, both in ecological and economic terms, over the past 20 years (Cumming 1989, Taylor 1990a). For practical convenience and client comfort the safari hunting season traditionally commences at the end of April or beginning of May, following the cessation of the rains. Consequently, most elephants shot on the safari hunting quota are taken from May onwards, during the dry season. There is, however, no legal restriction to hunting earlier. Quotas for elephant, based on a population estimate of 2,000, have not exceeded 0.8% of total numbers in any given year over the past 10 years (Table 8.1).

Table 8.1 PAC and Trophy Bull Elephant Offtake in Omay Communal Land, 1983–1992 (assuming an elephant population of 2,000)

Year	PAC Offtake		Trophy Offtake		Total	
	Number	%	Number	%	Number	%
1983	5	0.25	12	0.60	17	0.85
1984	8	0.40	12	0.60	20	1.00
1985	6	0.30	12	0.60	18	0.90
1986	10	0.50	12	0.60	22	1.10
1987	6	0.30	12	0.60	18	0.90
1988	9	0.45	16	0.80	25	1.25
1989	9	0.45	14	0.70	23	1.15
1990	8	0.40	12	0.60	20	1.00
1991	12	0.60	10	0.50	22	1.10
1992	8	0.40	12	0.60	20	1.00
Total	81	—	124	—	205	—
Mean	8.1	0.41	12.4	0.62	20.5	1.03

Elephants have also been shot to protect crops and people in Omay since the late 1950's following the relocation of the Tonga people displaced by the filling of Lake Kariba. In Northern Sebungwe 348 elephants were shot between 1955 and 1979 for crop protection measures (Cumming 1981). In Omay, probably less than 10 elephants were shot annually during the 1970's. Conflicts between people and elephants were minimal due to relatively low human and elephant numbers at the time. Furthermore, DNPWLM personnel probably considered elephants more important than people and avoided dealing with elephant-related depredation.

After 1980 the new government viewed conflicts between people and wildlife, especially elephants, more seriously and the DNPWLM was required to deal with problem animals in communal lands more diligently than it had done previously. Nevertheless, the number of elephants shot in Omay under the Problem Animal Control (PAC) program did not increase substantially and requests for eradication far exceeded the number killed. Although the number of trophy-hunted elephants have been strictly controlled, no limit has been set for animals shot under the PAC program.

With the acquisition of Appropriate Authority in 1989, the Nyaminyami District implemented a PAC monitoring program in Omay (Taylor 1990a). A comprehensive, yet simple data form was designed for completion by authorized control officers and other people involved with PAC. Between January 1989 and December 1991 some 1,000 PAC reports were filed at Siakobvu. Analysis of this data and DNPWLM records of the previous six years indicated that over 70% of reported incidents were elephant-related and occurred during the rainy season, between January and the end of April (Fig. 8.2). The reports

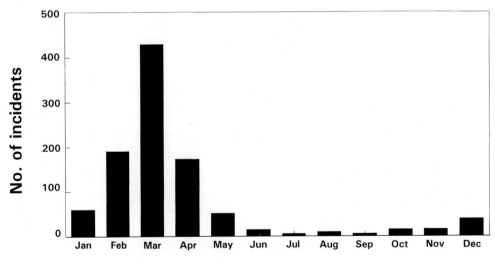

Figure 8.2 The monthly incidence of problem animal reports in Omay Communal Land 1989–1991 (n = 1,013 reports).

Courtesy of Stephen J. Thomas

Plate 8.1 Trophy animals are worth more than PAC animals which provide only meat.

Courtesy of Stephen J. Thomas

Plate 8.2 Biltong (dried meat) from elephants culled in the Gona re Zhou National Park, Zimbabwe, during the 1992 drought being distributed to communities neighboring the park.

peaked in March when growing maize, millet and sorghum are most attractive to crop raiding elephant. Despite the high number of incidents reported, an average of only eight male elephant were shot under the PAC program between 1983 and 1992 (Table 8.1).

Sustainable Trophy Elephant Hunting

To sustain good quality trophy elephant hunting, quotas should ideally not exceed 0.7% (Martin 1990). Based on a population of 2,000 elephant in Omay, the trophy offtake has been maintained at $0.62 \pm 0.049\%$ over the last 10 years (Table 8.1). But when the PAC male offtake was added to the trophy quota the total male offtake exceeded 0.7% every year and the long-term average offtake was 1.03%, an unsustainable level for maintaining high quality trophy animals. If Nyaminyami District wishes to continue offering competitive big game hunting on the international market, either the numbers of animals shot on safari or on PAC have to be reduced.

One possible solution for reducing both human-elephant conflicts and the number of elephants destroyed under the PAC program is to open a wet season "window" for safari hunting. By extending safari hunting of elephant bulls into the wet season, it would be possible to use PAC animals as safari trophies. To achieve this, the hunting season will have to be shifted forward gradually and several conditions will need to be met. Using the 10 year data set contained in Table 8.1, the following should apply.

- A combined PAC and trophy hunting offtake of 20 elephant bulls is equivalent to 1% of the estimated population of 2,000 elephant. Such an offtake is not biologically damaging to the population as a whole, but it will not sustain trophy elephant bulls in the long term. The desirable long term trophy quota should be less than 0.7%, and initially it should not exceed 0.6%, which is equivalent to 12 bulls per annum, if trophy quality is to be maintained.
- Setting an initial quota of only 12 bulls and hoping that this number will be sufficient for PAC offtake is unrealistic. Setting a quota of 20 bulls to cover both PAC and safari hunting and then reducing this number to 12 over time is more workable, especially if safari hunting can be extended into the wet season.
- An assumed quota of 20 bulls in year 1 can be allocated between PAC in the wet season and trophy hunting in the dry season. The safari operator should be allowed to market both the PAC and trophy portions of the quota. But, PAC elephant takings by safari operators should be subject to the conditions outlined below.
- The 20 bulls allocated to the combined PAC-trophy hunting quota in year 1 should be progressively reduced to 12 elephant over a 5 year period, by which time most should be marketed as trophy elephants for hunts in both the wet and dry seasons. The actual number allocated to wet season

Table 8.2 Suggested Allocations and Permutations for PAC and Trophy Hunting Quotas for Elephant in Nyaminyami District. (PAC = problem animal control quota; TH = trophy hunting quota; TQ = total quota)

Season	Year 1			Year 2			Year 3			Year 4			Year 5		
	PAC	TH	TQ	PAC	TH	TQ	PAC	TH	TQ	PAC	TH	TQ	PAC	TH	TQ
Wet	10			10			10			10			10		
Dry		10			8			6			4			2	
TQ			20			18			16			14			12
Wet	10			10			10			8			6		
Dry		10			8			6			6			6	
TQ			20			18			16			14			12
Wet	10			10			8			8			6		
Dry		10			8			8			6			6	
TQ			20			18			16			14			12
Wet	12			10			8			8			6		
Dry		8			8			8			6			6	
TQ			20			18			16			14			12
Wet	14			12			10			10			10		
Dry		6			6			6			4			2	
TQ			20			18			16			14			12

hunting would depend on hunter tolerance to such hunts and marketing success.

This approach would effectively deal with PAC problems and, by the end of the five year period, Omay residents should be sufficiently aware of the benefits of the approach to have greater tolerance for problem animals. Such animals could then be shot as trophies when they cause problems. Moreover, the financial return to the producer community would increase by converting a liability into an asset. Table 8.2 illustrates how such a scheme might operate. The combined PAC-trophy quota would be reduced gradually to the target figure of 12. Over 5 years a total of 80 elephant would be shot, representing an average annual off-take of 0.8% of the total population. Allocation between wet season PAC and dry season trophy hunting could vary between years and would be determined by the District Council after consulting with the safari operators. Five different permutations are illustrated in Table 8.2, but these are by no means exhaustive. Only even number allocations are illustrated because there are currently two safari operators in Nyaminyami District between whom the quota must be equally divided.

Conditions and Marketing

Clearly, a number of detailed conditions must apply for such a scheme to work properly but, since these would be area specific, only a general outline is given here. The total quota must not be exceeded, PAC must be undertaken only in the wet season, and the quota must be reduced to a sustainable trophy hunting quota over a specified time period. Elephants shot on PAC by a safari operator must be genuine problem animals destroyed as and where they cause problems. Prospective hunting clients would have no choice in the matter. If the PAC quota has to be exceeded, due to loss of life, then only the appropriate authority will be permitted to shoot elephants over and above the quota. Alternatively, if the PAC quota is not filled by the end of the wet season, the subsequent dry season's trophy elephant quota could be increased by the shortfall of animals shot under the PAC program.

Whether safari hunting can occur in the wet season is a question of good marketing. Since very good trophy elephants (80-100 lb tusk weight) have been shot in Zimbabwe during the wet season in connection with the PAC program, several safari operators are keen to market wet season elephant hunting. Some of the safari operators also recognize that local communities are more likely to tolerate elephants if they receive the larger economic benefits of trophy hunting compared with shooting problem animals.

Initially there may be market resistance to the concept of wet season hunting and safari operators may have to offer wet-season elephant hunts at a reduced price. Furthermore, the wet season "window" of 120 days (January-April) is too short for normal-length hunts for all elephants on the PAC quota. Safari operators should therefore be encouraged to market cheaper, shorter hunts, at least initially. Because of the restrictive conditions imposed on the client, a sliding price scale, based on the weight of PAC ivory, could be used for both the daily rate and trophy fee. Full trophy and daily fees would be charged for elephants with ivory greater than or equal to the average trophy weight for the district.

LAND USE

Fence Building

An electrified fence of 18 km encircling the 50 km^2 Negande settlement area was erected in September 1991. The fence is open along 12 km to the north where an abrupt, steep escarpment provides a physical barrier to elephant movement (Fig. 8.1). Fence erection followed protracted community debate which commenced in late 1988 and involved moving three villages which would not have been protected otherwise. Following completion of the fence, crop raiding incidents fell from 122 in 1990/91 to 42 in 1991/92, a 65% decrease (Mackie 1992). Although closing the open end of the fence might arguably increase its effectiveness, continued monitoring is necessary to determine the cost-effectiveness of such closure.

Prior to building the encircling fence, a smaller fence was installed around a 3 ha irrigation plot which produced green crops at the height of the dry season. This fence was severely tested by elephants during the first dry season but none entered the plot. After the crop was harvested, villagers returned to their traditional wet season fields and abandoned maintenance of the fence. Elephants and other animals breached and damaged much of the fence and some of it was swept away by the seasonal rains. Nevertheless both fencing projects were technically successful and construction defects were easily rectified.

No economic cost-benefit analysis has been undertaken for the Negande fences. Whilst the most important perceived benefit is reduced crop losses, its economic value has not been quantified, nor has its value been compared with fence construction and maintenance costs (Jansen 1992). Moreover the real economic benefit of fencing may well be decreased PAC-elephant eradication. Fencing programs are planned for the other major settlements in Omay (Fig. 8.1) but cost-benefit analyses are essential prerequisites to ensure economic effectiveness.

Land Use Planning

Long-term conservation of elephant depends on integrated land use which takes into account both the presence of elephants and their management and productive role in the economy of the district. Both district and ward-village level land-use planning and zonation pertaining to elephants and other wildlife management activities in Omay need to be considered. Under district level planning, communities have not been fully consulted while ward and village level planning has been conducted largely by agricultural extension officials who are largely ignorant of wildlife management requirements.

District level planning. The district has embarked upon a plan for developing wildlife-based tourism (Taylor 1990b). It includes proposals for zoning different uses and the following features.

- Establishment of a wildlife sanctuary within the existing Bumi Hills State Land where wildlife presently enjoys complete protection. Bumi Hills is an important international tourist destination with spectacular views of Lake Kariba amidst a full spectrum of wildlife. Elephants are an especially important tourist attraction.
- Zonation of the Mapongola hills as a Conservation Area. This would exclude human settlements and provide an effective link between Matusadona National Park on the Ume River in the east and Chizarira National Park on the Sengwa river in the west. The link is particularly important for the long term maintenance of genetic variability within the Sebungwe elephant population.

- Establishment of a number of sites with lakeshore frontage for small rustic camps (less than 20 beds). Commercial operators could lease these sites to make use of adjacent Lake Kariba Recreational Park and Matusadona National Park for walking, photographic, and game viewing safaris.
- Recognition of a number of key conservation areas. These include unique thickets (which constitute important elephant habitats), crocodile breeding areas on the lakeshore, and smaller areas of wetlands and minor escarpments in the Omay hinterland.

Much of the remainder of the area would be devoted to safari hunting which, in terms of consumptive resource use, is a low-pressure land-use option in which elephants are a key component. Areas designated for meat cropping (Taylor 1991b) would not conflict with other options such as tourism. Overall zonation would be linked to development objectives which are compatible and internally consistent.

Elements of this planned zonation are in the process of being adopted. For example, five 10 ha lease sites have been identified for non-consumptive tourism, advertised for competitive tender and private sector operators have been objectively selected. The district is now entering into joint venture partnerships with these operators to generate additional revenues for the district and provide local employment (Jansen 1990; Taylor 1992).

Agricultural planning at ward and village level. Approximately 80% of Omay is unsuitable for arable agriculture due to poor soils and broken terrain. Settlement presently extends over about 10% of the district but this is expanding due to illegal in-migration. There is consequently a need for participatory sub-district level landuse planning. Officers of the Department of Agricultural Technical and Extension Services (Agritex) are currently preparing residential, arable and grazing area plans for individual households at a ward and village level.

Whilst this involves greater community participation than does district level planning, Agritex has failed to recognize the increasing economic importance of wildlife in the district. Much of sub-district level planning is consequently being undertaken without due cognisance being given to wildlife needs. For example, grazing holdings are being allocated in anticipation of cattle introduction (tsetse fly having been eradicated only recently form most of Omay), rather than as holdings for wildlife. The main concerns over introducing cattle are the number of cattle that is ecologically sustainable, competition for resources between cattle and wildlife, and predation upon cattle.

WILDLIFE REVENUES

Over the three years 1989–1991, Nyaminyami District earned Z$1,273,503 (about US$467,000) from its wildlife. Moreover, revenue increased with each

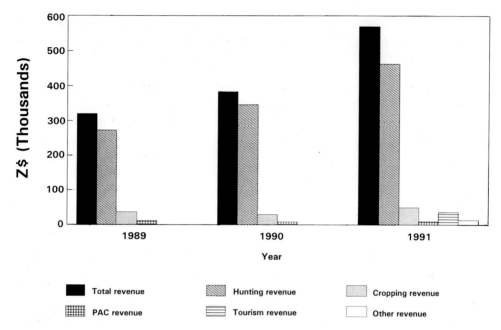

Figure 8.3 Revenues (Z$) from wildlife management activities in Nyaminyami District, 1989–1991.

successive year (Fig. 8.3), albeit only slightly in real terms. Earnings have come from a number of activities, including hunting, cropping for meat production, problem animal control and, more recently, from tourism. Elephants are central to these earnings, especially in the case of sport hunting. Trophy fees and hunting concession fees generated 85% of the total wildlife revenue (Fig. 8.3), while 38% of the three-year average hunting quota value was related to elephants (Table 8.3). Although PAC contributed only 2% to income, this was also derived mostly from elephants. These statistics emphasize the importance of minimizing PAC elephant killings in order to increase the number of elephants that can be offered for safari hunting. Transferring a PAC elephant to the safari-hunting quota increases its value nearly twenty-fold. This is likely to greatly enhance the prospects for conserving this ecologically, economically and aesthetically valuable resource.

Table 8.3 The Proportion of Revenue Earned from the Hunting Quota of Elephants in Relation to the Total Value of the Quota in Nyaminyami District

Year	Total Value of Quota (Z$)	Value of Elephants (Z$)	%
1989	189 400	83 000	43.8
1990	238 100	90 000	37.8
1991	223 100	75 000	33.6

DNPWLM guidelines (Anon. 1991) for the distribution of wildlife revenues earned under the CAMPFIRE program stipulate that the responsible District Councils retain no more than 15% of gross revenue as a levy; that up to 35% may be allocated to district-level capital and recurrent expenditure for wildlife management; and that at least 50% of revenue must be returned to wards, villages or households. Nyaminyami District has yet to meet these requirements. Only in 1989 did the ward dividend exceed 50% of revenue and of the total Z$1.27 million earned to date only 39% has been returned to the wards.

DISCUSSION AND CONCLUSION

Despite a growing human population in Omay, elephant densities have remained high (around $0.7/km^2$) and have probably increased in the Nyaminyami District over the past 12 years. Their continued existence, whilst ultimately linked to a limit in human population growth, depends upon human tolerance towards their presence. Such tolerance is being fostered by ascribing economic importance to elephants mainly through high-valued international safari hunting. To enhance this value the number of elephants killed to protect crops must be limited.

During 1992 a PAC elephant quota was set and four of the problem elephants were successfully sold by safari operators as trophy elephants. Moreover, the District Council agreed to return the revenues earned from these elephants to affected communities. The safari operator paid Z$13,000–Z$22,000 per elephant shot to the wildlife committee chairman of each affected ward at the end of the rainy season. In this way crop raiding elephants were effectively dealt with, affected people benefitted directly from associated hunting revenues, the safari operator was able to market more elephants and PAC was kept within sustainable limits.

Tourism based on game viewing, walking and photographic safaris are likely to become more important in Nyaminyami as joint venture partnerships increase during the next few years. Although this activity generated only 6% of the district's income in 1991, revenue derived from it is expected to exceed hunting revenue threefold during the next five years. Total earnings of around Z$6m per annum have been projected with non-consumptive tourism and hunting contributing Z$4.5m and Z$1.5m, respectively (V.R. Booth pers. comm.). Elephants together with a broad spectrum of spectacular wildlife and scenery characterizing the district are key components for such revenue generation.

The Nyaminyami District has demonstrated that earning money from wildlife, especially elephants, can be achieved very successfully. But this is only part of the requirement for ensuring the survival of elephants in communal lands. More importantly, the district must ensure that the beneficiaries of wildlife revenues are the rural poor and peasant farmers who live alongside wild animals. Returning such benefits to people who bear the cost of living with wildlife is at the heart of the CAMPFIRE program, but this has yet to be meaningfully achieved. Moreover, there must be greater participation by local communities in managing wildlife in order for them to become responsible and accountable for their wildlife and wildland resources.

Elephant conservation is as much an institutional problem as it is a technical one and its resolution lies in the hands of local people who will ultimately decide how their land is used. That decision will be strongly influenced by the extent to which wildlife benefits (especially those derived from elephants) accrue to individual householders and farmers. Only when elephants are perceive to be an asset will their conservation be voluntarily incorporated in locally developed land use, and become part of an economy that uses natural resources wisely and sustainably.

NOTES

1. This chapter is based on a paper first published in Pachyderm, Volume 17 in 1993 (pages 19–29) and which was presented at the IUCN/SCC African Elephant Specialist Group meeting, Victoria Falls, Zimbabwe 17-22 November, 1992. It is reproduced with the kind permission of the Editorial Board of Pachyderm.
2. The assistance of the Nyaminyami District Council and the Nyaminyami Wildlife Management Trust is gratefully acknowledged. Simba Hove, Elliot Nobula and Imke van der Honing were especially helpful with the collection of field data and the compilation of records. Rob Style, of Buffalo Range Safaris, provided much of the initial stimulus for reducing conflict between safari hunting and problem animal control and putting these ideas into practice. Doris Jansen and Ivan Bond kindly provided access to unpublished financial data and David Cumming made valuable editorial comments.

REFERENCES

Anon., 1987.
    The National Conservation Strategy: Zimbabwe's Road to Survival. Ministry of Natural Resources and Tourism. Government Printer, Harare.
Anon., 1991.
    Guidelines for CAMPFIRE. Department of National Parks and Wild Life Management, Harare.
Bell, R.H.V., 1985.
    The Man Animal Interface: An Assessment of Crop Damage and Wildlife Control. In: R.H.V. Bell and E. McShane-Caluzi (eds), *Conservation and Wildlife Management in Africa*. U.S. Peace Corps, Washington, D.C.
Booth, V.R., 1993.
    Personal communications, Price Waterhouse, Harare.

Cumming, D.H.M., 1981.
    The Management of Elephant and Other Large Mammals in Zimbabwe. In: P.A.
    Jewell and S. Holt (eds), *Problems in Management of Locally Abundant Wild Mammals*.
    Academic Press, New York.
Cumming, D.H.M., 1989.
    Commercial and Safari Hunting in Zimbabwe. In: R.J. Hudson, K.R. Drew and L.M.
    Baskin (eds), *Wildlife Production Systems*. Cambridge University Press, Cambridge.
Jansen, D.J., 1990.
    What is a Joint Venture? Guidelines for District Councils with Appropriate Authority.
    WWF Multispecies Project Paper No. 16. Harare.
Jansen, D.J., 1992.
    Economic Evaluation of Fencing Projects. In: R.E. Hoare (ed), *Fencing as a Manage-
    ment Tool in Zimbabwe's Wildlife Programs*. Proceedings of a Seminar. Department of
    National Parks and Wild Life Management, Harare.
Mackie, C., 1992.
    Implementing Electric Fencing Projects in Communal Lands of Zimbabwe. In: R.E.
    Hoare (ed), *Fencing as a Management Tool in Zimbabwe's Wildlife Programs*. Proceedings
    of a Seminar. Department of National Parks and Wild Life Management, Harare.
Martin, R.B., 1986.
    Communal Areas Management Program for Indigenous Resources (CAMPFIRE). Technical
    Report. Department of National Parks and Wild Life Management, Harare.
Martin, R.B., 1990.
    Elephant and Rhino Conservation in Zimbabwe. Paper presented to the Japan Wild-
    life Research Center, Tokyo, August 1990. Department of National Parks and Wild
    Life Management, Harare.
Parker, I.S.C. and A.D. Graham, (1989).
    Elephant Decline: Downward Trends in African Elephant Distribution and Numbers
    (Part 1). *International Journal of Environmental Studies*, 34:287–305.
Taylor, R.D., 1982.
    Buffer Zones: Resolving the Conflict Between Human and Wildlife Interests in the
    Sebungwe Region. *Zimbabwe Agricultural Journal*, 79:179–184.
Taylor, R.D., 1988a.
    The Indigenous Resources of the Zambezi Valley: An Overview. *The Zimbabwe Sci-
    ence News*, 22:5–8.
Taylor, R.D., 1988b.
    Elephant Numbers, Distribution and Movement in Omay Communal Land. Unpub-
    lished report, Department of National Parks & Wild Life Management, Harare.
Taylor, R.D., 1990a.
    *Ecologist's Report for 1989: Nyaminyami Wildlife Management Trust Annual General
    Meeting, February, 1990*. WWF Multispecies Project Paper No. 9. Harare.
Taylor, R.D., 1990b.
    Plan of Zonation for Wildlife and Tourist Development in Nyaminyami District,
    Kariba. Unpublished Report, Nyaminyami Wildlife Management Trust. Harare.
Taylor, R.D., 1991a.
    Aerial Census of Large Herbivores in Pilot Project Areas, October 1989. WWF
    Multispecies Project Paper No. 11. Harare.

Taylor, R.D., 1991b.
  Socio-economic Aspects of Meat Production from Impala Harvested in a Zimbabwean Communal Land. In: L.A. Renecker and R.J. Hudson (eds), *Wildlife Production: Conservation and Sustainable Development*. University of Alaska, Fairbanks.
Taylor, R.D., 1992.
  Ecologist's Report for 1990: Nyaminyami Wildlife Management Trust Annual General Meeting, February, 1991. WWF Multispecies Project Paper No. 28. Harare.
Taylor, R.D., D.H.M. Cumming and Mackie, C., 1992.
  Aerial Census of Elephant and Other Large Herbivores in the Sebungwe 1991. WWF Multispecies Project Paper No. 29. Harare.

9. SEEKING EQUITY IN COMMON PROPERTY WILDLIFE IN ZIMBABWE[1]

STEPHEN J. THOMAS

INTRODUCTION

Zimbabwe's Communal Areas Management Program for Indigenous Resources (CAMPFIRE) (Martin 1986) is widely regarded as an innovative approach to sustainable natural resource management. Since degraded 'open-access' characteristics are evident in many of Zimbabwe's communal areas, the program acknowledges the State's failure to effectively manage natural resources in these areas. Yet it is unlikely that rational human beings will deliberately destroy the environment upon which their livelihood depends. The reason for such degradation is attributable to the fact that people living in the communal areas only have rights of usufruct to such resources. Without giving these people clearly-defined proprietary rights, including rights of access and powers of exclusion, many of Zimbabwe's natural resources are in jeopardy. CAMPFIRE seeks to transfer proprietorship of these resources from the State to local communities and, hence, facilitate the conversion of open access to common property. It is evident that such conversion is occurring.

At present CAMPFIRE is a misnomer: It is not a Communal Areas Management Program for *Indigenous Resources* but rather focuses on wildlife resources in communal areas. The primary reason for this emphasis is that CAMPFIRE originated in the Department of National Parks and Wild Life Management (DNPWLM) which has no authority over resources other than wildlife outside of the Parks and Wild Life Estate. Furthermore, since Zimbabwe's safari hunting industry is internationally highly regarded and generates considerable foreign exchange, wildlife can provide significant, and immediate, financial returns to communities. This is an important factor in stimulating community participation in resource management. But wildlife can also inflict substantial costs. These include crop, livestock and property damage, human injury and death, and anxiety induced by living in close proximity to dangerous wild animals. Since such costs are borne by individuals, while benefits from wildlife accrue communally, they can be considered to be individual 'contributions' to the maintenance of the common property resource.

This begs a basic question of equity: "Do individuals get a reasonable and fair return on their contribution to a collective undertaking to regulate a common?" (Oakerson 1984:15). The answer is critical since equity and efficiency of resource-use are closely related. If individual benefits from wildlife do not outweigh the associated costs, any collective effort to manage wild animals is likely to fail. Free riding by disadvantaged people will lead to overuse of wild

animals and degradation of their habitats. Equity concerns are further complicated by the fugitive nature of wild animals. They can affect, and be affected
by, a variety of user groups. "These overlapping jurisdictions generate complex
management problems which require innovative institutional arrangements."
(Buck 1989:130).

This chapter examines the variety of approaches being advocated or adopted
by the different stakeholders in CAMPFIRE in order to address the question of
equity.

STAKEHOLDERS IN CAMPFIRE

Identity of Stakeholders

The CAMPFIRE concept grew out Zimbabwe's 1975 Parks and Wildlife Act
which designated land owners as the 'appropriate authority' for managing
wildlife resources on private land. A 1982 amendment redressed the Act's discriminatory nature by extending the term 'appropriate authority' to include
District Councils, the administrative authority for the communal areas and the
smallest, legally-accountable body to which such authority could be granted.
District Councils receiving appropriate authority status thus have the statutory
authority and the responsibility for managing wildlife in their communal areas
(Murphree 1991a:2). However, the DNPWLM remains responsible for wildlife
in Zimbabwe and, through its Minister, may withdraw appropriate authority
from a District Council not conforming to the conditions and objectives under
which authority was granted (DNPWLM 1992).

An ancillary objective of the government's decentralization policy is to increase local community involvement in the development of their areas. Village
Development Committees (VIDCOs) were identified as the fundamental planning units, each representing 100 households or approximately 1,000 people.
They were to submit annual development plans though their Ward Development Committees (WADCOs), which each represented six VIDCOs, for incorporation in integrated district plans by the District Development Committee
and subsequent approval by the District Council.

Although they comprise the elected chairpersons from the WADCOs in each
district, District Councils are effectively not autonomous local government entities because the District Administrator, the Council's Chief Executive Officer, is
appointed by the Ministry of Local Government, Rural and Urban Development. Although they are expected to act in an advisory capacity to District
Councils, District Administrators are accountable only to the Ministry. The potential for conflict is considerable.

A notable omission from this institutional structure is representation by traditional leaders. Their former role was significantly undermined by transferring the powers to allocate land and manage resources to the District Council.
The imposition of VIDCOs and WADCOs, whose boundaries were not necessarily aligned with traditional communal boundaries, further excluded them
from decision making. The transition from chiefly authority (i.e. local, heredi-

tary and long-standing) to elected bureaucratic authority (i.e. transient and possibly immigrant) has thus led to conflict.

Figure 9.1 shows the hierarchical structure resulting from the Prime Minister's 1984 directive to establish local government authorities (GoZ 1984a). This

CENTRAL GOVERNMENT

|

PROVINCIAL DEVELOPMENT COMMITTEE
Comprising Provincial Administrator [Chair], Provincial Heads of Government Ministries and Security Organizations, Chief Executive Officers of Rural/ District Councils, co-opted members of other organizations working within the Province

|

RURAL DISTRICT DEVELOPMENT COMMITTEE
Comprising District Administrator (chairman), Senior Executive Officer, District Heads of Government Ministries and Security Organizations, co-opted members of non-government organizations

|

RURAL DISTRICT COUNCIL
Comprising of WADCO chairpersons

|

WARD
WADCO — comprising Ward Councillor (chairman), chairperson and secretary of each VIDCO, 1 member each from ZANU-PF's Womens' League and Youth Brigades

|

x 6

|

VILLAGE
VIDCO — comprising 4 village members, 1 member each from ZANU-PF's Womens' League and Youth Brigades

|

x 100

|

HOUSEHOLD

|

INDIVIDUAL

Figure 9.1 Institutional structure established by the 1984 directive.

institutional framework presents the various stakeholders as points on a con-
tinuum along which questions of equity proliferate.

Perspectives of Stakeholders

At the national level, the two major stakeholders are the Ministry of Environment and Tourism, under which the DNPWLM falls, and the Ministry of Local
Government, Rural and Urban Development (MLGRUD). Initially there was
little liaison between them but, at the request of the MLGRUD, the DNPWLM
(1991) produced a set of guidelines for District Councils to follow in implementing CAMPFIRE.

The first principle was that "Councils are required to return at least 50% of
the gross revenue from wildlife to the community which produced it (i.e.
where the animal was shot)." It sought to forestall the bureaucratic impulse
encapsulated in what is colloquially known as 'Murphree's Law' which states
that: "There is an in-built tendency at any level in bureaucratic hierarchies to
seek increased authority from levels above and resist its devolution to levels
below." (Murphree 1989:4).

The second principle defined the size of the 'producer community'. "The
ideal size for a producer community is 100 to 200 households because this is
large enough for a wildlife program, and small enough that all households can
be involved in the program and accountable to it." This principle sought to
quantify the unit of proprietorship and is in accord with the view that: "The
conversion from open-access to common property will be facilitated in those
instances in which the size of the user group is small, the users are reasonably
homogeneous in important socio-economic characteristics, and the users reside
in close proximity to the resource." (Bromley and Cernea 1989:24). An implicit
assumption in quantifying the household constituency of a producer community is that the distribution of communal area settlements is uniform, which is
rare. The significance of the fugitive nature of wildlife for the equitable distribution of benefits is also disregarded.

The third principle stipulated that: "Producer communities must be given
the full choice of how to spend their money, including both projects and cash
payments ... Where communities value cash above projects, they should be allowed cash." This principle acknowledges the importance of livestock in the
rural African economy. Livestock and wildlife share vegetative and water resources in the communal areas but livestock is privately-owned while wild
animals are communally-owned. "(U)nless revenues from wildlife are translated into disposable individual or household benefits decisions on wildlife/
livestock options will be skewed towards livestock options even in situations
where it is apparent that the wildlife option is collectively more productive."
(Murphree 1991b:18).

These guidelines may appear prescriptive, but they counter perspectives, exemplified by the following two statements, which reinforced 'Murphree's Law'.

The first statement implies that the MLGRUD does not fully endorse the distribution of cash dividends; the second dismisses the concept of the 'producer community'. In his closing address to the first Annual General Meeting of the Campfire Association (which is comprised of District Councils with appropriate authority), the Minister of the MLGRUD stated:

> "The producer communities ... must be allowed a full choice of options whose aim is to improve the well-being of the people by providing direct benefits, through improved social services, like schools, clinics, infrastructural projects, like water, grinding mills, etc., or by paying cash dividends *where this is felt extremely necessary*." (MLGRUD 1991, emphasis added).

More recently, in his opening address to a seminar on CAMPFIRE, the Provincial Administrator for Matebeleland North Province, a senior official in the MLGRUD, made the following points:

> "Benefits should not be individual based but community based. Having given individual money, there is nothing to show for it the very next day. However upgrading the district through provision of infrastructure such as clinics, schools, play centers and community centers has long-term benefits.... The giving of financial handouts does not necessarily uplift standards but on the contrary creates a dependency syndrome. May I also point out that *the distribution of the benefits should be district oriented and not area based*. Those areas that do not have animals or game need not be left outside." (Mzilethi 1992, emphasis added).

It would be difficult to find two more contrary views. The DNPWLM has a 'microcosmic' view, believing that equity will be achieved only when each household receives its share of wildlife revenues derived from within their locality. It claims that only this will satisfy the principle that those who bear the costs of living with wildlife must benefit from its use. By contrast, the MLGRUD has a more 'macrocosmic' view and considers equity to arise through the diffusion of the same revenue to the district-wide constituency. Murphree has argued against this wider equity, stating that:

> "Communities which still possess good wildlife assets are those which subsist on lands marginal for cropping and which have largely been by-passed by the development process. This also is an historical cost to these communities and to argue on the grounds of equity that they should now share the benefits of the growing value of wildlife with their more affluent neighbours is highly tenuous." (Murphree 1991b:17).

This is a valid argument against the macrocosmic view, but it does not address the needs of the wider community which is affected by fugitive wildlife. The potential for costs to be incurred in one community whilst benefits accrue in another is very real.

The following case studies examine the assumptions of the national-level stakeholders and illuminate the perspectives of local-level participants. The first case examines intra-district variability in wildlife and agro-economic potential and questions whether equity issues are likely to be addressed satisfactorily if wildlife revenues are used for the benefit of the whole district.

CASE STUDIES

Guruve District

Guruve District straddles the Zambezi escarpment in the Mashonaland Central Province of northern Zimbabwe (Fig. 9.2). It is interesting for equity considerations because of the steep agro-economic gradient across it. Below the escarpment eight wards constitute the Dande Communal Land which is bordered by Mozambique to the north, the Chewore Safari Area to the west and the escarpment's Rukowakoona Mountains to the south. The area falls within Natural Region IV, which receives 450–650 mm of rainfall per year and is subject to periodic seasonal droughts and severe dry spells during the rainy season (GoZ 1984b). The incidence of tsetse fly (*Glossina spp.*) in the Zambezi valley has precluded cattle as a viable livestock option. Above the escarpment, 12 smaller

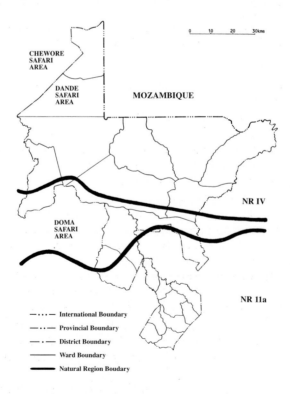

Figure 9.2 Guruve district.

Courtesy of Stephen J. Thomas.

Plate 9.1 Cash dividends earned from elephant hunting in 1991 being distributed to each of 140 households in Masoka Ward, Guruve District, Zimbabwe.

wards make up the Bakasa, Kachuta and Guruve Communal Lands. Bakasa and Kachuta together consist of only three wards and fall in Natural Regions IIa and III. Guruve Communal Land contains nine wards entirely within Natural Region IIa. This region receives 750-1,000 mm of rainfall per year with only rare severe dry spells in summer and is thus suitable for intensive cropping and livestock production systems (GoZ 1984b).

Fig. 9.2 presents the wards which comprise Guruve district. Since wards were delineated by central government to contain about 600 households, differences in ward size demonstrate variation in agricultural potential above and below the escarpment. In the Dande Communal Land the average area of the eight wards is 520 km². In stark contrast, the average area of the nine wards in the Guruve Communal Land is only 63,5 km².

The 1982 population census estimated that 40,000 people (70 km⁻²) inhabited Guruve Communal Land while only 18 000 people (4 km⁻²) were living in Dande Communal Land. These differences signify the positive correlation between population density and perceived potential economic well-being. It is clear that people considered their livelihood was more secure in Guruve Communal Land than in the valley below, at least, prior to implementation of CAMPFIRE.

Whilst appropriate authority for wildlife management is granted to District Councils, CAMPFIRE is generally being implemented in the most marginal areas within those districts. For example, in Guruve District it is being implemented only in the Dande Communal Land. In such circumstances, it is iniquitous to suggest that equity is better served by distributing wildlife-related benefits across the whole district rather than to those people coexisting with the wildlife resource.

The next case study examines two District Councils with relatively homogeneous characteristics. From the institutional perspective this example is interesting because it demonstrates that traditional relationships between communities often transcend imposed jurisdictional boundaries.

Bulilima Mangwe/Tsholotsho Districts

Bulilima Mangwe District is situated in Matebeleland South Province in the south-west of Zimbabwe and borders Tsholotsho District in the north and Botswana in the west. Tsholotsho District is situated in Matebeleland North Province and adjoins the Hwange National Park in the north and west (Fig. 9.3). The southern 30% of the Bulilima Mangwe District lies in Natural Region V while the rest of it and Tsholotsho District lie in Natural Region IV (Thomas 1992). Both districts experience only a short rainy season during the summer.

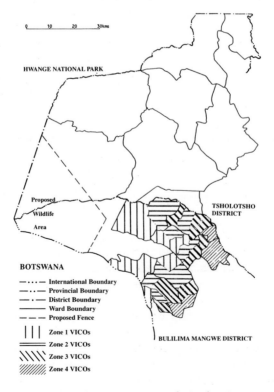

Figure 9.3 Bulilima Mangwe/Tsholotsho districts.

Unlike the Zambezi valley, this area is unaffected by tsetse fly and the rural economy has been based on extensive agro-pastoralism. The dominance of cattle and the traditional transhumance system (which long predates district and ward boundaries [Peterson 1991]) with winter grazing in the far west of Bulilima Mangwe, led to the development of a joint CAMPFIRE initiative between the two districts. Both District Councils realized that they would have to rationalize their land use in order to optimize benefits from cattle and wildlife. They planned to establish a wildlife area in the western portions of each district, along the Hwange National Park and Botswana borders. An electric fence was to be erected along the eastern border of the proposed wildlife area, traversing both districts, to prevent cattle moving into the wildlife area and constrain predation on cattle. The successful implementation of the plan required coordination of the grazing rights of communities in each district and a joint CAMPFIRE project.

Inhabitants of the western wards of Bulilima Mangwe suffer from significant crop damage by elephants during the wet summer season.[9] Yet the potential to generate revenues from these elephants through safari hunting is limited. In the wet season, when most of the damage is occurring, hunting is hindered by sticky black soils, high humidity and mosquitos. And during the preferred dry season, hunting is hampered by movement of cattle from both districts into the winter grazing area. Consequently, most of the hunting takes place in Tsholotsho.

In 1990, eight of the nine elephants shot under the joint hunting quota were taken in Tsholotsho wards, and the ninth in Bulilima Mangwe. At the end of the year, the joint wildlife committee decided that revenues from five of the elephants should accrue to Tsholotsho and the income from the other four to Bulilima Mangwe. This apparent recognition of the costs carried by Bulilima Mangwe and the need for inter-community equity was, however, reached only under strong DNPWLM pressure. Interestingly this position would have contravened the guideline that wildlife dividends should be returned to producer communities, which was later developed by the DNPWLM.

The seven Bulilima Mangwe wards involved in the CAMPFIRE project agreed that their elephant revenues should be divided equally amongst them. Their decision was based on the fact that rights of access to winter grazing were distributed widely and, since these rights were affected by the wildlife project, wildlife benefits should be distributed among the wards with grazing rights. The extent to which access to winter grazing is institutionalized in Bulilima Mangwe is not known, but the decision reached infers that the wards felt they have equal rights in this respect. Moreover, it seems that the benefits and costs of producing privately-owned cattle, through the use of common-property grazing resources, are more easily assimilated in decision-making than those associated with the 'production' of a fugitive resource like wildlife. This may account for the reason why the decision was reached even though inter- and intra-ward variability of wildlife damage was found to be so great that "simply to compare wards with reference to wildlife damage obscures the real differences by area" (Hawkes 1991:1).

Hawkes (1991) classified the 41 VIDCOs within the wards into zones (see Fig. 9.3); the first being adjacent to the wildlife area, the second only being reached by wildlife which had travelled through the first zone, and so on. Crop damage by elephants and the risk of predation by hyenas on cattle were found, not surprisingly, to be greatest in the first zone. The study concluded that:

> "Elephants are a serious problem only for the quarter of the households who live in the frontline area (Zone 1). However, the returns from safari hunting will go to the whole area covered by the seven wards. Aside from questions of fairness, is this enough return from elephants to give residents of the frontline the sense of proprietorship?" (Hawkes 1991:8).

The Tsholotsho District Council terminated the joint hunting concession with Bulilima Mangwe at the end of 1991. This is not surprising since, on balance, and in the terminology of the CAMPFIRE guidelines, they were the major 'producer' of wildlife revenue. Elephants, though, will continue to constitute a serious problem for those frontline VIDCOs in Bulilima Mangwe.

This example emphasizes the difficulty of determining guidelines which satisfactorily address wildlife-related equity issues. The term 'producer community' applies to all 13 frontline VIDCOs Bulilima Mangwe since they are providing the 'differential inputs' by tolerating the predations of large wild mammals, without which the wider community would not benefit as an 'end-user'. Yet the first principle of the CAMPFIRE guidelines states categorically that: "councils are required to return *at least 50%* of the gross revenue from wildlife to the community .... which produced it (*i.e. where the animal was shot*)." Murphree (1991b:6) proposed that for viable communal property regimes "differential inputs must result in differential benefits.... Wildlife assets are distributed unevenly in any national context; equally the cost of sustaining and managing these assets is unevenly distributed.... Policy must ensure therefore that benefit is directly related to input."

The final case study examines one way in which a District Council has attempted to address these complex equity issues. A fundamental failing of the CAMPFIRE guidelines is the assumption that the location where a wildlife asset is 'realized' is synonymous with the area where it was 'produced'. The previous example showed this assumption to be false for high revenue-earning species, such as elephant. Unless this deficiency is addressed, inequities will continue in the name of CAMPFIRE.

Nyaminyami District

The Nyaminyami District Council was the first to be granted appropriate authority status in 1988. The district borders the southern shore of Lake Kariba and comprises three communal lands, Omay, Gatshe Gatshe and Kanyati (Fig. 9.4). Some 75% of the district is in Natural Region V (Thomas 1992) and is affected by tsetse fly. In terms of wildlife resources, especially large mammals,

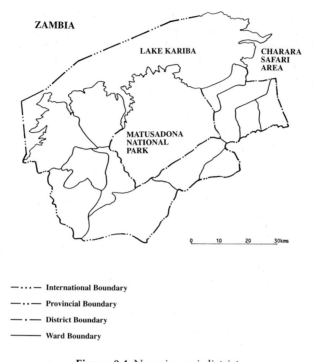

ZAMBIA

LAKE KARIBA

CHARARA
SAFARI
AREA

MATUSADONA
NATIONAL
PARK

0   10   20   30kms

—····— International Boundary

—··— Provincial Boundary

—·— District Boundary

———— Ward Boundary

Figure 9.4 Nyaminyami district.

Nyaminyami is probably the richest CAMPFIRE district in Zimbabwe (Child 1991).

In order to administer and implement its wildlife management programme, the District Council formed the Nyaminyami Wildlife Management Trust (NWMT). While the abundance of wildlife is potentially valuable, it also generates heavy depredation costs for proximate communities. Thus controlling problem animals had a high priority when the NWMT was founded. Unless local people could be assured wildlife-related costs would be met in some tangible way, the program was unlikely to be adopted. The Trust thus introduced a system for monitoring crop and livestock damage, compensated households suffering such damage, and discussed an insurance scheme to cover loss of life (Peterson 1991). Such compensation and insurance schemes are not only innovative but, arguably, more closely address equity issues than the attempts described in the preceding case studies, since those people that incur the costs of living with wildlife benefit from their involuntary inputs.

In 1989, the first year in which the crop compensation scheme operated, the NWMT paid Z$26,000 (ca. US$6,500) to 160 families (Nyaminyami District Council 1991; Peterson 1991). The following year it received 666 problem animal reports, related mainly to elephants and buffalo. Ninety percent of these reports were investigated, 14 elephant and seven buffalo were consequently killed, and Z$42,000 (ca. US$10,500) was paid in compensation to 360 families. But the total claims were more than double this amount and far exceeded the budget for compensation payments (Nyaminyami District Council 1990). The potential of the district-level compensation scheme to deplete wildlife revenues was recognized and the NWMT abandoned it in 1991. Instead it decided to provide wards with larger dividends since ward committees were better able to monitor crop damage and decide whether to compensate claimants. While some communities are confident that such compensation can be equitably allocated two wards expressed concern that such a responsibility could lead to dishonesty among their leaders.

By allocating wildlife benefits according to losses, compensation for crop and livestock damage theoretically solves the problem of defining a 'producer community'. But the rapid escalation of the number and scale of claims in this case study shows that the system is susceptible to abuse. The level of monitoring and adjudication necessary to prevent such abuse is likely to make the cost of compensation disproportionately high relative to the revenues available for distribution.

CONCLUSIONS

This chapter has provided only a brief overview of equity concerns in CAMPFIRE and has focused on the opposing perspectives of the 'national' stakeholders. The DNPWLM wants wildlife revenues to be distributed as cash dividends to households in producer communities while the MLGRUD wants them invested in district projects. The DNPWLM's view more closely reflects criteria necessary for successful common property resource regimes (see Bromley and Cernea 1989). The fugitive nature of wildlife resources, however, makes the producer community concept illusory, particularly when large species, such as elephant, are involved. CAMPFIRE has nevertheless been most successful where District Councils have returned revenues to the communities in whose locality the animals were shot. The MLGRUD's view that benefits from communal property resources should be distributed across the district, ignores the integrity of the unit of proprietorship. "Proprietorship .... cannot be separated from production, management and benefit and is a fundamental component in a communal resource regime" (Murphree 1991b:7).

The case studies examined the relevance of these national perspectives, and provided insights about some local concerns. The Guruve study demonstrated heterogeneous resource distribution within the district and consequently the inequity of the MLGRUD's view. The Bulilima Mangwe/Tsholotsho case study showed the difficulty of avoiding contradictions when considering equity issues. In this case the DNPWLM influenced the decision to distribute revenues more evenly between these two districts than would have been the case if their CAMPFIRE guidelines had been in force. The decisions made by the ward communities in Bulilima Mangwe to distribute revenues equally between them similarly ignored the concept of the producer community. Their decision was based on people's historic rights of access to winter grazing and it underscores the inter-relatedness of communal resources. In this case the benefits derived from wildlife offset the costs of reduced access to grazing land. Finally, the Nyaminyami example illustrated how a District Council sought to relate differential input costs (crop and livestock damage) to differential benefits through the medium of compensation. It remains to be seen whether the wards in Nyaminyami will develop their own compensation schemes.

The range of options for distributing wildlife benefits is clearly wide. CAMPFIRE is a dynamic program which demands that management be adaptive. Perspectives change; in the past two years the MLGRUD has increasingly accepted the need for devolution of authority below District Council level. In some districts cash has been distributed to wards, whilst in other districts ward communities have opened bank accounts. Such dynamics require that guidelines, incorporating sanctions and incentives, are not overly-prescriptive. The risk is that prescriptive guidelines are interpreted religiously leaving little room for District Councils or communities to be flexible in their application. It is, after all, local communities living with wildlife who will decide if the individual returns from investments in these resources make them worth conserving.

NOTE

1. The author is grateful to Rob Monro, James Murombedzi and Liz Rihoy for reviewing the text and making helpful suggestions to improve this chapter. The views represented, however, remain those of the author.

REFERENCES

Bromley, D.W. and M.M. Cernea, 1989.
    The Management of Common Property Natural Resources: Some Conceptual and Operational Fallacies. World Bank Discussion Paper No. 57. World Bank, Washington, D.C.
Buck, S.J., 1989.
    Multi-Jurisdictional Resources: Testing a Typology for Problem-Structuring. In: F. Berkes (ed), *Common Property Resources: Ecology and Community-based Sustainable Development.* pp. 127-47. Belhaven Press, London.
Child, B.A., 1991.
    CAMPFIRE Update: 1991 Report. Unpublished Report. Department of National Parks and Wild Life Management, Harare.
DNPWLM, 1991.
    Guidelines for CAMPFIRE. Unpublished Report. Department of National Parks and Wild Life Management, Harare.
DNPWLM, 1992.
    Policy for Wildlife. Government Printer, Harare.

GoZ., 1984a.
   Provincial Councils and Administration in Zimbabwe: Statement of Policy and a Directive by the Prime Minister. Government of Zimbabwe, Harare.
GoZ., 1984b.
   Zimbabwe 1:1,000,000 Natural Regions and Farming Areas Second Edition 1984. Surveyor-General, Harare.
Hawkes, R.K., 1991.
   Crop and Livestock Losses to Wild Animals in the Bulilima Mangwe Natural Resources Management Project Area. CASS/MAT Working Paper Series 1/91. Centre for Applied Social Sciences, University of Zimbabwe, Harare.
Martin, R.B., 1986.
   Communal Areas Management Programme for Indigenous Resources (CAMPFIRE). Revised Version. Department of National Parks and Wild Life Management, Harare.
MLGRUD, 1991.
   Speech by the Minister of Local Government, Rural and Urban Development at the Annual General Meeting of the CAMPFIRE Association of District Councils on 15th July 1991. Ministry of Local Government, Rural and Urban Development, Harare.
Murphree, M.W., 1989.
   Research on the Institutional Contexts of Wild Life Utilization in Communal Areas of Eastern and Southern Africa. Centre for Applied Social Sciences, University of Zimbabwe, Harare.
Murphree, M.W., 1991a.
   Preface to *CAMPFIRE in Rural Development: the Beitbridge Experience*, B. Child and J.H. Peterson Jnr. Working Paper 1/91. Department of National Parks and Wild Life Management, Harare.
Murphree, M.W., 1991b.
   Communities as Institutions for Resource Management. Centre for Applied Social Sciences, University of Zimbabwe, Harare.
Mzilethi, J.Z., 1992.
   Speech by the Provincial Administrator, Matebeleland North Province, on the official opening of the Provincial CAMPFIRE Workshop on 19th June 1992, Ministry of Local Government, Rural and Urban Development, Harare.
Nyaminyami District Council, 1990.
   Minutes of the Nyaminyami Wildlife Management Trust Board of Management meeting held on 23rd October 1990. Siakobvu.
Nyaminyami District Council, 1991.
   Minutes of the Nyaminyami Wildlife Management Trust Annual General Meeting held on 16th February 1991. Siakobvu.
Oakerson, R.J., 1984.
   A Model for the Analysis of Common Property Problems. Paper prepared for the Common Property Steering Committee, Board on Science and Technology for International Development (BOSTID). National Research Council, Harare.
Peterson, J.H.Jnr., 1991.
   CAMPFIRE: A Zimbabwean Approach to Sustainable Development and Community Empowerment through Wildlife Utilization. Centre for Applied Social Sciences, University of Zimbabwe, Harare.
Thomas, S.J., 1992.
   An Estimate of Communal Land Population Density in the Natural Regions of Zimbabwe. Zimbabwe Trust, Harare.

10. SCIENCE AND TRANS-SCIENCE IN THE WHALING DEBATE

MILTON M.R. FREEMAN

INTRODUCTION

The widespread call for an end to the consumptive use of whales presents certain political and societal problems. The reason this proposition remains troubling is because, even though whales are no longer used for food or trade in certain countries, in some societies whales continue to be of great importance. This importance is especially the case where traditional socio-cultural and economic institutions and dietary needs are sustained by harvesting marine food resources which customarily have included whales.

Such continuing dependence on harvesting and consuming whale products is especially pronounced in cold-sea regions or in those regions where land-based food production is constrained by geography or climate (Bockstoce *et al.* 1982; Akimichi *et al.* 1988; Joensen 1990; IWC 1992; Kalland and Moeran 1992; Pálsson 1992). Unfortunately for the whalers (and the consumers of edible whale products), these producer communities are small in both number and size, economically weak, and politically peripheral.

Campaigns to end the killing of whales for food are waged by organizations which, in comparative terms, are very large in number and size, economically and politically well supported, and situated close to political centres of power. These anti-harvest organizations tend to ignore the social and cultural costs borne by the traditional users of whale resources as a consequence of their campaign successes. Moreover, within rapidly urbanizing western societies, where the production of meat and the realities of making a living in small and remote coastal communities are beyond most peoples' knowledge or experience, such campaigns have succeeded to a remarkable degree. However, the success of such appeals depends upon a marked lack of relevant information available to the public and politicians who embrace these emotive animal welfare causes in order to demonstrate their "environmental awareness".

This chapter suggests that such zeal is misplaced. Though whales, and especially their habitat, are worthy of conservation concern, campaigns to stop the sustainable use of non-endangered whales are both ecologically unsound and environmentally unfriendly.

TRANS-SCIENCE AND THE SUPER WHALE PHENOMENON

The perceived special status of whales is derived in part from their biological characteristics, some of which cause these animals to occupy "anomalous" categories compared to other "animals" [i.e. mammals]. Anomalous animals, ones

143

that possess anatomical or behavioral characteristics that depart from more usual configurations and therefore challenge conventional norms, are those that frequently give rise to taboo and enliven myth (Kalland 1993; Peterson 1993). One such anomalous feature of whales is the enormous body size of some of the baleen whales, a feature which has, since earliest times, captured the public imagination.

Similar anomalies exist in regard to behavioral characteristics of cetaceans. However, the trainability of some dolphins, and indeed the affinity they may exhibit toward humans, is not shared by all species of dolphin, nor indeed by all individual animals within a particular species (Defran and Pryor 1980; Prescott 1981; Pryor 1981; Lockyer 1990). Thus generalizing becomes difficult and in the absence of unequivocal scientific evidence it becomes easier for both scientists and non-scientists to advance as science what is no more than pure conjecture unsupported by any critical data.

This indulgence in what has been termed "trans-science" (see Note 1) is by no means uncommon when scientists assume an advocacy stance on a popular or political issue. Given the public attention that the mass media can accord to scientists speaking out on topical issues, it may take little more than subsequent citing of a trans-science observation in a mass-circulation magazine or newspaper article for a scientifically-unsubstantiated opinion to become uncritically accepted by large numbers of people as an established, "scientific", fact.

With regard to creating and sustaining scientifically-unsubstantiated conventional wisdom about whales, there is no shortage of seeming authorities to cite in support of fashionable, yet fanciful, notions. Such ideas have been responsible for perpetuating the myth of the "super whale" advanced by whale protectionists (e.g. Barstow 1988) and popularizers of science (e.g. Carl Sagan, in Sagan and Druyan 1992). The super whale, a fanciful creation combining real and imagined characteristics of different cetaceans, though not existing in reality, nevertheless serves as a proxy for all whales (Kalland 1993, 1994; see also Gaskin 1982:115–6).

Perhaps in no other area of discourse has the imagined characteristics of the super whale taken hold so prominently and erroneously as in the matter of the presumed "intelligence" of whales, porpoises and dolphins. It should be noted that there is no testable, nor even generally agreed upon definition of "intelligence" so that it comes to mean almost anything the writer understands the term to mean. In commenting on this phenomenon, one cetacean behaviour specialist has warned:

> many of those who have observed and reported on cetacean behavior are not versed in or are not even cognizant of this area of biological discipline, being engineers, physicists or the like. Because of the unfamiliarity of such investigators with mammalian social behavior in general ....there seems to have arisen in both public mind and the attitudes of the scientific community a tacit assumption that cetaceans cannot be compared behaviorally to better

known social animals. Even people who do not subscribe to any wishful thinking about intelligence seem to treat cetaceans as if the whole group somehow fell from Mars....rather than being normal products of evolution.

(Pryor 1986:91)

Creative exercises of this nature, even if known to be fanciful and misleading or only intended to be metaphorical, are rarely considered as harmful by their perpetrators, and are often actively disseminated in the belief that such characterizations provide popular support for whale conservation and, by extension, other environmental causes. However, it is a serious mistake to believe that such false and misleading information causes no harm.

IMPRESSION MANAGEMENT AND HARM/BENEFIT TO WHOM?

It is now apparent that disseminating false and misleading statements about environmental issues can have extremely damaging impacts on some people, even as these same messages may bring profit to others who derive financial reward from a public willing to send donations in support of environmental or animal welfare issues they become emotionally concerned about (Sanderson 1990).

The best documented cases of this damage to others relate to the individual, family and community destruction occurring in various aboriginal and non-aboriginal societies in North America and Greenland due to the anti-sealing campaigns of the 1970's and 1980's (Henke 1985; Herscovici 1985; Wenzel 1991). These damaging campaigns were based upon the twin falsehoods that the target species was endangered and that the method of killing the animal was cruel and inhumane (Canada 1986).

At the same time as the economic destruction of small seal-hunting societies in the northern regions was occurring, a number of animal welfare and animal rights' organizations profited handsomely from widespread public concern about the fate of seal pups and a consequent willingness of large numbers of people to contribute financially to protest campaigns directed to stopping the killing by ending the trade in the products of the hunt.

The profitability of such protest campaigns can be illustrated by the size of financial contributions made by one of the principal campaigners, the International Fund for Animal Welfare (IFAW). During the 1992 national election campaign in Britain, IFAW contributed £68,000 (ca.$100,000) to the Conservative party election fund. To gauge just how large this contribution was in the U.K. context, the 30 leading British industrial corporations contributed an average of about £42,000 each to the Conservative party campaign. The IFAW also contributed £100,000 to the Labour party that year, and a further £20,000 to the Liberal Democrats. The IFAW contribution to the Liberal Democrats was twice as large as the contribution made to that party by the Marks and Spencer re-

tailing giant (The Independent, June 16, 1993, and Sissons 1993; see also
Schwartz 1991, Spencer et al. 1991, Bonner 1993).

Justifying their Opposition to Whaling

The damage caused, by environmental and animal welfare organizations' anti-
whaling campaigns, to whaling communities in Japan for example is documented elsewhere (e.g. Bestor 1989; IWC 1989; Manderson and Hardacre 1989;
IWC 1992) so will not be discussed further here. What will be examined however, are certain principles invoked by a number of governments to justify
their continued opposition to the consumptive use of whales and certain ecological implications of such actions.

The regulatory device used to prevent whaling from taking place is a year-
by-year continuation of the pause in whaling (commonly known as the "moratorium") imposed by the International Whaling Commission affecting all whale
stocks subject to commercial harvest. This pause in commercial whaling was
adopted at the 1982 IWC meeting, to come into effect in 1986 and 1987 and to
remain in effect until IWC scientists had completed various management-rel-
evant tasks by 1990 at the latest, after which time other catch limits were to be
considered (IWC 1983:21).

The IWC currently consists of about thirty-six contracting governments, of
which a majority justify support for continuing the pause in commercial whaling by invoking the "precautionary principle". However, this universal principle disregards the advice of the IWC Scientific Committee which has recently
concluded that some whale stocks can now be safely harvested. However, the
majority of members of the IWC are no longer influenced by the best scientific
advice. Indeed, the more extreme whale protection advocates at the IWC assert
that due to the depletion of several whale stocks in the historic past, humankind now has an obligation to "whalekind" to allow all depleted stocks of
whales to increase to their pre-exploitation levels of abundance, and until that
has occurred no stocks of whales anywhere, even if non-depleted, abundant or
expanding, should be subject to commercial harvest.

In the context of IWC discussions (and the public at large in many countries) it is seen as entirely reasonable to assume a preservationist position in
the interest of allowing depleted stocks to recover where there is significant
uncertainty about whale numbers and biology. Within the IWC there was, at
the time of the moratorium vote in 1982, widespread acceptance of the need
for extreme caution. However, even at that time some scientists doubted the
need to apply a uniform ban affecting all whale stocks irrespective of their actual status just because other stocks elsewhere were depleted. Despite several
such reservations being voiced, a pause in commercial whaling was agreed to
by all the whaling nations, subject to an understanding at the time by all IWC
member governments that by 1990 the best scientific advice would be used to
decide whether continuing the ban was justified.

The Precautionary Principle: How Valid is It?

The precautionary principle has a number of problems associated with its formulation, interpretation and application. For example, the notion that there is a particular pre-exploitation population level that could or should be attained and then maintained is an ecological absurdity. Such a notion assumes an ecological stasis that has never existed, and it also assumes a knowable pristine ocean community, the complexity and continuous variability of which are in fact unknowable. Moreover, despite a belief that its management decisions and actions are important and justified, in practice the IWC cannot manage ocean ecosystems because, *inter alia*, its mandate only extends to a minute component of the total ecosystem, namely to a few of the grazers (the baleen whales) and one predator (the sperm whale) at the top trophic level of this multiplex biotic community.

Far from acting in a manner that would demonstrate its serious intent to manage whale stocks purposively, the IWC behaves as though sound management can be carried out in the absence of information. For example, efforts by certain member countries to carry out research to determine whale feeding relationships within the marine biotic community are stringently opposed by anti-whaling countries because such research is held "to contribute nothing to whale management". The underlying belief expressed by leading anti-whaling governments at the IWC assumes that if human action (including whale-killing) can be stopped, then the cetacean component of the ocean biotic community will in time return to its pre-exploitation (and presumed optimum) size and composition.

Despite these various failings of what might be called the historico-romantic view of whale population ecology, there is an even more serious failing associated with the uncritical application of the precautionary principle to the management of whales and whaling. This is the construction and belief in false opposites: that policy options involve choosing between risk vs. no-risk, or destruction vs. safety (of populations). In reality however, doing nothing does not eliminate risk or damage, for the replacement of one set of actions by another set of actions (or inaction as in this case) results in particular risks or damages being replaced by yet other risks or damages (Bodansky 1991). According to the IWC charter, any such damages or negative consequences are to be expressly considered in relation to management actions taken by the Commission. However, since the ban on commercial whaling went into effect, majority decisions at the IWC have consistently failed to take account of such documented impacts (Gambell 1993:106).

WHY A RESUMPTION OF CONTROLLED WHALING DOES NOT POSE A THREAT

The historic excesses associated with over-exploitation of the great whales occurred when whales provided an abundance of needed and inexpensive oil for

food and various industrial purposes. Once petroleum-based products replaced
train [whale] oil (at the end of the nineteenth century) and, later, vegetable oils
could be hydrogenated for use in margarine manufacture (in the 1960's), the
main justification for continued large-scale whaling ceased to exist for many
whaling nations. Thus by the 1970's and particularly the 1980's, whaling continued principally to provide meat in those few societies where it had historically been an important part of the diet. Most other whaling nations abandoned the increasingly unprofitable whaling and trading in whale products.

The continued use of whalemeat as food is generally restricted to quite localized regions within a few countries. Meeting this need requires small-scale
fisheries compared to the earlier large, industrial-scale, whaling operations
which supplied international markets for whale oil and other whale-derived
products (see Note 2).

It is necessary to stress this singular and significant difference between past
and present whale-product demand, because the precautionary principle and
the need to maintain a total ban on commercial hunting of whales is continually justified by opponents of whaling by invoking the spectre of an uncontrol-
lable return of large-scale highly-capitalized and technologically-sophisticated
industrial whaling fleets to the ocean commons. Thus, it is warned, any retreat
from the present total ban on all forms of commercial whaling will allow
many large industrial operators to begin re-entering this once-lucrative fishery.
This prognosis ignores the historic decline (starting in the 1960's) of commercial whaling due to loss of markets and the lack of profitability attendant upon
the imposition of small quotas by the IWC. Such market losses are irreversible
because of the ready availability of cheap substitute commodities, not to mention trade restrictions placed on the importation of all whale products by most
of the major former whale consuming and trading countries.

Few countries have included whale meat in their national or local cuisine,
and those that do have the fishery capacity to adequately meet the demands
of their own domestic markets (Conrad and Bjorndal 1993; Ward 1992), and
thus have economic incentive to keep out foreign supplies. Recent research
carried out in some of the present and former whaling countries indicated low
levels of demand by the general population for whale meat as food. Thus
unacceptability of whale meat as food ranged from around 79% in Germany
and 88% in the U.S. to 93% in Australia and the U.K. Even in those countries
(Germany and U.S.) where up to 20% of the population views whale meat as
"acceptable", there appears to be no consumer demand. Even in those few
countries (including Norway and Japan) where whale meat is, for some, a dietary staple, nevertheless about 40% of the population expressed non-accept-
ance of whale meat as food (Freeman and Kellert, this volume) suggesting a
significant degree of consumer resistance to expanding market demand.

In a recent study based upon fish and meat consumption in a large number
of countries, it is argued that those countries that were whaling before the
pause in commercial whaling was instituted, can be quantitatively distinguished on the basis of high consumption levels of marine living resources

(mainly fish) compared to their consumption levels of meat from domesticated livestock. Some of these fish-eating nations consumed whale meat on a daily basis and others less often before the whaling ban, and it is these countries that now face foreign opposition to their traditional fishery and dietary practices from those countries where meat eating (and meat exporting) historically assumed and continues to assume far greater dietary (and economic) importance (Nagasaki 1993:10–11).

Therefore, not only is the size of the market demand for whale products quite limited, but contrary to earlier periods of whale exploitation these few remaining markets have recently been largely or exclusively supplied by their own national fishery. In the absence of serious international (or even national) competition, it is difficult to take as a practical reality the claim that there is any real possibility of a return to the destructively competitive and highly capitalized "Whaling Olympics" of earlier years, an eventuality that some whalers today would be in the forefront of opposing as it would undermine the sustainability of their community-based fishery (Broch, this volume).

Whaling and Environmental Protection

In some western countries opposition to the consumptive use of whales is largely based upon a widespread belief that cruelty is necessarily involved in methods used to catch whales. In addition, some government policies against whaling is based upon fears about harvest sustainability and attendant species endangerment and loss of biodiversity resulting from presumed future whaling-management deficiencies.

Discussion of the alleged cruelty and associated ethical issues involved with whale capture occurs both outside (e.g. IFAW 1991; RSPCA 1991) and inside the IWC, where a Humane Killing Working Group considers technical aspects of the topic. However, these objections to whaling based on animal welfare and animal rights issues are beyond the scope of this chapter and are examined elsewhere (e.g. Freeman 1992; Lynge 1992; Sandoe 1993).

To assess the environmental issues however, it is useful to return to considering the precautionary principle. During IWC debates, this principle appears to be understood to mean that whaling should be suspended in order to eliminate risk of extinction of whales and thereby safeguard ecosystem integrity and biodiversity. An appropriate assessment of these issues involves evaluating risk in an ecological manner, for it is germane to ask: if whaling as a food producing activity is suspended, what environmental costs are incurred to replace that lost food?

Whales are a component of a marine ecosystem which is subject to various external influences, most notably from food extraction (i.e. fisheries) and pollution (principally from land-based sources). One source of land-based pollution involves chemicals entering the oceanic sink from intensive agriculture. Can sustainable whaling, through its contribution to food production (no matter how small in global terms), help to reduce the pollution impacts on the oceans and their biotic resources? If the answer is affirmative, then the precautionary principle, if ecologically rational and environment-directed, would tend to support a resumption of whaling.

Two main areas deserve consideration, both subsumed in the question: how environmentally-friendly is catching whales for human food as compared with other means of producing protein/food for human use? In environmental terms the costs of land-based agriculture involves wildlife habitat loss and associated loss of biodiversity, soil erosion, pollution and often unsustainable use of groundwater reserves, heavy use of fossil fuels, release of greenhouse gases (especially nitrous oxides and methane), and drug/herbicide/pesticide additions to the environment and human diet. If the greatest threats to the global environment are exacerbated by poverty, as many now maintain, then it is also relevant to consider the additional costs to the environment resulting from the poverty caused in the developing world by the trade-distorting agricultural subsidies maintained by the advanced industrial nations in support of their own agricultural producers (WCED 1987).

Marine fisheries (and especially whaling) certainly do not contribute to loss of habitat, nor to soil erosion or groundwater contamination and over-use; furthermore, no fertilizer, pesticide, antibiotic or growth hormone applications are required for high-quality protein production from wild fish and whale catching.

Energy Efficiency in Food Production

In regard to energy use and greenhouse gas production, marine food production is generally more environmentally-favourable compared with land-based food production. For example, modern agricultural meat-production practices are generally more energy demanding than are fisheries. Thus compared to small-type coastal whaling in Japan where fossil-fuel derived energy input to protein-energy output ratios are 2:1, farm-produced chicken, pork and feedlot beef production ratios in North America are 22:1, 35:1 and 78:1 respectively (Freeman 1991). It therefore appears that coastal whaling can generate energy savings of between 90–95% when compared to some fin-fisheries (see Table 10.1).

Fishery-Caused Threats to Biodiversity

The last area to be considered relates to the issue of maintaining biodiversity. It is accepted by all whaling countries today that any truly endangered species or stocks of whale would continue to receive total protection from commercial exploitation. Indeed, this has been accepted practice for some years now and long before the pause in commercial whaling was adopted in 1982. For example, the blue whale and the humpback whale have been totally protected from

Table 10.1 Fossil Fuel Energy Demands of Selected Fisheries
(After Pimental & Pimental 1979 & Freeman 1991)

Type of Fishery	Fuel Energy Input to Food Energy Output
Japanese coastal whaling	2.1 : 1
U.K. overall	20.0 : 1
U.S. cod & tuna	20.0 : 1
Australian shrimp	21.9 : 1
U.S. shrimp	27.0 : 1
Adriatic (small vessel)	67.7 : 1
Adriatic (large vessel)	105.6 : 1

commercial hunting since the mid-1960's and right whales and gray whales since the 1930's (Gambell 1993:98,100).

However, most whale species are not endangered and several stocks of the minke whale can, according to the IWC Scientific Committee, with the utmost safety sustain stable levels of harvests at this time. These assurances stem from the past seven years work of the IWC Scientific Committee resulting in the development of a catch limit algorithm that controls for the uncertainty inherent in whale biology and management, guards against the danger of stock extinction, and allows depleted whale stocks to recover at the same time as allowing a safe harvest to be taken (Cherfas 1992; Kirkwood 1992; Gambell 1993: 100–101).

In comparison to whale fisheries, the question of ensuring the sustainability of many non-whale commercial fisheries is far more problematic, with continued fishing of severely depleted stocks of food fish being the norm rather than the exception in most countries. For example, in the U.S. 65 food fish species out of 153 assessed by the National Marine Fisheries Service were recently reported to be overfished, meaning that fish are being caught faster than they are reproducing. These endangered stocks include Atlantic cod, haddock, Atlantic halibut, redfish, flounder, red snapper, various species of coastal sharks, Atlantic swordfish, blue marlin and blue tuna as well as several shellfish species (Specter 1992).

One reason that stocks of so many species may be adversely affected is that under many commercial fishing practices, very few if any net-fisheries can be wholly selective and thereby avoid coincidently catching over-exploited non-target species. Thus, in the presence of a continued fishery, even endangered species that are not the target of the fishery may continue to be netted. In contrast to this ongoing reduction in biodiversity, highly selective whale fisheries can be considered exemplary. In whale fisheries the targeted whale can be identified by species and by size (in some species by sex) before being taken, and a prohibition on, e.g. the catching of females with calves or whales of a particular size can therefore be effected (see Broch, this volume). Such selective fishing activity stands in strong contrast to the comparatively indiscriminate net-based fin fisheries.

Associated with fin fisheries may be a large by-catch (the unintended and unwanted part of the catch) that is frequently discarded. Again, no such waste occurs in today's whale fisheries, where markets exist for all edible products (including all organ meats and cartilaginous parts of the carcass, Ward 1992; Manderson and Akatsu 1993).The magnitude of this by-catch in non-whale fisheries is sometimes very large with significant implications for local or regional biodiversity. For example, it has been estimated that in 1983, nets set for halibut and shark in California killed 1,000 sea lions, 100 harbour seals, 100 sea otters, 30 pilot whales, and numbers of gray, humpback and fin whales. In one study of 133 halibut nets in central California, there were more seabirds caught than halibut, with a total seabird mortality of over 25,000 reported (Specter 1992). Similar figures come from coastal nets employed elsewhere; e.g., in Australia, nets are deployed to make bathing areas safe from great white sharks. Unfortunately these nets whilst providing protection to bathers also constitute "walls of death" for more species than the great white shark. As sharks cannot swim backwards, any encountering the net are likely to be drowned rather than merely deterred. Over a sixteen-year period, shark-free bathing in Queensland resulted in the death of 20,500 sharks of different species (many of which are not dangerous), and in addition, 10,889 rays, 465 dugong, 317 porpoises and 2,654 sea turtles drowned (Farino 1990).

CONCLUSIONS

The rational discussion of whaling, a fishery which remains important for social, economic, cultural and dietary reasons in some societies, is emotionally clouded by the popular conception of whales as constituting a special class of animals. This presumed "special" nature of whales derives from a widespread belief that whales are intelligent, endangered, killed by methods that are cruel, and the products they provide are no longer needed.

This chapter has only touched on certain of these issues; in each case however, it is concluded that these ideas, in varying degrees, are not in accord with the facts.

The IWC Scientific Committee has now identified certain whale stocks that can sustain commercial exploitation, and has successfully developed a rigorous scientific procedure setting conservative catch quotas that guard against stock extinction whilst at the same time allowing a safe harvest to be taken and the stock, if depleted, to recover. The present pause in commercial whaling was understandable when adopted in 1982, given earlier excesses of industrial whaling operations and the ineffectiveness of the international whaling management regime operating in those days.

However, those historic circumstances and events no longer exist, and present circumstances (including a newfound political will to take effective action to safeguard whales against over-exploitation) are such that a return to the former uncontrolled situation is practically impossible since no large-scale international commodity markets exists today for whale products.

Whaling today is carried out to supply a comparatively limited demand for food in those few societies where whalemeat constitutes a traditionally-valued staple in the food culture. As these limited domestic fisheries occur largely within coastal waters, future whaling would appear to constitute a controllable and justified component of limited-access national fisheries. In response to the oft-heard argument that whales belong to all nations and not just the exploiting nations, this is also true for many other migratory living resource stocks, but in these other cases equitable sharing arrangements have been negotiated through international treaties (Young and Osherenko 1993).

The proclaimed "environmental" justification for opposing a resumption of controlled whaling is not sustained by the ecological evidence. Whaling today, is carried out exclusively to provide high quality human food; the environmental costs of providing nutritionally-equivalent land-based or marine non-whale food substitutes in place of whalemeat are extremely high. In addition, the loss of a specialized cultural adaptation from the progressively reduced inventory of human food-producing activities, is not in the best long-term adaptive interests of the global human community. Future problems of providing food for an expanding human population, and associated issues of social equity and environmental quality, are apparent to most responsible international resource management agencies. These problems, experts advise, are likely to grow more acute as increasingly uncertain and unstable times are experienced in the coming decades. Under such circumstances, the lack of leadership and responsible collective action evidenced in the international whaling conflict is not only regrettable, but in environmental and human terms, appears dangerously irresponsible.

NOTES

1. Weinberg (1976) coined the term "trans-science" in relation to statements made by scientists that are not supported by established scientific evidence. Such utterances usually take place in a public, rather than scientific forum and where consequently lower standards of "proof" are asked for or offered. Such fora often involve the interaction of science and society (environmental issues being a prime area for such trans-science discourse to occur; see Freeman 1989). However, the absence of pertinent information validated by science need not deter a scientist from entering a public debate: it merely requires the scientist to be honest and explicit in outlining where science ends and trans-science begins.
2. The contribution that edible whale products make to any national dietary is quite small, and at the present time only Iceland, Japan and Norway express interest in resuming commercial whaling. In each country, given the increasing diversification of dietary sources over the past decades, the consumption of whale products as a proportion of total protein intake has fallen. However, this overall reduction in

national per capita consumption obscures the fact that in several whaling districts or at particular times of the year in some of these countries, whale meat (and other edible parts) are considered important and cherished food for which a faithful market continues to exist.

REFERENCES

Akimichi, T., P.J. Asquith, H. Befu, T.C. Bestor, S.R. Braund, M.M.R. Freeman, H. Hardacre, M. Iwasaki, A. Kalland, L. Manderson, B.D. Moeran and J. Takahashi, 1988.
  Small-Type Coastal Whaling in Japan: Report of an International Workshop. Occasional Publication No. 27, Boreal Institute for Northern Studies, Edmonton.
Barstow, R., 1988.
  Beyond Whale Species Survival: Peaceful Coexistence and Mutual Enrichment as a Basis for Human/Cetacean Relations. Paper presented at Whales in a Modern World Symposium, Zoological Society of London, November 26, 1988.
Bestor, T.C., 1989.
  Socio-economic Implications of a Zero-catch Limit on Distribution Channels and Related Activities in Hokkaido and Miyagi Prefectures, Japan. Document IWC/41/SE 1. International Whaling Commission, Cambridge,
Bockstoce, J., M.M.R. Freeman, W.S. Laughlin, R.K. Nelson, M. Orbach, R. Petersen, J.G. Taylor, R. Worl and W. Arendale, 1982.
  Report of the Anthropology Panel. *Reports of the International Whaling Commission,* Special Issue 4:35–49.
Bodansky, D., 1991.
  Scientific Uncertainty and the Precautionary Principle. *Environment* 33(7):4–5,43–44).
Bonner, R., 1993.
  At the Hand of Man: Peril and Hope for Africa's Wildlife. Alfred A. Knopf, New York.
Broch, H.B., 1994.
  North Norwegian Whalers' Conceptualization of Current Whale Management Conflicts. In: M.M.R. Freeman and U.P. Kreuter (eds) *Elephants and Whales: Resources for Whom?*, pp. 203–218. Gordon and Breach Science Publishers, Switzerland.
Canada, 1986.
  Seals and Sealing in Canada. 2 volumes. Royal Commission on Seals and the Sealing Industry in Canada, Ottawa.
Cherfas, J., 1992.
  Whalers Win the Numbers Game: Whales can be Hunted without Risk of Extinction. *New Scientist,* July 11, 1992, p. 12–13.
Conrad, J. and T. Bjørndal, 1993.
  On the Resumption of Commercial Whaling: The Case of the Minke Whale in the Northeast Atlantic. *Arctic* 46:164–171.
Defran, R.H. and K. Pryor, 1980.
  The Behavior and Training of Cetaceans in Captivity. In: L.M. Herman (ed), *Cetacean Behaviour: Mechanisms and Functions,* pp. 319–362. John Wiley, New York.
Farino, T., 1990.
  Sharks: The Ultimate Predators. Gallery Books, New York.
Freeman, M.M.R., 1989.
  Gaffs and Graphs: A Cautionary Tale in the Common Property Resource Debate. In:

F. Berkes (ed) *Common Property Resources: Ecology and Community-based Sustainable Development,* pp. 92–109. Belhaven Press, London.

Freeman, M.M.R., 1990.
A Commentary on Political Issues with Regard to Contemporary Whaling. *North Atlantic Studies* 2(1-2):106–116.

Freeman, M.M.R., 1991.
Energy, Food Security and A.D. 2040: The Case for Sustainable Utilization of Whale Stocks. *Resource Management and Optimization* 8(3–4):235–244.

Freeman, M.M.R., 1992.
Why Whale? Do Ecology and Common Sense Provide any Answers? In: O.D. Jonsson (ed), *Whales and Ethics,* pp. 39–56. University of Iceland Press, Reykjavik.

Freeman, M.M.R. and S.R. Kellert, 1994.
Public Attitudes to Whales, Whale Management and Whale Product Use: Results of a Six-Country Survey. In: M.M.R. Freeman and U.P. Kreuter (eds) *Elephants and Whales: Resources for Whom?,* pp. 293–315. Gordon and Breach Science Publishers, Switzerland.

Gambell, R., 1993.
International Management of Whales and Whaling: An Historical Review of the Regulation of Commercial and Aboriginal Subsistence Whaling. *Arctic* 46:97–107.

Gaskin, D.E., 1982.
The Ecology of Whales and Dolphins. Heineman, London.

Henke, J.S., 1985.
Seal Wars: An American Viewpoint. Breakwater Books, St. John's, Newfoundland.

Herscovici, A., 1985.
Second Nature: The Animal Rights Controversy. CBC Enterprises, Montreal, Toronto, New York.

IFAW, 1991.
Commercial Whaling Humane? International Fund for Animal Welfare, Crowborough, U.K.

IWC, 1983.
Chairman's Report of the Thirty-Fourth Meeting. *Report of the International Whaling Commission* 33:20–42.

IWC, 1989.
Report to the IWC Working Group on Socio-economic Implications of a Zero Catch Limit. Submitted by the Government of Japan. Document IWC/41/21. International Whaling Commission, Cambridge.

IWC, 1992.
Norwegian Small Type Whaling in Cultural Perspective. Report Prepared by the International Study Group of Norwegian Small Type Whaling. Document IWC/44/SEST 1. International Whaling Commission, Cambridge.

Joensen, J.P., 1990.
Faroese Pilot Whaling in the Light of Social and Cultural History. *North Atlantic Studies* 2(1–2):179–184.

Kalland, A., 1993.
Management by Totemization: Whale Symbolism and the Anti-whaling Campaign. *Arctic* 46:124–133.

Kalland, A. and B. Moeran, 1992.
Japanese Whaling: End of an Era? Curzon Press, London.

Kirkwood, G.P., 1992.
    Background to the Development of Revised Management Procedures. *Report of the International Whaling Commission* 42:51–82.
Klinowska, M., 1989.
    How Brainy are Cetaceans? *Oceanus* 32(1):19–20
Klinowska, M., 1992.
    Brains, Behaviour and Intelligence in Cetacea (Whales, Dolphins and Porpoises). In: O.D. Jonsson (ed), *Whales and Ethics*, pp. 23–37. University of Iceland Press, Reykjavik.
Lockyer, C., 1990.
    Review of Incidents Involving Wild, Sociable Dolphins, Worldwide. In: S. Leatherwood and R.R. Reeves (eds) *The Bottlenose Dolphin*, pp. 337–353. Academic Press, San Diego.
Lynge, F., 1992.
    Arctic Wars: Animal Rights and Endangered Peoples. University of New England Press, Hanover, N.H. and London.
Manderson, L. and H. Hardacre, 1989.
    Small-type Coastal Whaling in Ayukawa: Draft Report of Research. Document IWC/41/SE 3. International Whaling Commission, Cambridge.
Manderson, L. and H. Akatsu, 1993.
    Whale Meat in the Diet of Ayukawa Villagers. *Ecology of Food and Nutrition* 30:207–220.
Nagasaki, F., 1993.
    On the Whaling Controversy. pp. 6–20 In: *Whaling Issues and Japan's Whale Research,* pp. 6–20. The Institute of Cetacean Research, Tokyo.
Pálsson, G., 1992.
    Modes of Production and Minke Whaling: The Case of Iceland. In: *The Report on the Symposium on Utilization of Marine Living Resources for Subsistence,* pp. 68–82. The Institute of Cetacean Research, Tokyo.
Peterson, J.H. Jr., 1993.
    Epilogue: Whales and Elephants as Cultural Symbols. *Arctic* 46:172–174.
Pimental, D. and M. Pimental, 1979.
    Food, Energy and Society. John Wiley, New York.
Prescott, J. H., 1981.
    Clever Hans: Training the Trainers, or the Potential for Misinterpreting the Results of Dolphin Research. *Annals of the New York Academy of Sciences* 364:103–136.
Pryor, K., 1981.
    Why Porpoise Trainers are not Dolphin Lovers: Real and False Communication in the Operant Setting. *Annals of the New York Academy of Sciences* 364:137–143.
Pryor, K., 1986.
    Non-acoustic Communicative Behavior of the Great Whales: Origins, Comparison and Implications for Management. *Reports of the International Whaling Commission, Special Issue* 8:89–96.
RSPCA, 1991.
    The Cruel Seas. Man's Inhumanity to Whales. Royal Society for the Prevention of Cruelty to Animals, Horsham, U.K.
Sagan, C. and A. Druyan, 1992.
    Shadows of Forgotten Ancestors. Random House, New York.

Sanderson, K., 1990.
   Grindadrap: The Discourse of Drama. *North Atlantic Studies* 2(1–2):196–204.
Sandøe, P., 1993.
   Do Whales have Rights? In: G. Blichfeldt (ed) *11 Essays on Whales and Man,* pp. 16–20. High North Alliance, Reine, Norway.
Schwartz, U., 1991.
   Umweltkonzern im Zwielicht. Geldmaschine Greenpeace. *Der Spiegel* 45(38), September 16:84–105.
Sissons, M. 1993.
   In the Fields of Fury: Concern for Animal Welfare has been Exploited by Animal Rights Groups on a Tide of Political Correctness. *The Daily Telegraph,* November 27, 1993.
Specter, M., 1992.
   The World's Oceans are Sending an S.O.S. *The New York Times,* May 3, 1992, p. E-5.
Spencer, L. with J. Bollwerk and R.C. Morais, 1991.
   The Not so Peaceful World of Greenpeace. *Forbes,* November 11:174–180.
Ward, S., 1992.
   Biological Samples and Balance Sheets. The Institute of Cetacean Research, Tokyo.
WCED (World Commission on Environment and Development), 1992.
   Our Common Future. Oxford University Press, Oxford and New York.
Weinberg, A., 1976.
   Science in the Public Forum: Keeping it Honest. *Science* 191:341.
Wenzel, G., 1991.
   Animal Rights, Human Rights: Ecology, Economy and Ideology in the Canadian Arctic. University of Toronto Press, Toronto and Buffalo, N.Y.
Young, O.R. and G. Osherenko (eds), 1993.
   Polar Politics: Creating International Environmental Regimes. Cornell University Press, Ithaca, N.Y.

11. WHOSE WHALE IS THAT? DIVERTING THE COMMODITY PATH

ARNE KALLAND

INTRODUCTION

At a conference on marine mammals held in Bergen, ca. 1980, participants esti-mated the "low-consumptive", that is the non-lethal consumption-value of cetaceans to be about US$100 million, which was about the same value as the commercial whaling industries of the day (Scheffer 1991:17–18). However, since that time commercial whaling has virtually ceased, at least temporarily, while the low-consumptive value of whales has increased. What we have wit-nessed during the last two or three decades is, to use a phrase taken from Appadurai (1986), the diversion of a preordained commodity path, or altera-tion in the route a commodity (loosely defined as goods and services having exchange value) used to travel from production through consumption. The pause imposed on commercial whaling by the International Whaling Commis-sion (IWC) in 1987 has left whale protectionists as the main economic benefici-aries of cetaceans, which, together with seals, have turned out to be a very im-portant source of income for several environmental, animal rights and animal welfare groups.

This paper explores how whale protectionists have tried to transform meat, oil and other whale products into products of no exchange value and thus to remove them from the "commodity state" (Appadurai 1986:13). A number of strategies have been used by these anti-harvest groups to achieve this end. For many years an ecological discourse dominated the rhetoric: whales were be-lieved to be endangered, and the moratorium was introduced in the name of conservation. More recently knowledge of the population status of some whale stocks has improved considerably however, and it is now evident that some whales can sustain a carefully monitored harvest (e.g. Barstow 1989:10; U.S. Marine Mammal Commission 1991:2; Butterworth 1992). This has caused some groups, but by no means all, to switch from an ecological discourse to one based on animal welfare, whilst at the same time embracing the image of a "super-whale" which combines traits from a number of different species of cetaceans as well as from human beings (Kalland 1993).

A second aim of this paper is to show how environmental and animal rights groups have created a demand for a "green" conscience or a "green" legiti-macy and how they have been able to bestow these highly desired characteris-tics on those who sponsor the super-whale myth. By using the ecological and animal welfare discourses as the cultural framework for exchangeability (Appadurai 1986:13), business corporations have acquired green legitimacy

(and partial immunity from environmentalist attacks) by economically supporting environmental and animal rights organizations, while government agencies
have obtained some degree of political legitimacy from the public by similar
strategies. Both exchanges have been wrapped in the metaphors of "super-
whale" and "goodness". Thus, the super-whale has taken on a life of its own
as a commodity. Decommoditization of meat and oil and commoditization of
the super-whale are simultaneous processes in the diversion of the commodity
path, and both find legitimacy through the annual IWC meetings and other
"tournaments of value", where "central tokens of value in the society" are
contested (Appadurai 1986:21).

Finally, the paper discusses concepts of rights *in* or ownership *of* whales in
order to show how it has been possible to redefine the rules for appropriating
nature. It is the symbiotic relationship between environmental and animal
rights group on the one hand and some national governments on the other,
brought about by skilful manipulation of the ecological and animal welfare
discourses, which has enabled this coalition to appropriate almost all whale
stocks, to the exclusion of whalers.

WHALE PROTECTIONISTS AND THE "SUPER-WHALE" MYTH

The Whale Protectionists

"Save the whale" organizations have proliferated in recent decades and 74
whale conservation and research organizations are identified in the U.S. and
Canada alone (Corrigan 1991). Some organizations, like the American Cetacean
Society, the Cetacean Society International (CSI), the Whale and Dolphin Conservation Society (WDCS), and the Whale Fund are aimed specifically to protection of cetaceans, while others, for example, Greenpeace, the World Wide
Fund for Nature (also known as the World Wildlife Fund, WWF), and the International Fund for Animal Welfare (IFAW), are more general in their scope
but nevertheless earn substantial parts of their incomes from sea mammal protection campaigns.

People are attracted to the "Save the Whale" cause for a number of reasons.
Some are concerned about the environment, and believing erroneous reports
claiming that all whales are endangered, they take part in anti-whaling activities or in other ways support environmental organizations. They might lose interest and turn to other issues if and when scientific data convince them that a
carefully monitored exploitation of certain whale stocks does not put these in
jeopardy. This is the attitude of the largest Norwegian environmental organizations, which are no longer opposed to whaling.

Others are concerned about the well-beings of individual whales, and a distinction needs to be made here between animal rights and whale rights advocates respectively. The former are concerned with all kinds of animal, at least

in theory, as any animal has an intrinsic value on its own (Regan 1984:264). However, as pointed out by Tester (1991:14), it has never been the intention of these advocates to extend rights to all animals: rights are extended to an inner circle of mammals (although seldom to wild rats and mice) with reptiles, fish, and molluscs on the margins. Insects and bacteria for example have no rights.

The whale rights advocates take their rhetoric from the animal rights philosophers and explicitly limited it to cetaceans. To these people whales are uniquely special, based upon claims of their biological, ecological, cultural, political, and symbolic specialness. Being uniquely special, they merit special moral and ethical regard from people (Barstow 1991:7). However, Barstow stops short of claiming whales have special rights, but in a lengthy article D'Amato and Chopra (1991) argue for whale rights while explicitly denying such rights to all animals.

There is a tendency among the whale rights advocates to indulge in the super-whale myth, and in this they join company with those whose interest in cetaceans springs from spiritual sources. The director of Sea Shepherd Society and co-founder of Greenpeace, Paul Watson, for example, claims that in 1973, after he had been initiated into the Oglala Sioux tribe at Wounded Knee, a bison approached and instructed him in a dream to save all sea mammals, and whales in particular (Scarce 1990:97). Since then he has embarked on an aggressive crusade against whalers and sealers, in what he envisions as the Third World War (Gabriel 1991:56). However, not all cetacean protectionists are violent, of course. In the New Age movement, whales and especially dolphins, are often seen as sacrosanct: pure agents for a higher existence and awareness. In a leaflet distributed by a group calling themselves "Sanctuary in the Pines", dolphins are depicted as a kind of messiah, having been sent to the Earth so that the "crystalline energy within their brain will activate the energy of the earth's group mind". Once this has been accomplished "their mission will be done", and they will leave our planet. This theme can also be found in science fiction (as exemplified by the film *Star Trek IV*; see also Adams 1984), and in the beliefs that dolphins and human infants have similar brain waves which facilitate telepathic communication between infants (or even unborn children) and dolphins (Dobbs 1990:181; Cochrane and Callen 1992:28). Equally speculative are the claims that we have much to learn from communicating with cetaceans (e.g. Lilly 1961).

Finally, there are those who see in the anti-whaling campaign the potential for profit-making as well as for influence and fame. Greenpeace's interest in whales and seals, for example, seems to follow such considerations. "We are strategic opportunists", says Harald Zindler, leader of Greenpeace-Germany (Schwarz 1991:105), a view fully supported by its international director, Steve Sawyer, who has stated that the philosophy of the organization is very pragmatic: the leaders choose issues they are able to win (Pearce 1991:40). Greenpeace employs public opinion polling organizations to uncover popular issues among supporters (Schwarz 1991:99), but also readily adopts issues proven popular by other organizations. The whaling and sealing campaigns are cases in point, as both issues were well developed by others before Greenpeace entered the campaign: others had prepared the public and Greenpeace profitted handsomely from others' early efforts (see Note 1).

The importance of selecting winning issues is because fund raising is enhanced and success attracts people to the cause. A large number of members and supporters makes the organizations more able to function as strong pressure groups while at the same time enabling them to play the role of stewards of nature and distributors of green legitimacy. Before a campaign is launched, the chances of winning must be considered to be good, the issue must be a timely one and the campaign activities must be able to reach a wide audience through mass media (Eyerman and Jamison 1989).

The Super-Whale

A leading whale protectionist and scientific advisor to IFAW, has given two reasons why it should be easy to save whales, i.e. to win and thus divert the commodity path. One such reason is that:

> [whales] are extremely attractive forms of wildlife; some of them sing, and many people have become familiar with their underwater performances on film and video. In many cultures whales and dolphins have an ancient and revered place as intelligent and benign companions of humans... (Holt 1985:192–193).

Cetaceans are indeed animals to which symbolic significance can easily be ascribed. Whales form an anomalous category of animals (Kalland and Moeran 1992:5–6) since they do not fit into simple classification of mammals and fish: whales, it appears, are "betwixt and between", and it is exactly those anomalous animals, the ones difficult to fit into our cognitive maps, that become the object of myths and taboos (Douglas 1966). Moreover, whales move in salt water, and given the difficulty of knowing what is going on in the oceans, this particular realm is open to mystery, manipulation and myth creation (Pálsson 1991:95; Kalland and Moeran 1992:7–8). Finally, the ocean, consisting largely of salt and water which are both important purifying agents used in religious rites throughout the world, becomes *the* symbol of purity, and thus stands in sharp contrast to the polluted soil on which we land mammals tread. Various environmental and animal rights groups have skilfully played on our susceptibility towards whales because it is precisely our predisposition towards these animals which forms the basis for these groups' success in commoditizing whales based on their constructed premises.

Whale protectionists tend to talk about *the* whale in the singular, thereby masking the great variety in size, behaviour and abundance that exists among the 75 or more species of cetaceans. By lumping together traits found in a number of species an image of a super-whale is created (Kalland 1993). Thus "the whale" is the largest animal on earth (this applies to the blue whale), it has the largest brain on earth (the sperm whale), a large brain-to-body-weight ratio (the bottlenose dolphin), it sings nicely (the humpback whale), it has nurseries (some dolphins), it is friendly (the gray whale), it is endangered (the blue and right whales) and so on. By talking about *the* whale, an image of a single whale possessing *all* these generalized (but often questionably accurate) traits emerges. Such a creature does not exist.

Whales are often anthropomorphised by being given human traits as well. They are depicted as living in societies similar to our own: they live in "families" rather than in pods (e.g. Cousteau 1988:191), the humpbacks "compose" music (Johannsen 1990:83), and they think and feel like humans (e.g. Nollman 1990). The super-whale is endowed with all the qualities we would like to see in our fellow humans: kindness, caring, playfulness. While commercialization has penetrated most contemporary human relationships, leaving many people with a nagging conscience for not taking care of aging parents and for not giving the children the attention they need due to career responsibilities, whales are depicted as the guardians of old values now lost. The super-whale cares for the sick and dying, baby-sits and runs nurseries, without charging anything for these services. Not only does it care for its own kind, but it has also been reported to rescue humans in danger.

In short, whales "represent the closest approach to civilization, not as defined in terms of machine or technology, but as realized among all intelligent beings, cetacean or human, where communication and social bonds transcended the mere exigencies of life" (Abbey 1990:80). What *Homo sapiens* is on land, cetaceans are in the sea (Barstow 1991:7). They are our brethren, they have become "the humans of the sea" (Gylling-Nielsen 1987). In Lévi-Strauss' terms, whale society has become a metaphor for the lost human paradise or utopian world and caring for whales has become a metaphor for kindness, for being "good".

TURNING THE SUPER-WHALE INTO A COMMODITY

This super whale, with all its cetacean and human qualities, has proved to have enormous economic and political potential. What has turned this image into a commodity, however, is the emergence of a new demand among individuals, corporations and governments to appear green. This demand has been created by the growing environmental awareness among people, fuelled by the crisis maximizing strategies of many environmental groups. In the environmental discourse, whales have come to play the role of a metonym for nature and the image of the "endangered whale" has become a symbol for environmentalists. "Saving whales is for millions of people a crucial test of their political ability to halt environmental destruction" (Holt 1985:192): if we cannot save whales, then what *can* we save?

The whaling issue has become a symbol to the environmental and animal welfare movements because this issue provides them with an easily identifiable enemy and a sense of urgency, two factors a consultant to Greenpeace identified as the requirements for raising money (Spencer et al. 1991:179). The creation of both an enemy and a sense of urgency are closely associated with the animal welfare and ecological discourses, respectively. Tournaments of value provide the arena and ecological and animal welfare discourses provide the cultural framework that protect transactions between corporations, governments and environmental/animal welfare groups from being regarded as either bribes or blackmail.

Creating an Enemy

The war metaphor is a favourite one among the "Green Warriors". At one level the metaphor is used to convey an image of a uneven fight between "defenseless" whales and "greedy" whalers, often ending in "massacres". In the protectionists' rhetoric whales are depicted as "lovely", "gentle", "peaceful", "graceful", "magnificient", "delightful", "beautiful", "playful", "loyal", "innocent", and so on. "Whales and dolphins are one-dimensional beings. They are only *positive!*" writes Paul Spong (1992:25), who brought the anti-whaling issue into Greenpeace (Brown and May 1991:32), and who goes on to claim that we love to watch whales because "they model such a perfect form of existence that they take us away from ourselves". Another of Greenpeace's founding fathers, Patrick Moore, says that to get people to "save the whale you have to get them to believe that whales are good" (Pearce 1991:27). That the killer whale is the largest predator on Earth is concealed, as is recent research showing that most species are promiscuous, suitors often engage in brutal fights over females, commit "gang rapes" and might secure mates by chasing and harassing females into submission (Carpenter and Schmidt 1992:60–61; Winton 1992:18).

The whalers are portrayed as the whales' opposites: they are "cruel", "brutal", "reckless", "barbaric", "insatiably greedy", "butchers", "savages", "sadists", and so on. They are "pirates" engaged in "evil" and "criminal" activities "defying" international law. Some even suggest that whalers are more likely to act genocidally toward "inferior" human beings (D'Amato and Chopra 1991:27), and the World Society for the Protection of Animals (WSPA Circular No. 881406) suggests that children exposed to hunting activities are more likely to show violent, criminal behaviour toward others. One of the more extreme expressions, which emphasizes the contrast between peaceful innocent whales and savage whalers, comes from the IFAW view of Faroese pilot whaling, which is seen as:

> a <u>savage harvest</u> ... the most *brutal festival* you can imagine. *Peaceful* pilot whales are herded together and then *lured* close to land through their *loyalty* to a captured comrade from their pod or *family group*. And there they are simply <u>hacked to death</u>. (Davies 1985:1, underlining in original text, italics added; however, see Note 2).

The above quote brings to light yet another aspect of the anti-whaling rhetoric, namely how the whales are killed. Some of the more moderate organizations claim that whaling methods are inhumane and not worthy of civilized nations without going into further detail. Others however, and here some mass circulation newspapers take the lead, think that the more cruel the killing can be made to appear, the better: thus in one newspaper report the Faroese whalers "smash gaffs into the flesh" of the pilot whales and "hack" them to death, and, in another report, the "gentle" minke whale is left "to thrash around for hours in its death throes" (*Daily Mirror*, June 25, 1992, p. 30).

At another level the war metaphor seeks to divide mankind into "good" and "bad" people. In the world view of the whale protectionists, the positive qualities ascribed to whales are extended to people who "defend" them and "fight" against the "bad" whalers and their supporters. Through this process the anti-whalers create a totemic dichotomy of mankind, with whales as the totem for themselves and with money as the totem for the whalers (Kalland 1993), a world view strongly opposed by the whalers.

One of the oppositions in this scheme, is that between the civilized (whale saver) and uncivilized (whaler). Whalers are uncivilized, and to Sir Peter Scott, who was chairman of WWF-UK for more than 20 years, "no civilised person can contemplate the whaling industry without revulsion and shame at the insensitivity of our own species" (quoted in Davies et al. 1991:2). Caring about whales is represented as a sign of personal and societal maturity (Scheffer 1991:19) and a qualification for membership in the "world community" (Fuller 1991:2). Envisaged is a convergence of national cultures (Holt 1991:8), where all people feast on the super-whale myth. Opposing whaling provides an inexpensive way of satisfying people's demands for being "civilized" member of the "global village".

Whalers make excellent enemies, and because there are few nations engaged in whaling the cost of the moratorium is born by the few and makes whalers easily identifiable and thus ideal scapegoats for environmental disasters and human cruelty to animals. Moreover, whalers tend to live in remote areas with only limited possibilities to influence central governments (see Note 3). Nor are their products regarded as "necessary" by the anti-whalers. This is the second reason why Holt (1985:193) believes it is easy to save whales.

The "needs" argument has been introduced by animal rights advocates in order to solve the contradiction between life and death, for as Albert Schweitzer observed (1950:189), life depends upon taking life. To Peter Singer (1978:9), because "animals should not be killed or made to suffer significant pain except when there is no other way of satisfying important human needs, it follows that whaling should stop". Though Singer had nutritional needs in mind, there are other needs to consider: e.g., subsistence and cultural needs were taken into account when the IWC authorized aboriginal subsistence whaling (Donovan 1982). Most of the anti-whaling groups have endorsed

whaling by aboriginal groups to meet subsistence needs, even though compared to modern commercial whaling it might be considered less humane due
to the simpler technologies employed.

Whereas a concession has been given to aboriginal peoples who are allowed
to catch whales to satisfy subsistence and cultural needs, this consideration has
been denied commercial whalers. Whalers' needs are evaluated in strictly material terms, and it is argued that commercial whalers have no need now to
catch whales because they share the general prosperity of the capitalist societies in which they live and work. They can afford to buy pork, beef and tur-
key. However, whaling is more than making a living, it is a way of life and
must be seen "as a process whereby hunters mutually create and recreate *one
another*, through the medium of their encounter with prey" (Ingold 1986:111).
In denying their cultural needs, the protectionists adopt an extreme materialis-
tic attitude, which may surprise many people who have taken the protectionists generally anti-capitalist rhetoric at face value. Nevertheless, by arguing
that there is no need to kill whales, anti-whaling campaigners represent whaling as a "senseless" activity which can only be understood in terms of "greed"
and "short-term profit", and the whalers as represented in a negative way,
making them easy targets for the protest campaigns.

Exaggerating Crisis

A recent survey conducted in four non-whaling countries (Australia, England,
Germany, U.S.) and two whaling countries (Japan and Norway) indicates that
knowledge about whale populations is poor, particularly in the non-whaling
countries where between 65 and 70 per cent believed (wrongly) that all large
species are in danger of extinction (Freeman and Kellert, this volume). But
only a few species are in fact endangered (Aron 1988). The minke whale,
which Greenland, Iceland, Japan, and Norway want to harvest, can hardly be
regarded as endangered with estimated stocks at around 100,000 in the North
Atlantic and 750,000 animals in the Antarctic (where it may well be more
abundant than ever before, Gulland 1988:44).

It seems to be easier for the animal and whale rights groups than for the environmental groups to accept the new estimates of whale populations since the
argument of the former is not based on conservation or ecological considerations but on the ethics of killing whales, such that their argument is thus not
endangered by higher numbers of whales. To those groups who profess to be
concerned with conservation issues however, the logical consequence of higher
estimates ought to be a feeling of relief accompanied by a transfer of effort to
more urgent matters. However, for the most part this has not happened,
though some anti-whaling advocates have changed their arguments from ecology to ethics, thus crossing the line between environmentalism and animal
welfare, while others stubbornly keep to the conservation discourse (claiming
that all whales are endangered) or argue both perspectives at once.

At the government level, the U.S. then-commissioner to the IWC is reported to have said in a 1991 interview (*Marine Mammal News*, 17(5):4, May 1991) that he would continue to defend the U.S. position (against commercial whaling) on ethical grounds since he could no longer object on scientific grounds. And the British then-Minister of Agriculture turned to the argument of inhumane killing methods after first admitting that minke whales are plentiful (*The Guardian*, May 27, 1991; *The Times*, May 29, 1991).

In the anti-whaling organizations, the president of CSI writes that "the science is now on [the whalers'] side. We can't even talk about extinction. Our arguments now focus on ethical, aesthetic, and moral reasons for the protection of the individual whale, not the population or the species" (Shields 1992). A WWF-USA spokesman admitted that the organization will "have a hard time continuing" to argue ecologically (Bright 1992:69), leading WWF to adopt a position that "even if the IWC ... could guarantee that whaling was only carried out on a truly sustainable basis, WWF would remain opposed to the resumption of whaling" (WWF 1992:1) and would seek changes to the Whaling Convention from one regulating whaling to one protecting whales (Sutton 1992:2). Greenpeace has adopted a similar position (Ottaway 1992:3).

Knowing that ecological arguments against whaling are more palatable for various reasons than are ethical and moral ones to a number of people, corporations, and government agencies, and realizing that the "terms of the [whaling] convention have required that this debate be conducted in a scientific guise" (Butterworth 1992:532), many protectionists are more than reluctant to change their rhetoric from an ecologial discourse to animal welfare or animal rights arguments. Thus the myth of the endangered whale is sustained by charging scientists producing new and larger stock estimates of being incompetent, biased and "bought" by governments of whaling nations, by refusing to accept these new population estimates or refuting their relevance, or by introducing new arguments into the ecological discourse.

Some anti-whaling groups continue to argue as if all whale species are close to extinction. Greenpeace launched a "Save the last whales" as late as 1992, just in time for the annual IWC meeting, and wrote that the Norwegian government "seems hell bent on waging a war of eradication on marine mammals" (Ottaway 1992:13). And in its "SOS Save the Whale" campaign WWF-Denmark recently rather emotionally appealed for support to save the last whale (WWF-Denmark 1990). Many of these organizations, as well as media, live by crisis maximization and, by giving the impression that the moratorium is about to be lifted (e.g. *The Mail on Sunday*, June 21, 1992), they exploit upcoming IWC meetings to launch fundraising campaigns. Such campaigns tend to exaggerate the potential scope of commercial whaling: "once again the blue whale ... will be ruthlessly hunted, although there are probably less than 1,000 remaining from 250,000 that used to roam the oceans" complained *The Mail on Sunday* (June 21, 1992, p. 8). Beside offering an incredibly low figure for the remaining blue whale population (see Baskin 1993), the newspaper fails

to mention that the blue whale has been protected since 1965 and that this total protection is not in any danger of being removed.

It is also claimed that the whale population size is irrelevant because commercial whaling will, by some asserted inevitability, lead to over-exploitation
and extinction. This argument is based upon the early history of pelagic, industrial-scale, whaling (e.g. Greenpeace International 1992:1), and ignores the
progress made in IWC's management procedures during the 1970s when the
IWC entered a short period of science-based whale management (Hoel 1986;
Freeman 1990). The argument further denies the possibility that mankind has
the ability to learn from past mistakes. Finally, the argument overlooks important differences in whaling regimes. To liken contemporary minke whaling
with historic industrial whaling distorts the issue because, among other reasons: (1) whereas the main product for industrial whaling was oil, the most
important product for minke whaling has always been meat for which there is
a limited market; and (2) the minke whaling boats are, with the exception of
Japanese minke whaling in the Antarctic, small and family operated under an
economic rationality different from that of the large pelagic expeditions (ISG
1992; Kalland and Moeran 1992). In reality there is thus little continuity between the two forms of whaling.

A recent strategy used by protectionists is to bring in new arguments into
the ecological discourse (Butterworth 1992), a strategy whaling nations perceive as foot-dragging. While the protectionists may accept the new stock size
estimates, they argue that knowledge about fertility and mortality is insufficient, that the impacts on whale resources from pollution, depletion of the
ozone layer, fishing and other human activities are unknown, that stock by
stock management is some distance away as definitive separation of stocks has
not been achieved, that inspection and enforcement measures of great
specificity must be written into the revised management procedure, and so on.

Tournaments of Value

The images of enemies and crisis are brought into prominence during tournaments of value, and the media have turned some of these encounters into major events. The most spectacular are the annual IWC meetings in which privileged participants compete for status, rank, fame, and reputation by contesting
central values in an attempt to diversify "culturally conventionalized paths"
(Appadurai 1986:21). Member countries send delegations ranging in size from
the large Japanese and U.S. contingents numbering more than thirty delegates
down to one-person delegations, with each country having one vote regardless
of delegation size. In addition to diplomats and bureaucrats, delegations may
also include scientists (mainly natural scientists but in recent years also social
scientists), environmentalists and animal rights advocates (in the case of anti-
whaling nations) and whalers (in the case of whaling nations). Their task is to
contest opposing values pertaining to whales and their exploitation (Moeran
1992b) and to give scientific legitimacy to their positions.

A large number of non-governmental organizations (NGOs) are allowed to attend the proceedings, without rights to vote or speak. The NGOs tend to form two major blocks, the largest composed mainly of environmental and animal rights groups, the smaller group composed of those supporting indigenous peoples or sustainable whaling in general. The NGOs have several tasks during the IWC meetings. First, they lobby the delegates and try to convince the general public through media interviews and written releases that their world view is the correct one. In addition to press conferences and demonstrations, about 20 protectionist groups jointly issue a paper called *Eco* during IWC meetings, while the High North Alliance (representing North Atlantic and other community-based whalers' interests) publishes *The International Harpoon*. The NGOs also report directly back to their followers without going through the mass media, thus enabling them to control the distribution of information regarding their own activities. Finally, and this is most important for the present analysis, the NGOs monitor the proceedings and report their interpretations and evaluations of the delegates' performances to their supporters and to the mass media that do not have direct access to the conference room. By this arrangement the NGOs are in a privileged position to manipulate the flow of information, to exert pressure on national governments and particular politicians, to endorse certain politician's statements thus enhancing their electoral prospects, and to rank and laud the most "progressive" delegations and nations.

The presence of the mass media at IWC meeting is important, for without journalists' participation the meeting would provide a much less attractive arena for contesting the commodity path. But the press is severely restricted in their work as it has no direct access to the conference room, and only parts of the proceedings are transmitted to the press room. To overcome this deficiency, pauses in the proceedings have been turned into intense press-briefing sessions where the media rely heavily on the services of the NGOs and some of the delegates. With anti-whaling NGOs in a majority and with most of the media coming from nations whose governments and public for the most part favour anti-whaling positions, it is not surprising that anti-whaling sentiments dominate the newspaper columns and news broadcasts. Anti-whaling demonstrations held outside the conference building also contribute to this situation. Moreover, the media willingly reports from the latest anti-whaling publications, particularly if they are sensational in character and grossly exaggerate crises. In short, the IWC meetings provide environmental and animal rights groups a rare opportunity to get their message inexpensively before millions of people.

Whale strandings offer other occasions for tournaments of value, with different sets of participants. Whalers and their supporters are usually not present, so these tournaments are left to protectionists, business corporations, and government agencies which may compete for leadership by suggesting solutions, and in getting credit for the outcome of the rescue attempt. For example, in a recent stranding of 49 false killer whales in Australia, rescue work undertaken by several animal welfare groups, National Parks officials, employees of marine parks and the Sydney Zoo, the Army and the Salvation Army, suffered from conflicting advice and priorities, culminating in the walk-out of the Organization for Research and Rescue of Cetaceans in Australia (Wheatley 1992:14-15).

At times rescuers form unexpected alliances, as when Greenpeace in 1988 joined with the Alaskan oil industry and other corporations, Alaskan Eskimo whalers, and U.S. and Soviet government agencies to free three gray whales trapped in Alaskan ice. The media may well have turned the case into "the World's Greatest Non-Event" (Rose 1989), but in reality they were a tool in the public relations contest waged between the rescuers, each trying to outdo the others and not be left behind (see Note 4). Importantly, the media attention and all the calls from people wanting to help made it possible to link this widespread sympathy for whales to the anti-whaling groups' boycott against Icelandic fish products in an attempt to bring an end to that country's whaling activities and thus further remove whale meat from the commodity state.

Direct confrontations during the hunt constitute another important arena which helps to bring about a diversion of the path. Such confrontations usually take place far from shore and therefore provide the activists with the opportunity to invent news or to monopolize news coverage (Pearce 1991:20). Although a nuisance to whalers these actions are not necessarily meant to produce an immediate end to whaling activities, but rather, they tell the world that the activists are concerned about the environment, that the issue is urgent and cannot wait, and that they fight against powerful enemies and great odds. The activists are always depicted as underdogs: small inflatable zodiacs against the big steel catcher boat, or swimming Greenpeacers in front of a Japanese factory ship. The situation is ideally suited for presenting a picture of David fighting Goliath. By skilfully manipulating the mass media, enormous sympathy to their cause is elicited, which is one reason behind their success in having whale meat removed from the commodity state and placing the super-whale in its stead. It matters little that the picture is false and that with environmentalism being a multi-million-dollar industry in most industrialized countries, it is the whalers who are the weak party in this confrontation.

Consuming the Super-Whale

Through the mass media, the public can participate as spectactors in such tournaments of value, which are, moreover, important marketing devices for low-consumptive use of whales. Tournaments and low-consumptive use of whales allow millions of people to partake in the super-whale myth, which provides the backdrop and *raison-d'être* for various commercial activities based on whalewatching, movies and books featuring whales and so on. But the super-whale is a symbolic type of commodity and what is consumed is not really the symbol itself, but human relations (Moeran 1992b), for which the super-whale is a metaphor. Therefore, the feasting on the super-whale does not exhaust the super-whale but adds to its economic and political power, at least up to a certain point. Whalewatching in particular is seen as a means to educate people in the "proper" way of appreciating the qualities of the super-whale (Hoyt 1992:1, Ris, this volume) and thus give further impetus to the mystification of whales and the anti-whaling campaign. However, as will be discussed in the last section of this paper, by becoming too successful, the super whale might put itself in danger of being removed from the commodity state.

Whale Tourism

One of the first low-consumptive uses of cetaceans was the appearance of dolphinaria which still are very popular and have probably done more than anything else to foster emotional warm feelings toward cetaceans. But dolphinaria are not completely non-consumptive because the mortality rate of captive dolphins is quite high. Moreover, to view dolphins in captivity is no longer regarded as authentic or as fulfilling as seeing them in the wild, and animal welfare concerns have caused many protectionist organizations to condemn dolphinaria.

The "real thing", of course, is to watch whales in the wild, and it is estimated that more than 4 million people reportedly spent more than US$300 million on whale watching activities in 1991 (Hoyt 1992:1). Many of the environmental and animal rights groups organize whale safaris, and those organized by the WDCS from London must be among the most exclusive: a tour to Alaska costs from £2,995, to Baja California from £2,070, to Galapagos from £3,260, and to the Antractic and Falkland Islands from £5,350. More than 200 commercial whale-watch operators offer more than 250 different tours in North America alone, ranging in duration from an hour to a fortnight with prices varying from $7 to $3,000 (Corrigan 1991:7). The tours all offer special excitement: one operator invites tourists "to reach out and touch nature" while "travel[ling] in safety aboard a comfortable cruise vessel, in harmony with nature and at nature's own pace" (ibid:182). Some tours seek to enhance this "one-ness" with nature by using small kayaks so as to be less separated from the water environment, while others prefer to observe nature through panoramic windows from a deluxe bar aboard a liner carrying 700 passengers.

Many of these tours also feature cultural attractions: one offers homemade Portuguese specialties and folk music, another announces that "the captain will sing for you and dance with you as part of the entire cultural experience" (ibid:255) or tempt potential tourists with such attractions as visits to Indian and Inuit villages along the route. In one Newfoundland tour the human/cetacean encounter has been turned upside-down by guiding "people into our spectacular marine environment so that the whales can watch them!" (ibid:250). To offer something for everyone, many operators advertized customized trips which "exactly suit the desires of the client".

It is interesting to note that so many of the operators claim that they operate in an "area where whales abound" and are able to guarantee sighting success or a new trip. This impression is underscored by Hoyt (1992:1), who can calm those believing the rhetoric of these same organizations that whales are on the brink of extinction, by stating that all "the large whale species and many dolphins and porpoises can be seen regularly on a wide range of tours".

These whale watching trips often develop a cult-like character, and those who have witnessed the collective *oooohhhhhh* from deck when a whale "waves" its tail in "goodbye" will be struck by the strong sense of community aboard the vessel.

Swim with a Whale

Not everybody is satisfied by watching cetaceans from the deck of a ship however, and more and more people want to actually swim with cetaceans. In Hong Kong "dolphin-lovers" have broken into the Ocean Park at night, in order to have a free ride (Carter and Parton 1992:5), and in Western Australia people are waiting in line to be in the water with a group of dolphins (Winton 1992). Babies born in close proximity to dolphins are believed by some to be more harmonious and to develop exceptional talents (Dobbs 1990:181), and (some believe) might even develop into *Homo delphinus* (Cochrane and Callen 1992:30). In the U.S. there are several licenced dolphinaria with "swim-with-dolphins-programs" (Hatt 1990:247), and these institutions claim that their programs are of therapeutic value for handicapped and distressed people (however, see Beck and Katcher 1984, Cochrane and Callen 1992:32–37).

Others, who dislike the practice of keeping dolphins in captivity, have taken these programs into the wild. The relatively few friendly dolphins become famous and attract large crowds of people, some of whom are seeking therapy. To the initiate, dolphins like Jojo, Fungie, Donald, Opo, Percy, Simo, Horace, Dorah, and Jean Louis have become close friends and cult objects, and apparently it matters little in which corner of the globe they appear.

One of the most celebrated "dolphin therapists" is Horace Dobbs who holds a doctorate in psychiatry. In a series of books and movies he describes encounters with dolphins and the reactions among his depressed patients. Several of Dobbs' patients testified that they felt relaxed among dolphins. Being with the animal released them from the anxiety of having to perform or to live up to other people's expectations. In the company of the dolphin the individual can behave "naturally", as the following testimonial clearly shows:

> I felt like a Princess being taken away to another land by her Prince... My Prince was taking me into his world beyond the realms of fantasy... We were together as one ... I was him and he was me. Complete harmony and love....

> I did not speak, we communicated with our hearts. I was totally and completely in love ... This beautiful dolphin loved me for what I was in my

heart. It didn't matter whether I was old, young, fat or thin. I didn't have to impress him with a string of degrees. I was loved and accepted for myself, for the person I was. Simo was far superior to me in every way in the water. I did not have to compete, all the stresses of human values and life no longer existed. (Dobbs 1990:82–83).

Testimonials, through which individuals can share experiences and receive emotional support from a small group of likeminded people, is a common feature in many New Age sects. Those who share the experiences are typically described as good and sensible people. The only person mentioned in the book *Dance to a Dolphin Song* who did not have a spiritual experience with the dolphin is described as "the very fat, rich, American lady" who got only fleeting attention from the dolphin (Dobbs 1990:97). To be rejected by a dolphin can thus involve rejection by the support group, but when good people meet in the water marvellous things can happen (Dobbs 1990:95).

Whales in Literature and Arts

Whale-watching guide books have started to appear to cater for whale tourists, and books about friendly dolphins help to satisfy the "searching minds". Such books might well be the latest genres of whale literature appearing on the market. But it is by no means the only genres. Melville's *Moby Dick* was probably the first best selling book to have a whale as one of the main characters. Other novels have followed: *The Last Whales* (Abbey 1990) takes the genre in a new direction in that all the characters are whales. In this novel interspecies communication is common, and the oceans would have been a paradise had it not been for human beings and killer whales.

Whales are also featured in other genres of literature. A steady flow of beautiful picture-books are being published, and *Whale Nation*, an odyssey to whales combining exiting photographs and emotional poems, has already become a classic for many whale lovers (Williams 1988). Some writers of science fiction are also intrigued by cetaceans: in *Startide Rising* (Brin 1983) for example, the space craft *Streaker* is crewed by humans and neo-dolphins that communicate in Japanese *haiku* verse and are the product of genetic engineering. "Writers of science fiction have often speculated about what it would be like to discover, on a planet in outer space, a much higher form of intelligence"; D'Amato and Chopra continue: "Stranger than fiction is the fact that there already exists a species of animal life on earth that scientists speculate has higher than human intelligence" (1991.21). This is the whale. The early scientific authority upon which this thesis is based (Lilly 1961) is taken as proof, and ignoring all later scientists holding quite different opinions D'Amato and Chopra seem to have taken the step from science to New Age. They are not alone in so doing.

A number of books can best be described as examples of dolphin cults: *Dolphin Dreamtime*; *Behind the Dolphin Smile, Dance to a Dolphin's Song, Dolphins and Their Power to Heal, Pictures in the Dolphin Mind* and *The Magic of Dolphins* are all telling examples. At the same time, "cetacean artists" have appeared on the stage painting pictures or recording music in praise of whales and dolphins. Books, movies, videos, "whale music", art objects, stickers, posters, photos, stamps, bags, T-shirts, soft toys, buttons, jewellery, computer games, and so on provide nourishment and visualize people's commitment to the cause and thus help build a community of believers.

Selling Green Images

Individuals may go on whale watching tours, pay for a swim with a whale, buy some of the many artifacts carrying whale symbols, or send a donation to one of the many anti-whaling organizations and get peace of mind believing they have done something for the environment. Government agencies may also bolster their green image by supporting, and thus giving legitimacy to, the anti-whaling movement.

In recent years there has also been a green marketing boom in which companies try to take advantage of the ecological discourse. This can be done through "totemic classification" by which a relationship between nature and product is established, or through "eco-commercialism" by which a company can create an image that it is aware of the environment (Moeran 1992a:197–198). One way to achieve this is to claim in their advertisements that they are green, by planting a tree for each car sold, for example. However, such claims will be more credible if they can be endorsed by a known third party. It is precisely here that the environmental and animal welfare groups have a role to play. In the 1988 rescue of the gray whales in Alaska, for example, oil companies and other industrial corporations worked side by side with Greenpeace (Rose 1989), and the work was transmitted worldwide for everybody to see. It is in this context that many of the transactions between environmentalist and animal welfare groups on the one hand and industrial concerns on the other can best be understood.

The WWF, for example, has on several occasions been willing to sell a green image to companies in need of one. In Denmark, WWF allowed the Norwegian Statoil company to use the WWF logo in advertisements and to announce that the oil company supported WWF's work for endangered species. One million Danish *kroner* (about US$150,000) was considered a reasonable price to pay for having WWF endorse advertisements for products that are harmful to the environment.

More companies than Statoil are in need of a green image, and these may be actively targetted by environmental and animal rights organizations (see Note 5). In a letter to Danish business leaders, WWF-Denmark, after claiming that the whaling moratorium was about to be lifted and all known whale species were consequently about to be eradicated for all time, wrote:

Therefore I send you this SOS for assistance in WWF's fight for an extension of the moratorium. Here your company can give a cash contribution by sponsoring a whale for 50,000 *kroner*. The sponsorship will in a positive way connect your activities with WWF... Through a sponsorship your company has the opportunity to show your associates that it takes the environment and 'the green wave' seriously... I am sure that you will see the opportunities which a whale sponsorship will imply to your business. (WWF-Denmark 1990, author's translation from Danish.)

The letter seems to have brought about the results expected, and one of the companies that decided to sponsor a whale was the Danish chemical company, Brøste. In order to celebrate its 75-year anniversary the company placed an advertisement in the *Børsen* (the Danish equivalent of the *Wall Street Journal*) stating: "We bought a giant sperm whale from WWF, World Wide Fund for Nature, as a birthday present to ourselves... We know from the seller that he has more whales, in many sizes". The advertisement asked those offering congratulations to the company for its anniversary to donate money to WWF rather than send flowers to the company.

WWF has taken the lead in developing whale adoptions as a fund raising measure: WWF-Denmark alone has earned more than £200,000 from individuals and corporations that have "adopted", or "bought" as it is frequently termed, sperm and killer whales from along the north coast of Norway during the second half of 1990. The purchase price is about £4,500 for a sperm whale and £450 for a killer whale; if this is too much money to give, people are invited to co-sponsor a whale. Other organizations have similar programs: for example, WDCS offers "peaceful" orcas off British Columbia (*Sonar* No. 7:4–5; the term "killer whale" is not used in the advertisements for understandable reasons). The Mingan Island Cetacean Study in Quebec offers 210 blue whales, each costing $100 per person or $1,000 for a corporate adoption and Allied Whale announces that "about 35 individual finback whales are available for $30; a mother and calf cost $50" (Corrigan 1991:311). Save the Whales International in Hawaii invites interested people to "welcome a Hawaiian humpback whale into their extended family" (ibid:312). Through these programs, people can adopt whales in distant waters, in the same way as many people in rich countries "adopt" children in the Third World. And, as with the adoption of a child, the whale's "adoptive parents" are provided with a picture of "their child" and annual "progress reports".

In all the activities being discussed a feeling of belonging is created through the partaking in the super-whale myth. This is most obvious on whale watching tours and in "swim with a whale" programmes, but participation in whale saving operations and adoption programmes can also be a boost to company morale (cf. Rose 1989:234). By displaying the proper buttons, T-shirts, jewellery, photos, bags and art objects, the fact of belonging to the movement is communicated to the world. Moreover, through books, movies and computer games, consumers are educated to appreciate the qualities of cetaceans in the "correct" way, which is a learning process that starts early in the childhood with picture books and toys. However, the very success of the super-whale poses also a threat to its exotism and its future. Although the super-whale may survive, the cultural framework in which it has been commoditised may not, for should all whaling activities cease, then a *de facto* appropriation of the whales by the anti-whaling movement leaves the ecological and whale welfare discourses meaningless.

WHOSE WHALE?

Appadurai (1986) talks about the path through which a commodity travels from production to consumption. From a whaler's perspective, *he* is a part of such a path, a step in a long chain of interconnected parts. The whaler's task is to hunt and capture whales and bring the carcasses to processors or merchants after which meat goes one way in the chain of exchange and usually money goes the other way, although a lot of whale meat is also bartered or given away as gifts. The money earned by the whalers is spent on various things including equipment and provisions to continue whaling.

The whaler possesses only a fraction of the total body of knowledge which is required in order to bring a whale from the sea to the dinner table. However whaler knowledge is crucial, and includes how to find whales, how to identify the species, how to pursue them and maneouvre the boat so that the gunner can aim and hit a moving target from an unsteady position, how to secure the animal once shot, how to winch it aboard and flense it, how to handle the meat so that it does not spoil, and so on. The necessary knowledge regarding whaling also includes rules of the game, and in this context the most important rules pertain to rights in whales.

In fisheries which are based on free access to marine resources, ownership rights to fish are usually obtained through investment of labor. Fish caught on hooks or in nets are no longer nobody's property but now belong to somebody, usually the owner of the gear. Rights in fish can also be acquired by spotting or initiating a chase, and in whaling all principles are at work. Striking a whale implies in many cases ownership, even should the whale subsequently be lost. Old Norse laws, for example, have detailed regulations about ownership of whales found drifting with marked arrows, spears or harpoons in their bodies (Martinsen 1964; ISG 1992:22). Today, a whale is rarely lost and hitting a whale implies *de facto* ownership. But what about sightings? In Japan and Norway, at least, sighting in itself is usually not enough to appropriate a whale, a chase has to be initiated. Only when a whale has been sighted *and* a chase has been started will the other boats recognize the claim to ownership. But the claim is void if the whale escapes.

If only one whale can be chased at the time, it makes little sense to lay claim on a school of whales. This has far-reaching consequences for communication between boats. In Japan, where boats operate near land and are obliged in most cases to land the carcass at a designated shore station before nightfall, a boat can seldom take more than one whale or two on any one trip. It might be optimal in such a situation to inform other boats when locating a school of whales. In Norway, however, boats often operate far from shore and they may stay at sea for three or four weeks at a time. The caught whales are winched aboard, where they are flensed while the crew pursues a second whale. The meat is laid on deck to cool for at least 24 hours; thus a limiting factor is deck space and the stability of the boat. Most boats may be able to keep 20 metric tons or more on deck. Only when the boat is "shot full", at which time the crew is forced to take a rest, might they inform others. Until then the crew might be tempted to keep silent in order to harvest other whales, either alone or together with a partner. Consequently, silence on the radio is often an indication that whales have been sighted.

Management information might be different in Japan and Norway as a result of different technologies and regulations, but in both countries whalers possess knowledge on how and when whales are appropriated. However, ownership does not imply that the whaler is free to dispose of the whale as he sees fit. In all societies there exist complicated rules as to how the animal shall be shared or the profit used. In the Faroe Islands, the pilot whales belong to the community and the distribution of the meat follows set rules (Joensen 1990), as did minke whales captured in bays outside Bergen in Norway until the beginning of this century (Østberg 1929). In Inuit culture, sharing the prey with others is fundamental to the successful continuation of relations between whales and Inuit and serves to legitimize hunting (Wenzel 1991). In Japan people obtain rights in the catch by giving gifts of *sake* (rice wine) to the owners and crews before the commencement of the whaling season (Akimichi et al 1988; Kalland and Moeran 1992). Moreover, whaling enterprises are obliged to support the communities from which they operate. These are "long term transactions" which are "concerned with the reproduction of the social and cosmic order" (Moeran 1992b). Such rules are also among the pool of knowledge required for a successful appropriation of natural resources.

Some person must own or have rights in an object for it to become a commodity having exchange value; but importantly, ownership does not signify a relationship between object and person but rather, a relationship between persons. "Any statement of property or of rights is a statement of what can be done by the owner to the non-owner if these rights are infringed" (Bloch 1984:204). Whalers' rights in whales are recognized only as long as the whaler is able, or his society is willing, to sanction infringements. In order to secure his society's favor and support, the whaler has to live up to the social obligation expected of him, which restricts his freedom to dispose of the carcass as he sees fit. In return for support and recognition of property claims, the whaler must forego some of his rights in the whale.

The whale protectionists use a different strategy in their attempts to deny whalers access to whale resources, one being to cause the resources to have severely limited value to the whalers by destroying their markets and thus removing their products from the commodity state through legal means. Norwegian catches of bottlenose whales, for example, came to an end when the UK banned imports of whale meat for pet food in 1972 (ISG 1992:32). A number of other restrictions have been introduced on the trade of whale meat: the Washington Convention on International Trade in Endangered Species (CITES), established in 1973, prohibits trade affecting several whale species, and in 1979 IWC prohibited imports of whale products into member states from non-members states. In 1986 the IWC passed a resolution recommending that products from scientific whaling should primarily be consumed locally, which means that no more than 49 per cent may be exported. In 1981 the European Community (EC) introduced licences for importing whale products into the community, and in the U.S. the 1969 Endangered Species Conservation Act and the 1972 Marine Mammal Protection Act both prohibit imports of whale meat into the U.S.

On the ideological level, the super-whale has made it seem barbaric, bordering to cannibalism, to eat whale meat (cf. *Daily Star*, May 11, 1992). Today more people in non-whaling countries appear to express greater adversion to eating whale meat than most other kinds of meat (Freeman and Kellert, this volume).

However, there are people living in Japan, Norway and elsewhere who cherish such meat; consequently a second strategy involves claiming that whales are not a free good owned by nobody until spotted, chased or captured. In furthering this strategy, attempts are made to turn whales from *res nullius* (an open access resource) into *res communis* (a communal resource; Hoel 1986:28), which means that nations without previous interests in whales, Switzerland for example, come to share property rights and management responsibilities with the others. The claim that whales are everybody's property is based on the notion that certain valued resources should be proclaimed "the common heritage of mankind".

Whalers see the policy of the anti-whaling movement as an attempt to close the whale fishery and appropriate the whales for themselves in order to "sell" them or give them away in "adoption" for a fee (see Note 6). The whalers are facing a new, and to them incomprehensible, regime of appropriation with other rules and sanctions. The whaling moratorium has, in many people's opinion, although not according to international law, given legitimacy to sanctions against the whaling nations; such sanctions are of two kinds. The U.S. has introduced legal measures which can be used against whaling nations. Under the 1971 Pelly Amendment to the Fisherman's Protective Act, the President is authorized to prohibit import of sea products from whaling nations, while the 1979 Packwood-Magnusson Amendment to the Fisheries Conservation and Management Act gives the Secretary of Commerce the option to reduce the fishing quotas of whaling nations within the U.S. 200-mile zone by 50 per cent and then cancel these allotments altogether.

The U.S. has certified, or threatened to certify, whaling nations under these two provisions. Moreover, these laws have given the environmental and animal rights groups the legal means to force the U.S. Government to impose sanctions on Japan (Sumi 1989). However, it is highly questionable whether these sanctions are in conformity with international law (Hoel 1992; Sumi 1989; McDorman 1991), and the U.S. has recently been critized by a General Agreement on Tariffs and Trade (GATT) panel for its embargo of Mexican "dolphin-unfriendly" tuna (GATT Panel 1991). It is therefore likely that extra-legal consumer boycotts organized by the environmental and animal welfare groups will pose the biggest threat to the whaling nations. Boycotts of sea products have already been used against Iceland (with questionable result) and will undoubtedly be tried again. Several organizations have so far tried to organize boycotts not only of Norwegian sea products but also of other Norwegian products and serrvices (including tourism and the 1994 Winter Olympics) because Norway resumed commercial whaling in 1993. These sanctions all have one aim in common: to close down whaling once and for all and turn the whale resources into a "common heritage of mankind".

It should be the case that low-consumptive use of whales, such as whale watching, and sustainable whaling could co-exist, particularly where tourists and whalers seek different species. The "swim with a whale" programmes, as well as the use of cetacean in literature and art, should be even more able to co-exist peacefully with whaling. However, this is not the case: the same cultural framework which has turned the super-whale into a commodity has also decommoditized whale meat and oil and created the image of "the evil whaler". Consequently the super-whale cannot coexist with whaling. Moreover, western urban dwellers tend to impose their totems on others, and it is this blending of totemization of whales and cultural imperialism which has caused the whale issue to stalemate. Having helped to create a powerful environmental and animal rights movement, anti-whalers have gained political and moral recognition. WWF with its many members and royal patronage has gained a position from where it can appropriate nature and dispense shares in it to those willing to pay for a good conscience or a green image. Greenpeace with large number of supporters has also gained power and international recognition sufficient to lay claim to being a steward of nature. Both organizations have been able to form coalitions with national governments, corporations and international bodies such as the E.C., the U.N.Environmental Programme, and the World Conservation Union (IUCN). In doing so these coalitions have redefined the whale as a commodity and managed to interrupt the commodity path.

Appadurai refers to this phenomenon as "diversion of commodities from their preordained path". One kind of diversion is theft; many whalers see the "sale" and "adoption" of whales (including the sale of whale images which have contributed so importantly to the finances of environmental and animal rights groups) as theft. In whalers' eyes the theft involves not just the whales, which they feel belong to them through several generations' involvement in whaling, but theft of their livelihood, their pride, and their culture.

NOTES

1. Greenpeace-Germany, for example, jumped on the anti-sealing bandwagon in order
   to grow:

   > "It was surely important to Greenpeace at that time to get bigger. Green-
   > peace was very small. At that time I found it completely legitimate to use
   > a cute animal with large eyes.... That I found OK. That was no issue."
   > (Wolfang Fischer, leader of the German section's anti-sealing campaign in
   > Bayrischer Rundfunk's TV program "Live aus dem Schlachthof", 15
   > January 1990. Translated from German by the author.)

   Today, Greenpeace has grown to one of the largest environmental and animal protection groups with an annual turnover of about US$ 142 million, with DM 57 mil-
   lion being collected in Germany (Schwarz 1991:88–89). This is only surpassed by
   the Greenpeace-US which has 2.3 million members and a budget of US$ 50 million
   (Gifford 1990:73). Only a few of the members have influence, however, for
   Greenpeace, like many of the anti-whaling groups, is centrally controlled: of the
   700,000 German members only 30 are entitled to elect the seven board members
   (Schwarz 1991:88–89), while in Sweden only 11 of the 150,000 paying members
   have the right to vote (Eyerman and Jamison 1989:106). Recently Greenpeace-Denmark has suffered loss of members and revenues which the newspaper *Politiken*
   (Dec. 4, 1991, p. 8) attributes to the "Leninist" leadership structure. It will be interesting to see whether some very critical coverage recently in the press (Fox 1991;
   Schwarz 1991; Spencer et al. 1991) will have a further negative impact on
   Greenpeace.
2. Not all anti-whalers partake in this rhetoric: the president of the WDCS, for example, asks his colleagues to stop accusing all whalers of cruelty (Payne 1991:22), and
   Robbins Barstow of CSI apologized for insulting remarks made by British politicians
   and others at an anti-whaling rally in Glasgow in connection with the 1992 IWC
   meeting.
3. Sealers make good enemies for the same reasons. Envirommental and animal rights
   groups campaigned against hunting of harp seals off Newfoundland, although this
   species, numbering in the millions, is far from endangered (Wenzel 1991). At the
   same time "international conservation organizations [including Greenpeace and
   WWF] are allowing one of the world's most endangered species [the Mediterranean
   monk seal] to slip silently into extinction" (Johnson 1988:5). There are only a few
   hundred monk seal left; this species is victim of massive environmental degradation
   due to military activities and high population concentration around the Mediterranean with millions of tourists flocking to its beaches every year. In the case of the
   monk seal it is almost impossible to identify the seal's "enemy", and the economic
   and political forces effecting monk seal depletion are powerful and influential. Furthermore, there is no "face-to-face" confrontation between the monk seal and the
   killer, no dramatic deaths and no blood. This situation contrasts strongly with the
   seal hunt off Newfoundland which is conducted during three short weeks in a very
   limited and uninhabited area where the white ice makes the perfect photographic
   background for slaughter and spills of blood, not to mention the pathos of the
   white pups with big, black eyes, helpless against the hunters with clubs and knives.
   These are the ideal conditions for making dramatic footage, while the Mediterranean setting is not.

4. In his book *Freeing the Whales*, Tom Rose (1989) vividly narrates the $5.8 million res-
cue operation and the prizes at stake: for Greenpeace the event meant the biggest
source of new money and members in its history, and for the U.S. National Oceano-
graphic and Atmospheric Administration (NOAA), which led Operation Breakout
and activited a new satellite ahead of schedule in order to provide the Soviets with
ice information, it meant its "coming of age". The oil industry improved its image
considerable (only to lose much of the newly won goodwill in the *Exxon Valdez* oil-
spill five months later). Through its involvement in freeing the whales, VECO
(Alaska's largest oil facility constructing company) was assisted in winning the
prime contract to clean up the mass after the *Exxon Valdez* oil-spill. The Alaskan
whalers benefitted by suddenly being pictured as good-natured humans and not as
greedy whale killers and, by showing personal interest in the affair, President
Reagan tried to shape up his environmental record, as did the Soviets who, by providing ice-breakers, presented a "human face" to their faltering regime and helped
the whale protectionists forget that Russia is the only nation hunting gray whales.

5. Such offers are not always accepted: in 1992 Greenpeace-Denmark approached the
association of supermarkets in Denmark with a suggestion that Greenpeace whould,
for a fee, give the shops' products a "green" stamp. The offer was perceived by the
association to be extortion and was consequently declined (*Politiken*, January 18,
1992, 3. section, p. 5).

6. When organizations appropriate whales to be used in their own "adoption" programmes, they not only contradict their own ideology (that wildlife is everybody's
property), but also face the problems of recognition of such claims and of sanctioning infringement of their exclusive "ownership". To solve these problems the organizations photograph whales and give them names; some individual whales can
be identified by their flukes or colour patterns, and it is precisely these species
which are appropriated by WWF and other organizations and offered for adoption.
The importance of photos for claiming ownership to individual whales surfaced in
a recent dispute between WWF-Denmark and the Center for the Study of Whales
and Dolphins (CSWD) in Sweden. CSWD photographs whales outside northern
Norway in order to identify and name individual animals, while WWF, who has
sponsored these activities, sent such photos to the "adopted parents" of whales. An
agreement was finally reached out of court when WWF-Denmark paid compensation for the use of the photos and promised to curtail the adoption program.

REFERENCES

Abbey, L., 1990.
   The Last Whales. Doubleday, London.
Adams, D., 1984.
   So Long, and Thanks for All the Fish. Pocketbooks, New York.
Akimichi, T., P.J. Asquith, H. Befu, T.C. Bestor, S.R. Braund, M.M.R. Freeman, H.
Hardacre, M. Iwasaki, A. Kalland, L. Manderson, B.D. Moeran, J. Takahashi, 1988.
   Small-Type Coastal Whaling in Japan. Boreal Institute for Northern Studies, Occasional
   Publication No. 27, Edmonton.
Appadurai, A., 1986.
   Introduction: Commodities and the Politics of Value. In: A. Appadurai (ed), *The*

Social Life of Things — Commodities in Cultural Perspective, pp. 3–63. Cambridge University Press, Cambridge.

Aron, W., 1988.
The Commons Revisited: Thoughts on Marine Mammal Management. *Coastal Management* 16(2):99–110.

Barstow, R., 1989.
Beyond Whale Species Survival: Peaceful Coexistence and Mutual Enrichment as a Basis for Human/Cetacean Relations. *Sonar*, No. 2, pp. 10–13.

Barstow, R., 1991.
Whales are Uniquely Special. In: N. Davies, A.M. Smith, S.R. Whyte, and V. Williams (eds), *Why Whales?* pp. 4–7. The Whale and Dolphin Conservation Society, Bath, U.K.

Baskin, Y. 1993.
"Blue whale population may be increasing off California". *Science* 260, 16 April, p. 287.

Beck, A.M. and A.H. Katcher, 1984.
A New Look at Pet-facilitated Therapy. *Journal of the American Veterinary Medical Association* 184:414–421.

Bloch, M., 1984.
Property and the End of Affinity. In: M. Bloch (ed), *Marxist Analyses and Social Anthropology.* Tavistock Publication, London.

Bright, C., 1992.
A Fishy Story about Whales. *Wildlife Conservation* 94(4):62–69.

Brin, D., 1983.
Startide Rising. Bantam Books, New York.

Brown, M. and J. May, 1991.
The Greenpeace Story. Dorling Kindersley, London.

Butterworth, D.C., 1992.
Science and Sentimentality. *Nature* 357 (June 18):532–534.

Carpenter, B. and K.F. Schmidt, 1992.
Whales. Demystified, the Watery Giant may come under the Harpoon Again. *U.S. News & World Report*, July 13, pp. 58–64.

Carter, J. and A. Parton, 1992.
Park in Flap Over Swims with Flipper. *Sunday Morning Post*, November 29, 1992, p. 5.

Cochrane, A. and K. Callen, 1992.
Dolphins and Their Power to Heal. Bloomsbury, London.

Corrigan, P., 1992.
Where the Whales Are: Your Guide to Whale-Watching Trips in North America. The Globe Pequet Press, Chester, Connecticut.

Cousteau, J., 1988.
Whales. Harry N. Abrams, Inc., New York.

Daily Mirror, 1992.
Having a Whale of a Time. June 25, 1992, p. 30.

Daily Star, 1991.
Sickest Dinner Ever Served: Japs Feast on Whale. May 11, 1991, p.1.

D'Amato, A. and S.K. Chopra, 1991.
Whales: Their Emerging Right to Life. *American Journal of International Law* 85(1):21–62.

Davies, B.D., 1985.
Letter to Friends, October. IFAW, Crowborough, U.K.

Davies, N., A.M. Smith, S.R. Whyte, and V. Williams, (eds), 1991.
Why Whales? The Whale and Dolphin Conservation Society, Bath U.K.

Dobbs, H., 1990.
Dance to a Dolphin's Song. Jonathan Cape, London.

Donovan, G.P., 1982.
The International Whaling Commission and Aboriginal/Subsistence Whaling: April 1979 to July 1981. In: G.P. Donovan (ed), *Aboriginal Subsistence Whaling*, pp. 79–86. Special Report No. 4, International Whaling commission, Cambridge.

Douglas, M., 1966.
Purity and Danger: An Analysis of the Concepts of Pollution and Taboo. Routledge and Kegan Paul, London.

Eyerman, R. and A. Jamison, 1989.
Environmental Knowledge as an Organizational Weapon: the Case of Greenpeace. *Social Science Information* 28:99–119.

Fox, J., 1991.
The Business of Greenpeace. *The Financial Post* (Toronto), January 7.

Freeman, M.M.R., 1990.
A Commentary on Political Issues with Regard to Contemporary Whaling. *North Atlantic Studies* 2(1–2):106–116.

Freeman, M.M.R. and S.R. Kellert, 1994.
Public Attitudes to Whales: Results of a Six-Country Survey. In: M.M.R. Freeman and U.P. Kreuter (eds), *Elephants and Whales: Resources for Whom?* pp. 293–315. Gordon and Breach Science Publishers, Switzerland.

Fuller, K.S., 1991.
Letter to the Ambassadors of Whaling Nations, dated May 24. WWF, Washington, D.C.

Gabriel, T., 1991.
Slow Boat to Trouble. *Outside*, September, pp. 54–60 & 128–113.

GATT Panel, 1991.
The United States — Restrictions in Import of Tuna. *International Legal Materials* 30:1594.

Gifford, B., 1990.
Inside the Environmental Groups. *Outside*, September, pp. 69–84.

Greenpeace International, 1992.
A Whale Sanctuary for Antarctica. (Paper distributed at the 44th IWC meeting, at Glasgow.)

The Guardian, 1991.
UK Battles to Keep Whaling Ban. May 27, 1991.

Gulland, J., 1988.
The End of Whaling? *New Scientist*, October 29, pp. 42–47.

Gylling-Nielsen, M., 1987.
Havets Mennesker. *Greenpeace Magasinet*, No. 1.

Hatt, J., 1990.
Enigmatic Smiles. *Harpers & Queen*, December, pp. 244–252.

Hoel, A.H., 1986.
The International Whaling Commission, 1972–1984: New Members, New Concerns. The Fridtjof Nansen Institute, Lysaker, Norway. 2nd edition.

Hoel, A.H. 1992
Internasjonal forvaltning av kvalfangst. Tromsø: Norwegian College of Fisheries Science Report Series.

Holt, S. 1985.
"Whale mining, whale saving". *Marine Policy* Vol.9(3):192-213.

Holt, S. 1991.
"The un-ethics of whaling". In *Why Whales* (eds. Nick Davies, Alison M. Smith, Sean R. Whyte, Vanessa Williams), Bath, U.K.: The Whale and Dolphin Conservation Society, pp.8-16.

Hoyt, E. 1992.
"Whale watching around the World". *International Whale Bulletin*, No.7.

Ingold, T. 1986.
The Appropriation of Nature. Manchester: Manchester University Press.

ISG (International Study Group on Norwegian Small Type Whaling). 1992
Norwegian Small Type Whaling in Cultural Perspectives. Tromsø: Norwegian College of Fisheries Science.

ISGSTW (International Study Group for Small-Type Whaling). 1992
"Similarities and diversity in coastal whaling operations: A comparison of small-scale whaling activities in Greenland, Iceland, Japan and Norway". The Report of the Symposium on Utilization of Marine Living Resources for Subsistence (Vol.II) held at Taiji, Japan. Tokyo: The Institute of Cetacean Research.

Joensen, J.P. 1990.
"Faroese pilot whaling in the light of social and cultural history". *North Atlantic Studies*, Vol.2(1-2):179-184.

Johannsen, W. 1990.
Hvalernes verden. Copenhagen: Skarv/WWF.

Johnson, W. 1988.
The Monk Seal Conspiracy. London: Heretic books

Kalland, A. 1992.
"Aboriginal subsistence whaling: A concept in the service of imperialism?". In *Bigger than Whales*, (ed. by G. Blichfeldt). Reine, Norway: High North Alliance.

Kalland, A. 1993.
"Management by totemization: Whale symbolism and the anti-whaling campaign". *Arctic*, 46(2). In press.

Kalland, A. and B. Moeran. 1992.
Japanese Whaling: End of an Era? London: Curzon Press.

Lilly, J. 1961.
Man and Dolphin, New York: Doubleday.

The Mail on Sunday, 1992.
Do We Really Want to Go Back to This? June 21, pp. 8–9.

Marine Mammal News, 1991.
Vol. 17(5):4, May 1991.

Martinsen, O., 1964.
Aktiv Hvalfangst i Norden i Gammel Tid., In: A. Bakken and E. Eriksen (eds) *Hval og Hvalfangst — Vestfolminne 1964*. Vestfold Historelag.

McDorman, T.L., 1991.
The GATT Consistency of US Fish Import Embargoes to Stop Driftnet Fishing and Save Whales, Dolphins and Turtles. *George Washington Journal of International Law and Economics* 24:477–525.

Moeran, B., 1992a.
   Japanese Advertising Nature — Ecology, Women, Fashion and Art. In: O. Bruun and
   A. Kalland (eds), *Asian Perceptions of Nature*, pp. 195–217. Nordic Institute of Asian
   Studies, Copenhagen.
Moeran, B., 1992b.
   The Cultural Construction of Value: 'Subsistence', 'Commercial' and Other Terms in
   the Debate about Whaling. *Maritime Anthropological Studies 5.*
Nollman, J., 1990.
   Dolphin Dreamtime — The Art and Science of Interspecies Communication. Bantam Books,
   New York.
Ottaway, A., 1992.
   The Whale Killers. Greenpeace-UK.
Pálsson, G., 1991.
   Coastal Economies, Cultural Accounts. Human Ecology and Icelandic Discourse. Manches-
   ter University Press, Manchester.
Payne, R., 1991.
   Is Whaling Justifiable on Ethical and Moral Grounds? In: N. Davies, A.M. Smith,
   S.R. Whyte, and V. Williams (eds) *Why Whales?* pp. 20–22. The Whale and Dolphin
   Conservation Society, Bath, U.K.
Pearce, F., 1991.
   Green Warriors. The People and the Politics Behind the Environmental Revolution. The
   Bodley Head, London.
Politiken, 1991.
   Greenpease muger ud i egne raekker, December 4, 1991, p. 8.
Politiken, 1992.
   Kaeder siger nej til miljpression, January 1, 1992, 3rd section, p. 5.
Regan, T., 1984.
   The Case of Animal Rights. Routledge & Kegan Paul, London.
Ris, M. (this volume)
   Conflicting Cultural Values: Whale Tourism in Northern Norway. In: M.M.R. Free-
   man and U.P. Kreuter (eds), *Elephants and Whales: Resources for Whom?.* pp. 219–239.
   Gordon and Breach Science Publishers, Switzerland.
Rose, T., 1989.
   Freeing the Whales. How the Media Created the World's Greatest Non-Event. Birch Lane
   Press, New York.
Scarce, R., 1990.
   Eco-Warriors. Understanding the Radical Environmental Movement. The Noble Press,
   Inc., Chicago.
Scheffer, V., 1991.
   Why Should We Care About Whales? In: N. Davies, A.M. Smith, S.R. Whyte and
   V. Williams (eds), *Why Whales?* pp. 17–19. The Whale and Dolphin Conservation
   Society, Bath, U.K.
Schwarz, U., 1991.
   Geldmaschine Greenpeace. *Der Spiegel,* 45(38), September 16, pp. 84–105.
Schweitzer, A., 1950.
   The Animal World of Albert Schweitzer: Jungle Insight into Reverence for Life. (Charles R.
   Joy, transl. and ed.). Beacon Press, Boston.

Shields, L., 1992.
  Be Careful What You Wish For. . . . *Whales Alive!* CSI Publication, Vol. 1(2), May, 1992.
Singer, P., 1978.
  Why the Whale Should Live. *Habitat* (Australian Conservation Foundation), 6(3):8–9.
Spencer, L. with J. Bollwerk and R.C. Morais, 1991.
  The Not So Peaceful World of Greenpeace. *Forbes Magazine*, November 11, pp. 174–180.
Spong, P., 1992.
  Why We Love to Watch Whales. *Sonar*, No. 7:24–25.
Sumi, K., 1989.
  The 'Whale War' Between Japan and the United States: Problems and Prospects. *Denver Journal of International Law and Policy* 17(2):317–372.
Sutton, M., 1992.
  Report on the 44th annual meeting of the International Whaling Commission, WWF, Washington, D.C.
Tester, K. 1991.
  Animals & Society. The humanity of animal rights. London: Routledge.
The Times, 1991.
  Beaching the Whalers, May 27, 1991. (editorial).
US Marine Mammal Commission, 1991.
  Letter to the US Commissioner to the IWC, John Knauss, dated December 5th.
Wenzel, G., 1991.
  Animal Rights, Human Rights - Ecology, Economy and Ideology in the Canadian Arctic. University of Toronto Press, Toronto and Buffalo.
Wheatley, J., 1992.
  The Three-day Miracle of New South Whales. *The Times Saturday Review*, August 22, pp. 14–15.
Williams, H., 1988.
  Whale Nation. Jonathan Cape, London.
Winton, T., 1992.
  Dolphin Mania. *The Independent Monthly*, September, pp. 14–19.
WSPA (World Society for the Protection of Animals), 1988.
  Circular No. 881406.
WWF (World Wide Fund for Nature), 1992.
  WWF Position Statement on Whaling and the IWC. WWF, Gland, Switzerland.
WWF-Denmark, 1990.
  Letter to Danish business leaders, dated June 19. WWF, Copenhagen.
Østberg, K., 1929.
  Gammelnorsk Primitiv Hvalfangst i Nutiden: (1) Fangst av Vaagehaval i Hordaland. *Norsk Fiskeritidende*, pp. 15–24.

12. *GRIND* — AMBIGUITY AND PRESSURE TO CONFORM: FAROESE WHALING AND THE ANTI-WHALING PROTEST

KATE SANDERSON

INTRODUCTION

"Cultural intolerance of ambiguity is expressed by avoidance, by discrimination, and by pressure to conform"

— Mary Douglas, *Implicit Meanings* 1975:53.

". . . if they want to kill whales in the traditional way, that's fine by us, if nothing else about their way of life, significantly anyway, has changed."

— Sean Whyte, Chief Executive, Whale and Dolphin Conservation Society, TV interview, 1991

The spontaneous, dramatic and bloody event of a *grind* — the driving and slaughtering of schools of pilot whales for food in the Faroes (see Note 1) — has made it one of the most colorful and exciting aspects of Faroese life, both for Faroe Islanders and for foreign visitors to the islands. In many accounts of the Faroes over the years, *grind* has inevitably been singled out for special mention as a characteristic feature of Faroese culture. In the nineteenth century it became an established symbol of Faroese national identity, along with other elements of traditional Faroese life which had lost their primary economic significance with the expansion and modernization of the fishing industry. The significance of *grind* in Faroese society today is, however, still an economic one, as catches of pilot whales, like other forms of local food production such as fowling, coastal fishing and some limited crop-growing, continue to supplement many household incomes. Until not long ago, the national symbolic significance of *grind* was actively and explicitly fostered, with scenes from the kill depicted on postcards and posters designed for sale to foreign visitors to the islands.

However, since 1985 *grind* has been the object of organized protest by a number of animal welfare and environmental groups in Europe and North America, and once proud descriptions of pilot whaling as a dramatic and positive element of Faroese life have now been replaced by equally dramatic but highly critical descriptions in many media accounts and campaign newsletters generated by such protest groups. The reasons for this change of focus derive from widespread changes in attitudes to the exploitation of the natural environment and its wildlife. Whales, in particular, have come to be seen by many

Courtesy of Ole Wich

Plate 12.1 The dramatic and bloody event of a *grind*.

in the urbanised Western world as potent symbols of the way in which modern industrialised forms of production and exploitation have destroyed natural habitats and decimated natural resources. Large-scale industrial whaling operations earlier this century led to the depletion of several stocks of large whales, a fact that has been transformed into the mistaken but commonly-held belief in many urbanised societies today that all whales are threatened with extinction.

Another aspect of modern urban perceptions of nature is the emergence of concerns for the welfare of animals (see for example Thomas 1983; Löfgren 1985 and Tester 1991), which have found practical application in most Western nations in the form of animal welfare legislation governing the slaughter of domestic livestock. Such concerns have also been extended to the hunting of some wild animals and particularly with respect to whales, whose perceived intelligence, social cohesion and beauty make the manner in which they are hunted, or indeed in some cultures the fact that they are hunted at all, morally unacceptable:

> "From an ethical perspective, the most blatantly insupportable aspect of whaling is the slow and painful death inflicted on whales" (Singer 1978:9)

> "Australia's present community attitudes therefore dictate that the whale should not now be regarded as a natural resource available for exploitation. We are confident that, in the light of all the facts put to the Inquiry, reasonable Australian citizens would conclude that, now there is no necessity, it is wrong to kill an animal of such significance as the whale" (Frost 1978 Vol. I:208)

Such views have been most clearly formulated and reinforced through a now well-established discourse of environmental and animal welfare campaigning, which is largely reliant on the mass media for the dissemination of its ideologies. *Grind* has been effectively incorporated into this discourse, and as a result, many external representations of whaling in the Faroes today have taken on a new dramatic function as agents of political and ideological persuasion. The reasons for the persistent and aggressive campaigning to stop Faroese whaling can be found in the nature of *grind* itself and the ambiguities it presents in relation to predominant cultural perceptions of nature and human society found in the urbanized western world.

Grind

Long-finned pilot whales (*Globicephala melas*) belong to the group of whales known as the toothed whales, which, with the exception of the large sperm whale, are much smaller than the baleen whales. An adult pilot whale measures between four and six metres in length and weighs around eight hundred kilos (Bloch *et al.* 1990:43). Pilot whales migrate in schools numbering from around 50 to 1,000 or more animals. Analysis of recent sightings surveys of the species estimates a population size in the central and eastern North Atlantic of 778,000 (Buckland *et al.* 1992).

Each year, and usually several times a year, Faroe Islanders drive schools of long-finned pilot whales into bays and kill them with knives on the shoreline and in the shallow water. A total of 1572 pilot whales were killed in this way

Courtesy of Ole Wich

Plate 12.2 The whales are assessed according to the traditional value of *skinn*.

in the Faroes in 1992 in fourteen separate drives. The annual average catch over the ten years from 1983 to 1992 was 1,550 whales. Almost continuous unbroken annual catch records kept since around 1600 show a long-term annual average catch of around 1000 whales. Catch levels, while fluctuating a great deal, have not decreased over time. Pilot whales are killed to provide food for the local community, and both the meat and blubber are eaten.

Whale drives only take place when a school of whales is sighted close to land, and when sea and weather conditions are suitable for driving the whales. Catches can occur at any time of the year, but are most frequent in the summer months. It is a communal, non-commercial form of whaling which provides local households with a fresh supply of meat and blubber (Bloch 1990; Sanderson 1991, 1992a). Depending on the number of whales in the school and the location of the drive, the meat and blubber is distributed free among participants in the drive and kill, residents in the district where the whales were beached, and anyone else registered with the local authorities at the time of the catch. The whales are assessed according to the traditional value of *skinn*, which derives from the valuation of land-based produce. One *skinn* is equivalent to ca. 72 kg of meat and blubber (ca. 34 kg of blubber and 38 kg of meat). A catch of 61 pilot whales on the island of Suðuroy in April 1993 was valued in total at 411 *skinn*. According to the tradition on Suðuroy, the catch was distributed to the entire island. The 411 skinn were divided into 6118 shares for all the registered residents of Suðuroy, with each individual share consisting of 3.81 kg of meat and blubber. In the capital of Tórshavn, with a population of 15,000, a small catch such as this would most likely only be divided among the participants.

Plate 12.3 There is no factory processing.

Those who receive shares of a catch butcher their meat and blubber themselves. There is no factory processing or export of any part of the catch and most of the meat and blubber is divided without the exchange of money. No-one in the Faroes makes a living from killing pilot whales. Some of the catch may be auctioned off after the assessment and calculation of shares and purchased by those who are not otherwise entitled to a share, the proceeds of which go towards covering the costs incurred by the local community, such as claims for damages and cleaning afterwards. Thus it is possible to purchase whale meat and blubber in stores, but the maximum retail price is regulated by governmental order and is about half the price of other meats such as beef and pork. The meat and blubber of pilot whales have long been an important part of the diet of Faroe Islanders and a catch of pilot whales provides many households with an important source of highly-valued foods which are regarded as healthy and nutritious (Bloch & Hanusardóttir 1993; see Note 2). The blubber and meat are eaten together, and the blubber is also eaten with other food such as dried fish. Both the meat and blubber are preserved in modern and traditional ways, either by freezing, salting down (meat and blubber) or wind-drying (meat).

The earliest specific account of the organised driving of pilot whales in the Faroes is from 1587 (Storm 1881), although archaeological evidence suggests that pilot whales have been used for food in the Faroes since the earliest days of Norse settlement in the Middle Ages. The organization of *grind* was first codified in written regulations in 1832 (*Reglement for Grindefangsten paa Færøerne* 1832), but organized methods of message-sending, driving and killing have existed since at least the late seventeenth century (see for example Debes 1673; 1676). Today *grind* is regulated by an executive order which is issued by the Faroese Government. The regulations stipulate the responsibilities for the supervision of *grind*, the manner in which the driving and killing should be conducted, and the procedure for the assessment and division of the catch (*Kunngerð um grind* 1986). Locally elected whaling officials, known as *grindaformenn* ("*grind* foremen") are in charge of the immediate conduct of the drive and kill, and the district sheriff, or *sýslumaður*, supervises proceedings in general and is responsible for the division of the catch.

AMBIGUITY

A *grind* is always an exciting and very public spectacle, and it is these very elements which have also made its incorporation into international environmental and animal welfare discourses particularly easy and effective (Sanderson 1990). The essentially dramatic and, in many ways, ambiguous nature of *grind*, can be best understood in terms of its position in relation to different cultural perceptions of nature and society.

The many accounts of *grind* over the centuries have almost always under-
lined the dramatic spectacle represented by a pilot whale drive in the Faroes,
transforming the otherwise calm of the daily routine into a hectic few hours of
great activity and excitement. American anthropologist Jonathan Wylie has explained the excitement of a *grind* in terms of three essential anomalies inherent
to it, namely, the unpredictable nature of the arrival of a *grind*, the combination of fishing and farming techniques used to drive and kill the whales, and
the fact that a successful catch provided "an "imported" food that did not
have to be bought with exports" (Wylie 1981:107).

Expanding on Wylie, these anomalies can be seen to be largely conditioned
by the very nature of the target species itself, the pilot whale. Pilot whales
travel in schools, and can be seen close to the Faroe Islands at any time of the
year. Their unpredictable, yet frequent occurrence has meant that the means by
which they are exploited is neither completely systematised, nor entirely uncontrolled. It is, strictly speaking, neither just a hunt of wild animals, nor a
slaughter of domestic livestock, but combines elements of both. *Grind* can be
said to be a hunt in the sense that, as in other forms of hunting, the target
animal is wild. But the inclusion of *grind* in the general category *hunt* is complicated by the fact that the whales are not sought out but are *driven* (like
sheep) only when sighted quite close to land, and this only happens when a
school is encountered, by chance, in the course of other activities such as
coastal fishing. There is no exclusive season for *grind*: the initiation of a drive
depends upon the movement of the whales close to land as well as prevailing
weather conditions.

In considering farming and fishing in relation to *grind*, the parallels between
the terminology of *grindadráp* and the language used in sheep raising and fishing are apparent: both sheep and whales move around in flocks — *flokkar*, and
both are herded in a drive - *rakstur*. The drive is led by a *grindaformaður*, which
in fishing terminology is simply the *formaður* of the boat crew. Individual pilot
whales are sometimes referred to as fish in compounds such as *finningarfiskur*
(finder's fish), the largest whale in the catch, which goes to the individual
who first sights the *grind* (*Kunngerð um grind* V, 18a, p. 93; Wylie 1981:
109–110).

The apparent anomaly represented by the combination of the techniques and
functions of farming and fishing characteristic of *grind* is worth considering in
greater depth, since it manifests itself in many different ways in representations and perceptions of the distinctiveness and excitement of *grind*. These concepts can be related to the examination of perceptions of nature and the environment in the work of two contemporary anthropologists, both of whom have
written about Iceland. In a study of Icelandic society from 1400 to 1800, the
historical anthropologist, Kirsten Hastrup (1990), sees a clear conceptual distinction between the human, social system within which farming and domestic
animals were incorporated, and the wild, untamed environment of the sea
(and outfields), where wild creatures existed and forms of hunting such as
fishing were conducted. The activity of farming, as a harvest, involved plan-

ning and social organization, while fishing was, as Hastrup explains "based on an instantaneous relation between man and prey" (ibid:274).

Applying this conceptual model to the Faroes, *grind* blurs the otherwise clear distinction between fishing and farming, for killing pilot whales takes place on the shoreline, in the marginal area where the two quite distinct realms of the social and the wild meet. According to Hastrup's model, these realms were mutually exclusive and hunting (*veiðar*) was, for pre-modern Icelanders, external to the social system. A similar conceptual model for the Faroes would have to allow a meeting of the two spheres of activity in relation to their respective environments in order to accommodate *grind*, thereby providing a different conceptual representation of the relationship between the social and natural environments for Faroe Islanders. This meeting of spheres could also accommodate the equally dramatic and exciting Faroese method of fowling, in which men attached to lines venture down cliff-faces to catch seabirds or collect their eggs, and which, like *grind*, is also an activity which occupies a somewhat marginal position in relation to the environment(s) in which it is carried out (on land and in the air).

When schools of pilot whales are herded into a bay and made to beach, men and whales meet at the intersection of the social and the wild at a place where the whales have been driven partially from their own environment into another. They are stuck between these two realms, unable to turn or dive. The depth of the water will determine both their chances of escape, and the speed with which they can be killed. The fate of the animal is thus partly determined by its own natural environment and the way in which it has adapted to it. The whale is an anomalous creature, a fish-like mammal, air-breathing, but one which, unlike other marine mammals such as seals and walruses, lives entirely in the ocean environment and cannot survive for long outside it. Like fish, pilot whales, in particular, move around in relatively large schools. But unlike fish, they cannot be caught in nets or on lines with hooks from boats. Due to the large amount of food they provide for humans they must be brought (in whole groups) as close as possible to the human environment, an event facilitated by the strong herd instinct of the species, a characteristic of the pilot whale and its behaviour in the hunt which has often been commented on in many accounts of *grind*, from as early as 1673 (Debes 1673:75).

The human participants join the whales in the borderline region, and, like the whales, they are in their own and an alien environment at the same time. Knives are used to sever the major arteries of the whales, in a similar way to the slaughtering of sheep (although without prior stunning), but the further the whales are from dry land, the deeper the men must wade into the water (and into the realm of fishing), and the more necessary it is for them to use a large iron hook attached to a strong rope to secure the whales for killing. The Faroese word for the hook used in the hunt is *sóknarongul*, a compound of the elements *sókn* (from the verb *søkja*, to attack) and *ongul* (fishhook). The *sóknarongul* functions in this context as an instrument most closely related to the technology of fishing.

The inherently dramatic and exciting nature of *grind* derives from its ambiguous position between two very different environments, resulting in a spectacle of mass killing and large quantities of blood. The nature of the spectacle also derives from its location in this marginal area, on the shore of a bay. The geographical character of the Faroese shoreline is such that the bay or bottom of a fjord forms an intimate arena, where the action can be observed at close range from the shore and surrounding hillsides. It is a public slaughter, similar in its slaughtering methods to the killing of land animals, but distinct in the fact that it is carried out *en masse*, with many slaughterers performing their work simultaneously. The anomaly of the whale itself, a warm-blooded, sea-dwelling animal, results in the anomalous sight of a large slick of bright red blood in the water.

Icelanders of the sixteenth and seventeenth centuries saw nature as both power and resource (Hastrup 1990:271). Its power manifested itself through its wild and untamed elements, and the fact that it could not be fully controlled. Until the expansion and industrialisation of fishing in the nineteenth century, when the sea was increasingly appropriated by large fleets of trawlers, fishing still took place in the conceptual realm of the wild. "To fish was to engage temporarily in an appropriation of the wild, but this was ... a dangerous activity. By stepping into the wild, the Icelanders became vulnerable to forces beyond their control..." (ibid:275). Farming, on the other hand, was an activity which belonged to the human social and cultural realms, beyond which nature was seen to contain a wealth of potential resources. However, in the pre-modern period (1400–1800), although the wild could be tamed and exploited to some extent, it could not be fully controlled (ibid:272). Gradually, as technology became more sophisticated, ideas of nature as constituting unlimited and controllable resources began to dominate. As more and more of nature was appropriated for human exploitation the distinctions between wild and domestic became less clear. This newly appropriated area included the sea, whose previously feared inhabitants came to be seen increasingly as marine resources as their environment was able to be used (Pálsson 1990:130).

The same can be said for Faroe Islanders, for whom, as the occupants of an archipelago in the middle of a vast ocean, the sea was an even more dominant controlling factor of life in earlier centuries, prior to the development of an export economy based on an expanding fishing industry. In the nineteenth century, fishing, and to a certain extent also *grind*, came to be controlled and managed, as the sea was increasingly appropriated for large-scale human exploitation. The difference was that for various social and economic reasons, *grind* did not develop into a commercial industry in the way that fishing did, although it seems to have come close to doing so in the abundant pilot whale years of the mid-nineteenth century, when train oil from pilot whales was a fairly significant export commodity (Petersen 1968:47).

Despite the social and technological changes which have taken place around and within it, pilot whaling in the Faroes still functions as a means of providing food which is independent of the commercially-based export economy that has arisen since the nineteenth century. The occurrence of a catch remains sporadic and unplanned, and depends largely on the movement of the whales themselves past the Faroes. As such, it could never conceivably be controlled and systematised to the extent that land-based food production can be. But the apparent paradox of its "unpredictable regularity" has also allowed for the development of detailed organisation on the community level, codified in formal written regulations, both for the conduct of the drive and kill, and the distribution of the catch.

The survival of a subsistence hunt for food in what is now, in most other respects, a modern technological society requires a different model to explain perceptions of the human/nature relationship than those suggested by both Hastrup and Pálsson. The necessarily spontaneous and unpredictable nature of a *grind* still allows room for a view of nature as power as well as resource. In concrete terms, the danger to participants of thrashing tails, often mentioned in accounts of the kill, is still present. In this respect it retains the characteristics of the precarious exploitation of an untamed environment of pre-modern times, as opposed to the controlled and industrialised exploitation of nature characteristic of modern Western societies today. The initiation of a drive is no guarantee, even with modern technology, that the chase will end in a successful catch. The whales must be killed like land mammals, but to do so, the men must partially enter the whales' own environment. Like the pre-modern activity of fishing/hunting defined by Hastrup, *grind*, by definition, is "based on an instantaneous relation between man and prey" (1990:274).

It has been suggested that changes in Icelandic fisheries have come full circle, meaning that as a result of many decades of over-exploitation and abuse of nature, modern attitudes now advocate a return to respecting nature: "Now as in a primitive society, people are induced to cooperate because of a threat to nature; those who violate the new prohibitions are considered guilty of creating disorder in nature" (Pálsson 1990:130). Thus the distinction between the social and the wild is once again very clear, but with the important difference that nature's previously perceived power over human society has been completely reversed, and nature is now seen as the vulnerable party which must either be kept entirely external to the realm of the social, or fully incorporated into it.

The ambiguity of *grind* today is contained in its form and function as a subsistence hunt for food, largely independent of the development of the industrial exploitation of nature and still beyond absolute human control, yet conducted with a large degree of social organisation in a modern Western society. However, prevailing urbanized perceptions of the human/nature relationship in western societies, most clearly manifested in, and accentuated by wildlife protection and animal welfare discourse, cannot accommodate congruence between the "social" and "wild". This distinction is manifest, for example, in terms of British attitudes to animals and animal welfare:

It is no coincidence that aesthetic and moral sensibilities towards animals
have almost exclusively emerged in the highly urbanised societies which practice a far-reaching freedom from, and over, the natural world. The urban way
of life sees animals as experiential and geographical outsiders or, at least, sub-
ordinates who need to be looked after. Relationships with animals would
have other characteristics if the urban had not been quickly and firmly established as the home and garrison of the fully social in eighteenth- and nine-
teenth-century Britain. (Tester 1991:197).

PRESSURE TO CONFORM

The ambiguous position of *grind* between the spheres of the social and the
wild has been appropriated and transformed by an international anti-whaling
discourse into pressure to conform. The pressure has manifested itself in a
number of ways since the mid-1980s, when directed campaigns against Faroese
whaling by a number of environmental and animal welfare groups in the U.K.
and U.S. began in earnest and continue to provide several such groups with a
reliable source of revenue. The end result of such campaigns has been a constant stream of protest letters addressed to the Faroese Prime Minister, as well
as countless television and press reports of the "issue" which have for the
most part relied heavily on campaign material for their information. There
have also been some incidents of physical confrontation in the Faroes between
anti-whaling activists and Faroe Islanders.

The easy public access to a *grind* and the actual killing and blood involved
has made it a prime target for international scrutiny and media exposure
through the camera lens, a scrutiny which has provided an already well-established anti-whaling lobby with the raw material necessary to maintain public
interest and outrage at a steady level. Now, however, nearly 10 years after the
first campaigns began, the anti-*grind* lobby has still not managed to "claim a
victory" by forcing the Faroes to stop whaling. Nor are recent renewed attempts to increase the pressure by organising a consumer boycott of Faroes
fish products likely to have any significant effect on a Faroese economy already severely depressed for other reasons. The lobbying and campaign tactics
of certain environmental and animal protection groups, and the way in which
they tend to abuse and exploit public concern for the environment with misin-
formation for fund-raising purposes, has also become the subject of scrutiny
and criticism in recent years (e.g. Bonner 1993; Spencer et al. 1991). This is
hardly surprising, given the often aggressive and confrontational nature of
much of the anti-whaling rhetoric, including that directed specifically against
the Faroes. A boycott campaign newsletter from the spring of 1993 is headed:
"Frighten the Faroes Whale Eaters. Join our campaign and hit them where it
hurts" (WDCS 1993). Recipients of this newsletter are urged to buy "Whale
Bonds" which will provide their owners with, among other things, a "Wage
War on Whale Eaters" window sticker.

This form of campaigning in turn incites the general public to write abusive letters to Faroese authorities, not infrequently expressing the hope that Faroe Islanders are physically punished in some way, or comparing Faroese whaling with atrocities perpetrated against human beings:

> ... If it was left to *me* and *many more*, we would *drop* a *Very Large Bomb* on the lot of *you*, or pray for an *Earthquake* to *Destroy* every *Man Woman* and *Child* on your Godforsaken, wicked cruel islands. May *you* all *perish*, none of you are *fit* to *live* and should be *slaughtered* as you do to these wonderful creatures that do you no Harm... (From the UK, 1990).

> Can you help me? I have recently conducted a survey in an effort to determine what people generally considered the most debase and despicable act perpetrated by humanity for the year 1992. It seems opinion is divided between the systematic and mass rape of Bosnian women by Serbian forces, and the systematic and mass slaughter of pilot whales by the Faroese. The question is do you think your country should receive the greater acclaim rather than Serbia? ... (From the UK, 1993).

Even the most cursory glance at the many protest letters received by the Faroese Government over the years leaves no doubt about the strength of feeling against pilot whaling aroused amongst members of the public in France, Germany, the U.K. and U.S. As already suggested, the perceived ambiguities of *grind* have been transformed in the anti-whaling discourse with the deliberate purpose of outraging the public and maintaining momentum in, and support for, a conflict of their own creation. With reference to the three essential anomalies represented by a *grind*, as outlined by Wylie (1981:107, see above), and drawing in general on the campaign material directed against Faroese whaling, some of these transformations can be summarised as follows:

1. The unpredictable nature of the arrival of a *grind*.

This has been transformed into the claim that *grind* is uncontrolled, unregulated and indiscriminate. That modern Western man has no control over the occurrence of the resources he uses is not consistent with a world view in which most natural resources have been fully appropriated into the social and cultural sphere, and where it is consequently man, not nature, who determines the frequency and extent of use.

2. The combination of fishing and farming techniques used to drive and kill the whales.

This anomaly has served to provide the dramatic film and pictures of the actual killing on which campaign material is almost exclusively focused. Not least the fact that a whole school is usually driven and killed at once, and that this is done by a number of men simultaneously, is transformed into claims of "mass slaughter" and the "largest whale kill in the world". The "barbarism" and "cruelty" as it is so often described in campaigns against Faroese whaling seems primarily a reaction to, and transformation of, not only the public nature of the activity, but also the simple technology of the whaling knife used in the kill (see Note 3). This is seen to be inconsistent with the use of modern motor boats to drive the school, and the use of crane trucks and modern cars to collect the shares of the catch.

That *grind* represents a mixture of hunting and farming technologies disturbs the preconceived view of these activities as completely separate. The sight of fair-skinned men, fully clothed in woollen sweaters and boilersuits entering the water with knives in their hands to kill whales is transformed into "sport" or "ritual slaughter". In the simplistic urban view of hunting for food, the hunter is at one with the unspoiled wilderness in which he hunts and must not therefore display any of the incongruous trappings or influences of "modern civilisation".

3. A successful catch provides an "imported" food that does not have to be bought with exports.

In the rhetoric of the anti-whaling campaigns, the fact that the pilot whale catch in the Faroes is non-commercial is not necessarily in its favor, despite the fact that particular criticism is made of the commercial nature of whaling elsewhere (e.g. in Norway and Japan). Rather, that pilot whales, a locally occurring resource, are freely shared, without financial considerations, among participants and residents in a modern, largely cash-based society is transformed into an "unnecessary indulgence". Free food distributions belong to the realm of the Third World or elsewhere where extreme poverty and starvation prompt handouts from the wealthy West. Campaign material inevitably points out that "the Faroe Islands have one of the highest standards of living in the world" and that they therefore "do not rely on pilot whales for their survival". In response to this interpretation of the issue, a recent letter-writer complained "It is surely only justified to people who are starving with hunger".

Simplistic interpretations of Faroese whaling continue to dominate the way in which it is represented today. Confronted with consumer boycott campaigns, a constant series of TV and press reports depicting the Faroese as "savage barbarians", the resultant stream of protest letters to their Government, and even threats of physical violence, Faroe Islanders find themselves in the often frustrating position of having to defend what for them has always been a source of food and a way of life. People who take part in *grind* do all the work themselves, from the driving and killing of the whales to the butchering of the whale carcass from which they receive their shares. As Wylie (op. cit) pointed out, the anomaly of a *grind* is that it represents an "imported" food that doesn't have to be paid for with exports. If forced to stop making use of pilot whales as they have done for centuries, Faroe Islanders would have to import other meat in its place, not only paying dearly for it with exports, but also having it transported over long distances from industrialised livestock production in continental Europe and elsewhere. Thus, not only would Faroe Islanders lose a degree of self-sufficiency in food production on which they have been able to rely for centuries, but in environmental terms in regard to the wise use of natural resources including non-renewable fossil fuel, the opposition to whaling appears misguided and counterproductive. It is, after all, nature that produces the renewable pilot whales eaten in the Faroes.

The confrontation between the ideologies embedded in the current discourse of anti-whaling protest and the needs and wishes of various coastal communities, such as the Faroes, to use whales for food, reflects very clearly the divergence in attitudes to nature and the degree of intolerance associated with attempts to force others to conform to a particular world view. This raises questions of culture and identity which have even greater resonance in the present context of increasing European centralisation, and the growing environmental responsibilities felt by industrial nations for the problems facing developing countries. Pilot whaling in the Faroes is a particularly interesting case, as it challenges the basis of such historically-based, yet still used, categories such as "commercial" and "aboriginal subsistence" applied to the exploitation of fish and wildlife.

Pilot whaling represents a meeting and merging of the boundaries between land and sea, between the social and the wild, between culture and nature, between the pre-modern and the post-modern, between the historical continuity and modern function of a traditional form of food production and prevailing perceptions of modern society. As a result it also challenges us to rethink our all too rigid definition of what it is to be modern and civilised, and our increasingly artificial relationship with nature. Pilot whaling in the Faroes provides Faroe Islanders with food; for others, it may also provide some food for thought.

NOTES

1. The Faroese word *grind* denotes both a school of pilot whales, as well as the activity of driving and killing pilot whales. Since the word has so frequently been used in English texts and is listed with its Faroese meaning in the *Oxford English Dictionary* (1991), it is used in this paper, albeit in italics for clarity.
2. It is known however, that pilot whale meat contains elevated levels of heavy metals and the blubber is contaminated with organochlorines. This led the Faroese Food and Environmental Institute to issue recommendations in 1989 for the maximum monthly intake of pilot whale meat and blubber. International studies are also being conducted to determine the effects of these pollutants on the development of Faroese children (Grandjean et al. 1992). Ironically, certain environmental groups, whilst attacking *grind* and Faroese society and culture, express concern for the well-being of Faroe Islanders by using this contamination as an argument for banning

grind, rather than addressing the problem of the source of the pollutants and their effect on the marine environment.

3. Alternatives to the whaling knife, such as guns, have been discussed, but none are considered a suitable alternative to the simple technology of a knife for speed and efficiency given the conditions in which the killing takes place and the anatomy of the whale. By cutting across the back of the head to the spinal cord, the major blood supply to the brain is severed in a matter of seconds.

REFERENCES

Bloch, D., G. Desportes, K. Hoydal and P. Jean, 1990.
   Pilot Whaling in the Faroe Islands July 1986–July 1988, *North Atlantic Studies* 2(1–2): 36–44.
Bloch, D. and M. Hanusardóttir, 1993.
   Marine Mammals. Whales as Source of Food. In: *Encyclopedia of Food Science, Food Technology & Nutrition*, pp. 2902–2907. AcPress, London.
Bonner, R., 1993.
   At the Hand of Man: Peril and Hope for Africa's Wildlife. Alfred A. Knopf, New York.
Buckland, S.T., Bloch, D., Cattanach, K.L., Gunnlaugsson, Th., Hoydal, K., Lens, S. and Sigurjónsson, J. 1993.
   Distribution and abundance of long-finned pilot whales int he North Atlantic, esti-mated from NASS-87 and NASS-89 data, in G.P. Donovon, C.H. Lockyer and A.R. Martin (eds), *Biology of Northern Hemisphere Pilot Whales. Report of the International Whaling Commission*, Special Issue 14, Cambridge.
Debes, L., 1673.
   Færoæ & Færoa Reserata, Einar Joensen 1963 (ed.) Einars prent og forlag, Tórshavn.
Debes, L., 1676.
   Færoæ & Færoa Reserata: That is a Description of the Islands and Inhabitants of Foeroe, Englished by John Sterpin, printed by E.L. for William Iles, Flower-de-Luce in Little Brittain, St.Bartholomews Gate.
Douglas, Mary, 1975.
   Implicit Meanings - Essays in Anthropology, Routledge & Kegan Paul, Boston.
Frost, S., (ed.) 1978.
   Whales and Whaling, vols I & II, Australian Government Printing Service, Canberra.
Grandjean, P., P. Weihe, P.J. Jørgensen, T. Clarkson, E. Cernichiari, T. Viderø, 1992.
   Impact of Maternal Seafood Diet on Fetal Exposure to Mercury, Selenium, and Lead. *Archives of Environmental Health* 47(3):185–195.
Hastrup, K., 1990.
   Nature and Policy in Iceland 1400–1800, An Anthropological Analysis of History and Mentality. Clarendon Press, Oxford.
Kalland, A., 1993.
   Management by Totemization: Whale Symbolism and the Anti-Whaling Campaign. *Arctic* 46:124–133.
Kunngerð um grind, May 12, 1986.
   Føroya kunngerðasavn fyri árið 1986, Føroya Landsstýri, Tórshavn: 90–96.
Löfgren, O., 1985.
   Our Friends in Nature: Class and Animal Symbolism. *Ethnos* 3(4):184–213.

Pálsson, G., 1990.
   The Idea of Fish: Land and Sea in Icelandic World View. In: R.G. Willis (ed), *Signifying Animals: Human Meaning in the Natural World*, pp. 119–133. Unwin Hyman, London.
Pálsson, G., 1991.
   Coastal Economies, Cultural Accounts, Manchester University Press, Manchester.
Petersen, P., 1968.
   Ein føroysk bygd, Egið forlag, Tórshavn.
Reglement for Grindefangsten paa Færøerne, 1832.
   J. H. Schultz, Copenhagen.
Sanderson, K., 1990.
   Grindadráp — the Discourse of Drama. *North Atlantic Studies* 2(1–2):196–204.
Sanderson, K., 1991.
   Whales and Whaling in the Faroe Islands. Department of Fisheries, Tórshavn.
Sanderson, K.,1992a.
   Grindadráp — A Textual History of Whaling Traditions in the Faroes to 1900. Unpublished Master of Philosophy Thesis, University of Sydney, Australia.
Sanderson, K., 1992b.
   Grind — Næringstof til Eftertanke. *Norðurlandahúsið í Føroyum - Árbók 1991–92*: 48–52. Norðurlandahúsið, Tórshavn.
Singer, P., 1978.
   Why the Whales Should Live. *Habitat* (Australian Conservation Foundation) 6(3): 8–9.
Spencer, L., J. Bollwerk and R.C. Morais, 1991.
   The Not so Peaceful World of Greenpeace. *Forbes Magazine*, November 1991:174–180.
Storm, G. (ed), 1881.
   Samlede Skrifter af Peder Claussøn Friis, A.W. Brøgger, Kristiania.
Tester, K., 1991.
   Animals and Society. The Humanity of Animal Rights, Routledge, London.
Thomas, K., 1983.
   Man and the Natural world - Changing Attitudes in England 1500–1800, Penguin, Harmonsworth, U.K.
WDCS, 1993.
   Whale and Dolphin Conservation Society Newsletter, April 1993, Bath, U.K.
Wylie, J., 1981.
   Grindadráp. In: J. Wylie & D. Margolin (eds), *The Ring of Dancers - Images of Faroese Culture*, pp. 95–132. University of Pennsylvania Press, Philadelphia.

13. NORTH NORWEGIAN WHALERS' CONCEPTUALIZATION OF CURRENT WHALE MANAGEMENT CONFLICTS.

HARALD BEYER BROCH

Since 1987, there has been an international moratorium on commercial whaling imposed on the whaling nations by the International Whaling Commission (IWC). The moratorium was the result of prolonged and determined international campaigns by various environmental and animal protection organizations for the purpose of saving threatened cetacean populations from the possibility of extinction. To Norwegian whalers this resulted in at least a temporary halt to minke whaling, which activity had come to be important in North Norway, where a reliance on small scale minke-whaling emerged in the 1920's and gradually became an integral part of ecological and cultural adjustments securing the foundation of livelihood in several fishing communities (Jonsgård 1955; Kalland 1990; ISG 1992). However, whaling and the consumption of whale meat have ancient roots in Norwegian coastal culture (Risting 1922; ISG 1992).

The importance of minke whaling during recent decades in the Lofoten regions of Northern Norway derived from the fact that whaling provided a highly predictable source of income. Thus, from an ecological perspective the minke whale can be regarded as a key resource, one that insures the viability of many local communities when other resources constituting the ecological niche of the human population cannot be harvested (see Broch 1985, 1991 for discussion of the niche concept). The key nature of the minke whale therefore results from its role as a buffer during those difficult periods when the cod (or other northern fish stocks) provide poor returns. Whaling, at such times in particular, consequently provides an important measure of adaptational security.

Although the IWC allows for non-commercial whaling, under the category of aboriginal-subsistence whaling as carried out by Inuit whalers for example, this particular exemption from the moratorium on commercial whaling is not applied by the IWC to the traditional Norwegian minke whaling operations.

The focus of this paper is on the whalers' interpretation of various groups' statements on the future of minke whaling, how they view the recommendations of marine biologists, and how they understand the intentions and ideology of various environmental and animal rights' groups. Most importantly, the paper will examine how whalers consider their profession within an ecological perspective (see Note 1).

To appreciate the whalers' view of their own profession it is necessary to pay attention to how they present themselves both in their local communities and to external authorities. It is argued that the whalers share common identity elements and a reputation as brave entrepreneurs within Norwegian coastal fishery operations. The first part of this paper examines how this identity and reputation is constructed. Following this examination, the paper seeks to describe how whalers are regarded in their own local communities by their non-whaling neighbors and how they are represented in the rhetoric of non-local anti-whaling activists.

Doubt has been expressed concerning the degree to which Norwegian whalers can be considered as competent ecologists who understand the intricate interplay between various species interacting within the marine ecosystem. Pål Bugge, involved with Greenpeace Norway, claims that both Norwegian whalers and top ranking politicians are ecological ignorants. In Bugge's opinion Nature itself has to restore ecological balance where humans have destoyed it; the only way that balance can be restored is by leaving Nature to itself with no interference from people (Teigland 1990).

However, in the present discussion this is not the main issue. Whalers are without doubt competent hunters and fishermen with extensive knowledge of marine life and the environment upon which they base their living. Indeed, such specialized and discriminating knowledge is necessary in order to utilize the resources upon which they depend. Many whalers have higher education through the secondary school level and they know biologists and their science-based discourse through cooperation and discussion with fishery scientists. Thus they are quite aware of the current terminology and the scientists' perspectives when matters such as resource management are being discussed and negotiated. The whalers are also very well informed regarding ongoing debates reported in national and local newspapers and the other media, and they participate in these debates both on an individual level and through their regional organizations and professional associations.

One consequence of the whalers' increased sophistication in utilizing ecologically-sound rhetoric has probably been the need of those opposed to whaling to shift the focus of discussion away from arguments based upon conservation concerns or ecologically-sound resource harvest levels, to arguments favouring animal rights and, more particularly, cetacean rights. The ecological argument that whales are endangered is losing salience, as scientists newly report robust stocks levels, and so is now rarely advanced. Even though endangerment or extreme depletion may still be the case for some few species (such as the blue whale) the minke whale, the object of interest and discussion in North Norway, is definitely not an endangered species according to current scientific understanding (Aron 1988:104; Conrad and Bjørndal 1993:165).

This paper will demonstrate how two fundamentally different forms of rhetoric, that of the whalers and of those opposed to all whaling activities, clash in verbal communication. The term "rhetoric" is not used in any negative sense, but follows Oliver (1965) who says that rhetoric reflects a mode of thinking and a mode of activating all available information and means for the achievement of a particular end (see Note 2). What people perceive and how they experience their life-world is largely dependant upon their perceptual frame of reference. Rhetoric is thus not universal but rather is culture-specific, and may vary from time to time within a particular culture. Logic (in the popular sense of the word), which is the basis of rhetoric, also evolves out of a culturally-influenced worldview and is no more universal than rhetoric (Kaplan 1988).

"WE ARE HUNTERS"

Though many whalers present themselves primarily as hunters, they see themselves quite differently from the way most urban citizens (who often have minimal or no direct experience with game management and hunters) conceptualize hunting and hunters. To some people, hunting is a morally dubious activity connected with inhumane sports and entertainment where people kill to amuse themselves. For those individuals advancing the notion of animal rights, all killing of animals poses an enormous moral problem. Thus, the perception and standing of professional hunters in these circles is evidently not favorable: professional hunters are, according to their opponents, uncivilized bloodthirsty brutes, either because they are misled and unfortunately do not know any better, or because they suffer some personality disorder (Tester 1991). In either case, hunters should be helped to deconstruct their basic values in order to develop more humane attitudes towards animals and to accept animal rights. From an anthropological and relativistic point of view, the disparate valuations about hunting held by hunters and their opponents represent a classical and profound confrontation of ideologies where, despite the magnitude of the differences, neither party has any right to impose their own ideas upon the other with brutal force.

Obviously no hunter thinks of himself, other hunters, or hunting activities, in such negative terms. All hunters are doubtless aware of the possibility that there may be some who violate hunting ethics by making their prey suffer unnecessarily and that some individuals may even experience deviant satisfaction from mistreating animals. Such exceptions are just that, exceptions, and emphasize the fact that, in general, hunting activities are indeed governed by ethical norms. Further, many professional hunters, including subsistence hunters and small scale commercial hunters who both depend in significant degree upon hunting for their livelihood, strongly disapprove of many forms of sports-hunting where animals are killed merely for entertainment. For such professional hunters, there is a significant difference between saying that hunting is "just for fun", and appreciating an occupation that at times is thrilling.

When whalers in many contexts present themselves as hunters, this implies that they regard whaling as a specialized form of hunting. In the present context it is relevant to note that those fishermen who engage in whaling and those who do not whale, all generally relate fishing for cod, halibut, haddock, or coalfish as being more similar to hunting than any other form of livelihood. Thus one whaling skipper stated that it should be obvious to all with some hunting experience that fishing activities actually are specialized hunting techniques.

In the case of fishing one never knows exactly where the fish can be found; the fish are somewhere, but it takes experience and skill to discover them and then knowledge about the movement and habits of the fish in order to make a good catch even after they have been discovered. During fishing operations, the fish may suddenly disappear, only to reappear at some other location. In fishing too, you must know how to handle the relevant technology, about the nature of the sea floor, how the currents run etc. Stamina and faith in one's luck are prerequisites in order to succeed, for one day the fisherman may be lucky, the next day, even the best among fishermen may fail. Whalers and fishermen agree that in whaling, as well as in other kinds of "fishing" or hunting, it is "the hunting instinct" that motivates and nurtures the fascination of the activity and moulds the successful outcome. Whilst gathering information on minke whaling in Northern Norway, a fisherman, who had never participated in whaling, stated:

> [The Lofoten cod fishery] is much better than working here on shore. This year there is so much fish in the ocean that all men around here are just turning crazy to get out on the sea. There is some kind of hunt going on out there too, you know.

(Field Notes. Lofoten, 1991)

None of the whalers interviewed during fieldwork in Lofoten participated in small or large game hunting (for ptarmigan, grouse, ducks, hare or moose etc.) on land. However, a few fishermen did hunt, and some whalers have reported that they indeed went ptarmigan hunting when they were young and regarded that as training for future hunting and whaling (Kalland 1992). It appears most probable that whalers are not particularly interested in small game hunting, and this accords with what many other large-game hunters state, that when you first get engaged in the hunt of large animals, small-game hunting gets less interesting. However, all whalers agree that whaling is the most fascinating, most thrilling and exciting of all kinds of hunting. Although whalers generally do not engage in hunting for land animals, many have successfully hunted for fresh meat on Spitsbergen while whaling off the coasts of that archipelago. These whalers enjoy talking about the hunting aspects of their profession and many stated that they felt privileged to have such an engaging occupation: most people have less interesting work, with more repetitive routines to earn their living than they do themselves, the whalers claimed. The whalers felt proud of their profession where they can prove themselves as men in skills and courage, and they find their job meaningful and valuable.

HARVESTING RENEWABLE RESOURCES TO PROVIDE HIGH QUALITY FOOD

The whalers consistently legitimatized their activities by referring to the fact that they harvest renewable marine resources and supply society with high quality food. This is done in a world that seems to lack food and where many others exploit less robust resource stocks in an unsustainable manner. Some whalers also referred to recent research proving the preventive value of whale meat with regard to cardiac diseases and stressed its superior food value (see, e.g. Innis and Kuhnlein 1988; Dyerberg 1989; Nordøy et al. 1991). According to the whalers, they are conservationists by mind and heart and have to be so if they are to continue as marine hunters: over-hunting the minke whale is seen to be against their own self-interest as that would remove their source of live-lihood. One whaler/skipper said that the nature of small scale whaling, utiliz-ing family-financed boats of not more than 50 to 60 feet (15–18 m), in itself is a guarantee against the extinction of the minke whale, for using this technol-ogy it becomes unprofitable to continue searching for whales once they be-come scarce, given the high operating expenses for fuel and payments on the boats. This evaluation of the safety of the whale stocks is dependent upon the scale of technology remaining at its present level, and the skipper added that this indicates just how important it is that renewed whaling remains in the hands of the current whalers and is not opened to more highly capitalized en-trepreneurs who could afford to gamble with the introduction of new technol-ogy. In the event of such technological and ownership changes, this skipper feared the situation could possibly be altered in a negative fashion.

It may well be argued, and with some justification, that Norwegian fisher-men have not been particularly foresighted in their recommendations and claims for quotas in various fisheries, and in recent times overexploitation for short term profit seems to have been the rule rather than the exception. How-ever, with modern and highly efficient fishery technology in widespread use, many fishermen have come to accept a need for regulation and scientifically-based resource management in order not to deplete economically important fish species. Experiences with the seal invasion in Northern Norway during the 1980s come firmly to their minds, and fishermen relate that disaster to two factors: over-fishing and reduced commercial sealing. The whalers compare that situation with the present ban on whaling; one whaler thus made the point by asking rhetorically: what would happen to newly-planted conifer woodlands if all moose hunting was brought to an end? Some whalers agree that unlimited whaling may represent a danger to their own adaptation and risk the extermination of various whale species. However, they do not see that they themselves represent a significant threat to any whale population; the threat would come from others, especially if new methods for whaling were introduced or more people were allowed to enter the hunting grounds. This way of analyzing the situation comes remarkably close to Brox's discussion of common property theory where he predicts the collapse of many natural resource stocks where large capital interests come to control wildlife and fishery management decisions. Those utilizing intermediate technology backed by limited economic resources will not survive competition with enterprises backed by national or international large-capital interests (Brox 1991).

Whalers, who are also fishermen, also worry about what will happen if no whales are hunted and the whale population multiplied without this limitation through hunting being placed on its growth. Will not an ever-increasing whale population threaten the future of the commercial fisheries? The minke whale, for instance, consumes large quantities of fish (Nordøy 1991). When some animal rights' supporters claim that indeed both whales and ultimately all fish have the same rights as humans to live, the rhetorical standpoints are so far apart that fruitful discussion between the two sides becomes almost impossible. When Norwegian politicians seem to agree on the importance of adopting a policy allowing sustainable exploitation of resources, this is in order to secure an income for future generations based on maintaining viable populations of renewable plants and animals. When no Norwegian politician has ever promoted the view of a future society based on vegetarianism, whalers do not understand how animal rightists' ideologies can provide any acceptable premise upon which to base current or future Norwegian resource management policies. The whalers can see no rational arguments supporting a continuing ban on minke whaling, not from an ecological, or local community viability, or labour market point of view. Many whalers feel that environmentalists and those living off the sea should cooperate in order to save threatened marine life: whalers are, for instance, concerned about the ongoing pollution of the oceans, with some whalers playing an active role in opposing pollution. These whalers are especially careful about what they dump into the sea and they actively promote the notion of the importance of a clean sea.

At a personal level, there is more to whaling than just the chase. Many of those who present themselves as hunters, describe and indeed stress the beauty of the Arctic regions. The coastal areas of Bear Island and Spitsbergen are especially mentioned for their aesthetic values. During whaling the individual needs to remain alert; however, the whale hunt also involves long periods of waiting when little happens, and such periods give the whalers time to contemplate the beauty of nature. One steersman/gunner explained:

> We have time to reflect about life, about our [humans'] place in the world. No one feels particularly great or important in a 50 to 60 foot vessel on the open ocean in rough weather, or in bright sunshine among drifting and collapsing

icebergs. When there is little to do, it is impossible not to be impressed by the beauty of the Arctic summer; the colors of the water, ice and clouds have a great impact.

(Field notes, Lofoten, 1991).

How the shape and sizes of clouds and wave-shaped icebergs give inspiration to imagination was mentioned by many. The time spent on whaling expeditions thus give the participants varied emotional experiences and enrichment.

It should not be forgotten that no man is only a whaler and hunter: minke whaling has always been carried out as a complement to other tasks, such as fishing, in order to improve living standards in Norwegian coastal communities. In addition to the reputation earned as a whaler and by other part-time occupations, all men, of course, play several other roles in their local communities. In order to understand whalers, it is important to remember that their statuses are composed of a variety of different elements: some may be members of the parish council, active in a church congregation, a member of the volunteer fire brigade, or the community school board for example. Additionally, whalers' values are influenced by their family concerns and attachments: they may have parents who need to be cared for, they may have wives and children of their own to support; in short, they are concerned about the well-being and reputation of their families. Not only are the whalers proud of whaling, but so too are the members of their families proud of this activity, for in the local context it represents honourable man's work, where men prove themselves as brave and responsible individuals.

When the whaling aspect of the families' adaptation is selected as the crucial identity marker, this is indeed because whalers at times are envied, admired and respected in their local communities. The emblematic importance of whaling is precisely because whaling represents the best part of life, where work and existential meaning merge. Additionally, women — that is mothers, wives, and daughters — seem proud of their masculine men and most boys hope to be allowed to participate in the adventurous life on the whaling vessels when they are old enough.

In the ongoing controversy about whaling, whaling families say they are firmly supported in their claim for continued whaling by their neighbors, including those who have never participated in the activity. This may be because some hope to be able to participate in whaling in the future, but more importantly it is because residents appreciate the positive effect whaling has on the viability of their local communities. The reality of this support was indeed affirmed by the interviews carried out among non-whaling residents in Lofoten during fieldwork there. However, despite this overwhelming local-level support, people in Lofoten in 1991 asked: "But where is the support from our government and politicians?" (See Postscript).

"ENVIRONMENTALISTS GIVE A FALSE PICTURE OF US AND MINKE WHALING"

Environmental and animal welfare and animal rights organizations were far from popular in the Lofoten communities when this investigation was carried out in 1991. All whalers, their families and their supporters felt they had been cheated by prominent members of such organizations. When viewing the picture of whalers and whaling that environmentalists and animal rightists construct and advocate through their protest rhetoric, this should surprise nobody. The protest movements' goals are indeed to create negative attitudes towards all aspects of whaling (Ris, this volume). To succeed in this strategy they make selective use of research on whales and whaling, confuse different whaling techniques and create the image of a "super whale" by lumping together traits found in different species of whales (Kalland 1993). The super whale appears to be constructed from diverse images best suited to foster the greatest emotional involvement in whales by large numbers of people in western urban societies. The same rhetorical tactics have probably been in use when trying to establish the negative image of the whale hunters (ibid:129–130). Research in the U.S. shows that peoples' attitudes toward and willingness to work for protection of endangered species are connected to several factors (Kellert 1988). First, the animal species must indeed be believed to be endangered, which the super whale of course is. The question of the aesthetic image and the purported phylogenetic relatedness to humans are also important: the more the animal appears to be similar to human beings, the greater the likelihood of public support for the species. People also pay heed to how many and what kind of people are threatening the animal in question. In this context aboriginal rights and cultural survival may moderate some of the willingness to otherwise provide absolute protection for an animal. Finally, how much information people have on the species and the perceived humaneness/inhumaneness of the activity threatening it, are also relevant considerations. In this American study it became clear that the general public overwhelmingly approved of the two most pragmatically justified kinds of hunting, that is subsistance hunting as practiced by traditional native Americans and hunting exclusively for meat regardless of the identity of the hunter (ibid.:159–161).

   At the beginning of the controversy leading to the moratorium on whaling, the whalers hoped they could gain sympathy and understanding for their cause and continue their profession. Later, after the moratorium was in effect they believed they would improve their position if they fully cooperated in the search for sound information on the population status of the minke whale. Indeed, whalers wanted to cooperate with marine biologists and with conservation organizations. The whalers felt sure that if it was established that the minke whale was indeed not an endangered or threatened animal species, commercial whaling should and would recommence. All this time, given their own environmental knowledge of the region, they were themselves quite certain that the minke whale population was large enough to allow the hunt to continue.

However, as time went by they came to understand that they had not realized how their opponents' rhetoric is constructed. The whalers had opened their homes to scientists and anti-whaling activists alike, and had let them join in whaling expeditions before the ban was introduced. This strategy had been adopted because it was rumored that an end to whaling could become a reality due to strong negative world opinion. The strategy of openness worked relatively satisfactorily with the scientists, but according to the whalers, the activists misused the whalers' hospitality and willingness to cooperate, for not only did some activists claim they were marine biologists and thus gain access to information and hospitality on false premises, but they subsequently wrote completely false and blood-curdling reports in which whalers were represented as criminal murderers lacking moral obligations and humanity. Moreover, the activists did not reciprocate, in an open information exchange, the hospitality and insights into whaling provided them by the whalers by disclosing their own intentions and project goals.

The following dramatic statement by a former leader of Greenpeace Denmark indicates the level of misrepresentation that occurred:

> I have never been so afraid in my entire life. Sitting at the table hour after hour with a crew for whom Greenpeace were the lowest thing they could tread on. Playing the role of an unbiased marine scientist fully aware that they might stab me with their knives if I revealed my true identity.

> (Claudi 1988:91).

This obviously says more about the Greenpeace leader than about the whaler/fishermen he writes about. However, it would be naive to dismiss these observations as merely the result of a powerful imagination. Rather these views should be taken to illustrate very well an individual and a representative of a subcultural group which is quite distant from and ignorant of people who earn their living from the sea. It is indeed a common social process that people separate very completely their own ways of behaviour and values from those of members of groups they oppose. Much of an individual's self legitimation of the right to morally over-rule the values and traditions of others is based on such dichotomization. Tight and closed ingroup discourse confirms and elaborates such inter-group differences and make it difficult to alter the opinions being expressed. More or less explicit categorization of others serves to stereotype them in order to give clues about how interaction should be carried out. Stereotypes are based on exaggerated beliefs either built on misunderstandings or on purposive socio-cultural constructions. These sets of belief are strongly associated with the in-group members, and have the function of justifying group members' conduct, and in this particular case, their moral outrage. The function of stereotyping is thus to justify an individual's (or group members') conduct in relation to those being stereotyped (Simmons 1988).

The above quoted statement from the Greenpeace leader also has great rhetorical potential for mobilizing resources for use against the "brutes" who
would, unless stopped, violate the whales. The animal rights' defenders an-
thropomorphize the whale: it becomes at least as intelligent as humans and it
is nurturant and has developed a sense of humour (Gylling Nielsen 1987), and
is believed to want to befriend and protect humans (Watson 1985). In a way
the whale represents innocent, good, unspoiled human qualities. The whalers,
on the contrary, are made inhuman, animal-like predators. The portrayal
ulimately represents yet another version of the classical confrontation between
good and evil, but in this particular case additional poignancy pertains because the good forces cannot defend themselves, indeed cannot survive, unless
urgent help is provided (by various anti-whaling organizations).

During this fieldwork, the whalers and their wives and children were not
unlike most Norwegians encountered in small rural communities: they were
open, informal, and thoughtful of others. People in Lofoten, and elsewhere
along the coast, are used to accepting strangers with hospitality and curiosity.
Now however, they are learning to be more careful because of incidents like
those described above and because of a growing tourism that brings more
strangers to their doorsteps during summer (for some aspects of tourism in
Lofoten, see Ris, this volume).

ANIMAL RIGHTS AND "HUMAN" WHALES

It is difficult for whalers to accept the logic of animal rights. Whalers are concerned about the survival of whales, but that does not imply that they are
against killing of all animals. Whale protectionists urge people to believe:

> "... that whales are at least as intelligent as we are, they are more skilled in
> handling social relations, they deserve "whale rights" and there are talks
> about "whalekind" as a counterpart to mankind".

(Kalland 1993:128).

To whalers however, the animal rights ideology represents misplaced involvement in a brutal world where people of many different regions are suffering badly. Nevertheless, those hunting whales are sometimes impressed by
the "intelligence" of some whales, for it takes skill to outsmart them during
the hunt, even though with experience whalers come to learn whale behaviour.
Whalers generally insist that whales, being animals, are significantly different
from humans; however, they respect whales, as most of them respect many
other living organisms, but it is also a firmly established part of their religious
faith that animals are placed on Earth for the benefit of people. As whales are
not humans, the whalers regard it as false to claim that whales have human
feelings and human thoughts. However, no serious hunter finds any pleasure

in causing unnecessary pain to the animal he hunts, and whalers agree that the modern penthrite whaling grenade is superior to the non-exploding "cold harpoon" used in earlier times. After first having distanced himself from the view that whales have human feelings, one whaler offered the following comment:

> A few years ago I asked a biologist at a conference I attended, about the nervous system of whales. He told me that there is still lots to learn in these matters, but that all indications point towards the conclusion that their nerves are not working so fast and efficiently as those of humans. I believe that even if whales should have feelings such as we do, there is still little need to worry about the pain inflicted by modern harpoons. I was myself shot by a German during the war and I did not feel any pain before several hours after the bullet hit me. It never takes that long before a whale is killed after it is harpooned, he would never have the time to understand what is going on. Our present way of whaling is not a brutal affair.

> (Field notes. Lofoten, 1991)

Obviously, animal rights' activists and people living by any occupation involving the killing of animals, will never agree on these issues.

DIFFICULT YEARS TOWARD A BETTER FUTURE?

This paper provides examples of how Norwegian minke whalers discuss and express their frustration through ecologically-based rhetoric reflecting their own particular environmentally-based worldview. Their argumentation is in most ways opposed to those advocating animal rights or the more limited goal of universal whale protection. It would be quite unrealistic to believe that people possessing such conflicting interests and understandings should reach a *modus vivendi*. The whalers are fighting for a lifeway and their communities' socio-economic survival, their opponents are fighting for the life of large marine animals and for the economic and political enhancement of their organizations. The whalers agree that some regulation of their activities should be put into effect, and they acknowledge such measures to be in their own interests. They explicitly state: "If the minke whale is threatened with extinction, we stop whaling" and "If the number of (minke) whales is small it is not economically possible to hunt for them".

It is apparent that effective mediation is required to end the controversy and conflict. The members of the local communities in Northern Norway are not in a sufficiently strong position to stand up against international pressures generated by well-financed international organizations. The whalers receive moral and symbolic support from non-whaling fishermen, who recognize the benefits to their communities provided through whaling, but unfortunately for the whalers such support helps little. Further, the fishermen say they are worried about what might be the environmentalists' next actions, aware as they are of the hardship imposed on North Norwegian society through the anti-sealing and anti-whaling campaigns. Hopes for the future, by both whalers and fishermen in northern Norway, is very much in the hands of the politicians who have to be both rational and courageous in order to secure a varied and viable occupational basis for continued livelihood in the coastal areas in the North.

The whalers find the situation especially frustrating because they believe they have promoted their view according to the rules of officially approved discourse: whalers have not organized major demonstrations, neither have they complained unnecessarily, but rather, they have waited patiently and cooperated on all levels, behaviour that had at that time apparently resulted in little support and understanding from the politicians. Whalers' patience was running low in 1991, and they felt they had the right to demand to know when commercial whaling will resume: the year and date! (see below). One of the fishery advisors in Lofoten expressed the current situation in this way:

> I must say I am impressed by how well many whalers have handled the difficult situation after the whaling moratorium was introduced. Most of them have indeed kept spirits high. Many have learned new methods of fishing, such as the use of Danish seine. The first year they had little luck, the second not much better. But by now most manage well and are doing OK. But some, and in particular young people, who invested in new boats just before the ban, are depressed. If whaling is not reopened in a few years time these young men will have serious problems in paying the debts on their vessels. A few have also gone broke. Time is indeed running out. Whaling is part of one of the most important specializations we have to make a living in Lofoten. That does not say that the whalers know no other ways, and that they know no other techniques. But here it is indeed urgently important that we do have specialists on several sectors. We have specialists on sealing and whaling, on long lines, jigging (juksa), nets and Danish seine, etc. This knowledge and competence is built on experiences gathered through decades. It is of the utmost importance to the overall fishing communities that there be someone with whom you can talk and share the same type of experiences. Recruitment is also about having someone for sharing new ideas and, not the least, to discuss previous experiences in order to develop new strategies.

> (Field notes, Lofoten, 1991)

Finally, a whaler's ecologically-based rhetoric provides a summary of the whalers' understanding of the present controversy:

> Our politicians tell us that we must take ecological considerations into account when discussing the future of whaling. Of course we must, we understand this. If we do not, there will be no whales to hunt. Actually if there are too few whales it will not be profitable for us to hunt them. But how dare the politicians criticize us in these matters? Are they themselves led astray or are they short of knowledge? Why did they, for instance, support and give licences to those scallop-trawlers that were bound for the coast of Spitsbergen a few years ago? High cost expensive boats, three seasons only and then they

went broke. What was the foundation for those boats? What did the authorities know about the ecological consequenses of that activity? As far as I know, many marine biologists claimed that the activity should never have been allowed.

We, the whalers, have an occupation, a profession that is ecologically sound. Why do the politicians give in to external pressure in this matter? Why should we be prohibited from carrying out an economically successful enterprise? The question is even more frustrating when we consider the high rate of unemployment both locally and nation wide.

(Field notes, Lofoten, 1991)

POSTSCRIPT

During the summer of 1992 (one year after the fieldwork reported above was completed) a new dimension was added to the public discourse on whaling. First, the Norwegian Government decided to allow a catch of 110 minke whales for research purposes, with the meat from the whales made available for commercial distribution after the biologists carry out their studies. Suddenly the whalers and national politicians joined in a common rhetorical effort, and utilized the same arguments to gain general acceptance in Norway for eventual renewed commercial minke whaling. As a result of now widely accepted scientific estimates of a North Atlantic minke whale population numbering about 100,000 (with ca. 86,000 in the region hunted by Norwegian minke whalers), a regulated harvest can now be resumed without threatening the survival of the species. The outcome of future confrontations between antiwhaling supporters and the whalers (now assisted by the Norwegian Government) is not entirely certain: there still remains the threat of organized trade boycotts of Norwegian fish and fish products in certain other countries for example. However, the present study illustrates that an ecologically-based rhetoric has been useful in mobilizing support for a resumption of whaling.

The arguments have many general implications for both practical and ethical issues connected with wildlife and domesticated animal management. Controlled utilization remains an essential element of policies allowing sustainable exploitation of animal resources for human consumption. The issue emerging now seems to be, which parties are to control when, where and how animal resources should or may be harvested. There can be no doubt concerning the need for protection of certain animal species against the threat of extinction, but who, if not the scientific community, can offer rational advice as to which particular stocks or species require protection? Thus the whaling controversy may reflect and indeed trigger more important and wide ranging consequences for resource management than those merely related to minke whaling.

In spring 1993 the Norwegian Government decided that commercial minke whaling should resume in early summer that year. The decision was based upon an evaluation carried out by an international group of marine biologists who determined that the number of minke whales was sufficient to allow the safe resumption of whaling. The quota for the 1993 season was set at 160 whales, to be shared among the 36 boats. Although the whalers claimed that the quota was too small, they were nevertheless relieved to find themselves able to set out for whales once again.

The Norwegian government's decision was met by loud protests from various anti-whaling groups and individuals, and there were renewed threats of trade boycotts; representatives of the European Economic Community claimed that Norway could never join that organization as long as whaling continued. Although some industrial firms claimed they had encountered problems in marketing their products in certain foreign countries, the general view in Norway was that traditional whaling is both ecologically and morally justified.

The quota of 160 whales was filled by mid-July. However, it is interesting to note that the whalers expressed dissatisfaction with the government whaling regulations: whalers wished the regulations specified the tonnage of meat, rather than a set number of animals, allowed to be taken. The whalers' reason for this was that under a quota specifying number of animals, the largest animals are targetted and this necessarily includes the large pregnant female whales. In the whalers' opinion, it would make more ecological and economic sense to hunt the smaller young males, whose meat tastes best, and to avoid catching so many of the pregnant adult females. Thus the whalers continue to argue their case in ecological terms even following the resumption of whaling.

NOTES

1. The data in this paper were gathered during fieldwork at Lofoten in April 1991 during the author's membership in the International Study Group on Norwegian Small Type Whaling. This study group consisted of ten researchers most of whom were social scientists. Thus this paper draws on information obtained by other members of the group and from our discussions held during the preparation of our collective report (ISG 1992). I am indebted to Arne Kalland, Nordic Institute of Asian Studies, and Inger Lise Lien, Department of Social Anthropology, University of Oslo, for their many stimulating comments concerning this particular chapter.

2. Oliver (1965:x–xi) writes:

> Accordingly, rhetoric concerns itself basically with what goes on in the mind rather than with what comes out of the mouth. . . . Rhetoric is concerned with factors of analysis, data gathering, interpretation and synthesis. . . . What we notice in the environment and how we notice it are both predetermined to a significant degree by how we are prepared to notice

this particular type of object.... Cultural anthropologists point out that given acts and objects appear vastly different in different cultures, depending on the values attached to them.

REFERENCES

Aron, W., 1988.
The Commons Revisited: Thoughts on Marine Mammal Management. *Coastal Management* 16:99–110.

Broch, H. B., 1985.
Resource Utilization at Miang Tuu, a Village on Bonerate Island in the Flores Sea. *Contributions to Southeast Asian Ethnography* 4:5–29.

Broch, H. B., 1991.
Ecological Adaptation on Bonerate with Emphasis on the Niche Concept and Cultural Symbiosis. In: R. Grønhaug, G. Haaland, G. Henriksen (eds), *The Ecology of Choice and Symbol. Essays in Honour of Fredrik Barth*, pp. 210–224. Alma Mater Forlag, Bergen.

Brox, O., 1991.
The Common Property Theory: Epistomological Status and Analytical Utility. In: R. Grønhaug, G. Haaland, G. Henriksen (eds), *The Ecology of Choice and Symbol. Essays in Honour of Fredrik Barth*, pp. 426–444. Alma Mater Forlag, Bergen.

Claudi, E., 1987.
Greenpeace — Regnbuens krigere. Vol. 1. Tiderne Skifter, Copenhagen.

Conrad, J. and T. Bjørndal, 1993.
On the Resumption of Commercial Whaling: The Case of the Minke Whale in the Northeast Atlantic. *Arctic* 46:164–171.

Dyerberg, J., 1989.
Coronary Heart Disease in Greenland Inuit: A Paradox. Implications for Western Diet Patterns. *Arctic Medical Research* 48:47–54.

Gulland, J., 1988.
The End of Whaling? *New Scientist* October 29:42–47.

Gylling Nielsen, M., 1987.
Havets Mennesker. *Greenpeace Magazine*, p.11.

Innis, S.M. and H.V. Kuhnlein, 1987.
The Fatty Acid Composition of Northern Canadian Marine and Terrestrial Mammals. *Acta Medica Scandinavia* 222:105–109.

ISG (International Study Group on Norwegian Small Type Whaling), 1992.
Norwegian Small Type Whaling in Cultural Perspectives. Norwegian College of Fisheries, Tromsø.

Jonsgård, Å. 1955.
Development of the Modern Norwegian Small Whale Industry. *Norsk Hvalfangst-Tiendene* 44(12):409–430.

Kalland, A., 1990.
Whaling and Whaling Communities in Norway and Japan. *North Atlantic Studies* 2(1–2):170–178.

Kalland, A. 1992.
Personal communication.

Kalland, A., 1993.
  Management by Totemization: Whale symbolism and the anti-whaling campaign. *Arctic* 46:124–133.
Kaplan, R. B., 1988.
  Cultural Thought Patterns in Inter-Cultural Education. In: J.S. Wurzel (ed.), *Toward Multiculturalism. A Reader in Multicultural Education*, pp. 207–221. Intercultural Press, Yarmouth, Maine.
Kellert, S. R., 1988.
  Human-Animal Interactions: A Review of American Attitudes to Wild and Domestic Animals in the Twentieth Century. In: A. N. Rowan (ed.), *Animals and People Sharing the World*, pp. 137–175. University Press of New England, London.
Nordøy, E. S., 1991.
  Vågehvalens økologiske plassering og betydning i norske farvann. *Ottar* 184:13–16.
Nordøy, A., E. S. Nordøy and V. Lyngmo, 1991.
  Hvalkjøtt — delikatesse med helsemessige fortrinn. *Ottar* 184:53–59.
Oliver, R. T., 1965.
  Foreword. In: M. Nathanson & H. W. Johnstone, Jr.(eds), *Philosophy, Rhetoric and Argumentation*. University of Pennsylvania Press, University Park.
Ris, M., 1993.
  Conflicting Cultural Values: Whale Tourism in Northern Norway. In: M.M.R. Freeman and U.P. Kreuter (eds) *Elephants and Whales: Resources for Whom?* pp. 219–239. Gordon and Breach Science Publishers, Reading U.K.
Risting, S., 1922.
  Av hvalfangstens historie. W. Cappelens Forlag, Kristiana.
Simmons, O., 1988.
  Stereotypes: Explaining People Who are Different. In: J. S. Wurzel (ed), *Toward Multiculturalism. A Reader in Multicultural Education*, pp. 57–65. Intercultural Press, Yarmouth, Maine.
Teigland, A., 1990.
  Uredd natur-elsker. *Magasinet* (August 10.) (An interview vith Pål Bugge)
Tester, K., 1991.
  Animals and Society. The Humanity of Animal Rights. Routledge, London.
Watson, L., 1985.
  Whales of the World. Hutchinson, London.

14. CONFLICTING CULTURAL VALUES: WHALE TOURISM IN NORTHERN NORWAY

MATS RIS

INTRODUCTION

This paper examines the case of a whale tourism project in northern Norway. In recent years, along with its development, a cultural conflict has arisen between the foreign tourist entrepreneurs and the whalers in the region. The project has caused conflict between marine biologists, whalers and international financial sponsors, as well as controversy surrounding a new whaling museum connected to the tourism project. By introducing whale-watching tourism as a substitute for whaling, the explicit purpose of the project is to "slowly but surely change the image of the whale in northern Norway" (Ostrowski 1989:17).

Why do these entrepreneurs find it necessary to change the local people's attitudes about whales and whaling? This paper addresses the questions of how this situation has emerged and its ideological background. Before approaching the specific case, a brief outline is given of the international background and the surrounding concepts of tourism, Norwegian small-type coastal whaling and the cultural significance of whales in urban Western societies.

One of the most distinct cultural conflicts of post-modern Western society is the open clash between various hunting communities and the animal-rights/welfare movement. Numerous anti-hunting campaigns have created serious problems for indigenous peoples in the Arctic and thus put emphasis on the crucial issues of land and hunting rights as well as the future of hunting societies (Keith and Saunders 1989; Wenzel 1991).

A noteworthy situation has emerged wherein environmental organizations by adopting animal-rights positions and endeavouring to bring indigenous trappers off their hunting territories have thus taken the most crucial step toward clearing the land for pipelines, dams, mining and other industrial projects. Instead of being natural and close allies, these organizations have shown unusual insensitivity in their relationship with hunting communities and have unconsciously served the interests of large-scale industries by undermining the economies of those who still live in close contact with nature (Herscovici 1985).

Parallel to these successful public campaigns against indigenous peoples, the very same organizations have been involved in similar, and equally successful, anti-sealing and fund-raising campaigns against fishermen in the North Atlantic, thus causing as much economic damage to local people as profit for themselves (Henke 1985; Herscovici 1985). Though aimed primarily at Newfoundland fishermen, the consequences for Inuit hunters have been at least as serious. The attacks of the animal welfare movement and the European Economic Community are carried out in light of their own culture specific values and without respect to either the ecological or socio-economic nature of subsistence hunting (Wenzel 1991).

The same can indeed be said about anti-whaling campaigns during the last two decades. Animal-rights and environmental groups are also in this case heavily backed up by several governments, thus creating very strong joint political pressure on whaling nations (Freeman 1990; Suter 1981). The history of international regulation of whaling has gone from incapability and mismanagement of the world's whale stocks, through the United Nations (UN) Conference on Human Environment in 1972 and the International Whaling Commission (IWC) moratorium on commercial whaling in 1982, to the present political situation (Asgrímsson 1989; Hoel 1990), with its goal the total protection of all cetaceans, irrespective of scientific findings, sustainable development principles and social and cultural considerations (Freeman 1990).

TOURISM

Tourism is linked to many aspects of modern urban life. The most important one is the creation of varied leisure-time pursuits, which has, in turn, given

Courtesy of Mats Ris.

Plate 14.1 Typical tourist photograph.

rise to a variety of ways of satisfying these newly-created demands. But this development does not only imply positive effects. Nash (1978), among others, focuses on an important contradiction in the tourist industry: despite the negative consequences of tourist promotion in alien regions, native peoples often 'choose' to take the responsibility upon themselves to make the necessary physical and social adjustments to suit the needs of the tourist (Nash 1978:41–42). For instance, charter tourism to New Guinea and Amazonia has, due to the introduction of money and the visible presence of outsiders and their demands, reinforced the total process of culture change among indigenous peoples (Smith 1978:14).

For other than indigenous peoples, especially in the Western world, the situation is more complex, since there is an element of cash economy involved from the beginning. The tourist industry creates jobs, but on certain conditions. As Nash further points out, tourists from highly industrialized countries expect, even demand, that their vacation abroad meets expectations they take for granted at home. And the fact that the tourist entrepreneurs guarantee that their expectations are met makes it up to the hosts to adjust to the guests, not the opposite (Nash 1978:35). As many tourist resorts are located in rural areas or at the periphery of the industrial world, the pressure on the hosts to make a living becomes so great that they often willingly accept their subdued role vis-à-vis the guests.

Quite often though, dramatic culture change takes place when tourist entrepreneurs more directly influence the inner structure of another culture. Then it is not necessarily a question of uneven centre-periphery relationships, but rather a case of focused manipulation of important key symbols. In such cases, the commoditization of culture does not require the consent of the participants. Tourism simply packages the cultural realities of a people for sale along with their other resources. It can be done by everyone, and once set in motion, it very subtlety prevents the affected people from taking any clear cut action to stop it (Greenwood 1978:137).

This tendency of tourism to move very rapidly within other cultures without necessarily having explicit and outspoken motives towards them nourishes a special kind of cultural imperialism: "tourism is like Coca Cola; it is not a plague in itself, but if it is not handled carefully it can bring about irremediable damage" (Rossel 1988:19). The structural power of the tourist industry to regard other peoples' cultures as a common property available to be exploited and, in so doing, to penetrate almost every corner of the world and change the inner meanings and dynamics of other cultures does indeed give a new meaning to Hardin's (1968) concept 'tragedy of the commons'.

Norwegian Small-Type Coastal Whaling

Contemporary Norwegian whaling is a case of small-type coastal whaling for minke whales (*Balaenoptera acutorostrata*). It was developed by fishermen on the

Courtesy of Mats Ris.

Plate 14.2 Family owned whaling boat.

coast of Møre in the late 1920s and was at first performed from small boats of only 7–12 meters (Jonsgård 1955). Today minke whaling is characterized by bigger, mostly 12–24 meters fishing boats and takes place mainly in Vestfjorden in Lofoten, Barents Sea, and off Spitsbergen during about six weeks in the summer. The boats are equipped with winches and cooling facilities, since the meat and blubber is primarily flensed on deck and then stowed onboard before it is delivered to and processed by local fish plants mostly in northern Norway as food for human consumption (Foote 1975; ISG 1992).

Ownership of the whaling boats is family-based: the boats are often owned by two or three brothers, with their sons and other close relatives included in the crew of 5–7 members. The whalers are in fact fishermen by definition, since whaling is only one of the fisheries they are engaged in, and boats are therefore equipped for both whaling and fishing. The income from whaling is slightly lower than from other fisheries but is regarded as more stable. Being a relatively minor economic sector in northern Norway, whaling as a primary economy (with important secondary effects) is most vital to some small and isolated communities in Lofoten, such as Reine and Skrova. Here, 20% of the work force is directly involved in whaling, and in Skrova another 30–40% in the processing industry (Mønnesland *et al.* 1990).

In the last eight years, the situation has changed drastically. In 1982, the IWC adopted a moratorium on commercial whaling, which came into effect in 1987. In 1983, the last year with "normal" activity, the quota was 1690 whales. Today, only a few whales are taken for scientific programs. There are only about 35 whaling boats left in Norway (with about 200 whalers), and this

Courtesy of Mats Ris.

Plate 14.3 North Norwegian fishing/whaling community, Lofoten Islands.

number is steadily declining due to personal bankruptcies (Norwegian Small-Type Whaling Association, pers. comm. 1992).

In contrast to this development, recent research on the status of the minke whale stock in the northeast Atlantic has shown that the present size of the stock is 86 700, with a 95% confidence interval of 60 700 to 117 400, as agreed by the IWC Scientific Committee in 1992 (IWC 1993).

The negative social and economic effects of the present moratorium, the non-endangered status of minke whales, the antipathy to whaling among anti-whaling organizations and a majority of government members of the IWC and the until recently undeclared intentions of the Norwegian government has left the whalers and their families with a strong feeling of despair and abandonment. On top of this, they feel misused by foreign organizations who have destroyed their livelihood and ridiculed their culture. They see themselves as pawns in a political game and that has little to do with real conservation (ISG 1992). In light of this development, as will be shown, the introduction of a whale tourism project with clear anti-whaling incentives, is seen by the whalers as another attempt to attack their culture and deprive them of their livelihood.

Whales and Key Symbols

A most distinguished trait of the modern save-the-whale movement is its creation of an image of "the Whale". This image is build up by real or imagined traits found in several species of whales and by desirable human characteristics. Kalland (1993) shows how traits found in different whale species are put together to create a veritable "super whale", a non-existing mythical creature. Since real whales have several ambiguous marks and features compared to most other animals, they are difficult to place in a cultural category and therefore become culturally charged. An invented super-whale is even more powerful than real whales, since it comes to possess a whole set of human-like characteristics. Such a whale is perceived as at least as intelligent as humans, friendly and caring, fond of music, able to effect inter-species communication, with a huge repertoire of accumulated knowledge and stories, etc., and holding all these traits in one imaginable body. The Super Whale has now become a totem, since symbolic association between a sacred animal (the whale) and its creators (the animal-rights movement) has been created. The totem is not only a way of integrating various like-minded social groups, but also reinforces a common opposition to others. If there is no appropriate opposition, then one must be created. The result has become a single and powerful symbol, or rather an *elaborating key symbol* (Ortner1973). Such a symbol holds a central status in a culture, due to its ability to sort out experiences, feelings and ideas, "making them comprehensible to oneself, communicable to others, and translatable into orderly action" (1973:1340).

By communicating the Whale as a totem, it is possible to distinguish visible and comprehensible opponents, i.e., whalers. As a way of actively defending the Whale, the self-appointed guardians have introduced an extensive discussion about the ethics of harvesting whales for so-called "consumptive" use (Barstow 1991). A central question here is whether it is morally acceptable to kill whales, regardless of motive. As animal-rights philosophers point out, the ethics of killing an animal is philosophically equal to the killing of any other animal of any species (Singer 1975). The problem of resisting whaling without getting entangled in a perpetual discussion of human relations towards animals is solved by separating cetaceans from all other animals:

> I am not arguing for the sanctity of all life on earth. I am not advocating equal rights for all animal species. I am seeking to set forth a rational and moral basis for a future determination by one, specialised, international, human agency that one order of marine mammals should be managed in this manner. Why whales? My rationale most simply is that whales are uniquely special. They really are in a class by themselves
>
> (Barstow 1989:12).

By this move it is easier to implement the Whale in practice, i.e., to set up an effective plan of action on its behalf and especially outside one's own social group. The goals can be achieved by, for instance, creating a new management regime of permanent protection from within the IWC or, as shown in this paper, introducing whale tourism in whaling countries. In both these cases, the defenders have developed a special strategy to impose their specific cultural views on those holding different cultural views.

NON-CONSUMPTIVE UTILIZATION

The strategy of developing whale-watching facilities in a whaling region like northern Norway follows an increasing international trend in the 1980s. The key concept is *non-consumptive utilization of whales*. This idea signifies a fundamentally different view about whales as a natural resource. It rests on opposition to regarding whales as food or a fishery resource, in favour of an emotion-based affective relationship between humans and whales. The term applies to "any use of cetacean resources which does not involve deliberate killing or critical harming of whales . . . in contrast to the . . . whaling industry which has been to kill and consume whales by processing their bodies" and involves, among other issues, benign research, recreational whale-watching and cultural valuation (Barstow 1986:155–156). The latter issue focuses on "aesthetic, educational . . . and even religious values of whales alive", and the explicit challenge is to support through educational and cultural channels on a global basis the optimum utilization of the world's whale resources by non-consumptive means as a unique part of the common heritage of all humankind (ibid:163).

It is argued that the abundant occurrence of mass media features has provided the general public with facts and imagery about whales, knowledge that has brought humankind to the threshold of a profound moral transformation, since the world is turning from valuing whales dead to valuing them alive (Barstow 1989:13). Or as stated by the U.S. branch of World Wide Fund for Nature (WWF): "*not* killing whales is evolving as the norm among the nations of the world" (Fuller 1991). In this respect, whales are seen as an earthly good, offering humans moral and material support. Attitudes towards them are expected to be sensitive and proportionate, based upon feeling, as well as knowing: "Caring about whales is a mark of personal and societal maturity; and it is good practice in caring: the most difficult assignment of Homo Sapiens climbing toward humanity" (Scheffer 1991:19). Thus, the understanding of non-consumptive utilization of whales not only seems to be based on a culture-specific and evolutionary view of cultural change, but also includes an ambitious plan of action for propagating the views of moral progress to a much wider audience, i.e., other cultures, and especially whaling communities, as will be shown below.

However, modern whaling represents a problem. It is now admitted among whale-protection advocates that there is little scientific doubt that some whale stocks can sustain limited and regulated catches in the future without endangering the species (Barstow 1989:11; Gummer 1991; Marine Mammal Commission 1991:3–4). It is said that the argument against the killing of whales can no longer be based on preventing extinction, so a different rationale is required. Such a rationale is a "new world moral and ethical standard". It is thought that the past 40 years of highly significant changes in attitude and ethics with regard to whales will permanently protect them from consumptive exploitation in the future (Barstow 1989:11–13).

This is not an isolated view among the whale-protection movement alone. It is also supported by some of the leading nations against whaling in the IWC. The government of New Zealand recognizes the "economic contribution that can be provided by live whales, in the form of whale-watching and eco tourism" and regards whales as "fellow denizens of planet Earth, with perhaps much to teach us, rather than as potential steaks or pet food" (Government of New Zealand 1991:13). And since "many United States citizens remain opposed to commercial whaling on moral and ethical grounds", the U.S. Marine Mammal Commission recommends that the United States government adopt the position that non-consumptive value of whales may be equal or greater value than their consumptive value, and that science alone is not sufficient to dictate that commercial whaling should be resumed (Marine Mammal Commission 1991:4). The U.S. commissioner to the IWC has also confirmed this position by stating that he will continue to defend a ban on whaling,, but that it cannot be done in the face of scientific evidence (Anonymous 1991:4).

It is also argued, that the perception of whales as merely one class among other classes of exploitable living marine resources is naturally held by those who are involved in whaling or live in whaling countries, but despite "continuous global changes in ethical attitudes", many people do not yet accept these changes (Holt 1991:8). The reason for the delay is said to be that they have had much less exposure than people in non-whaling or ex-whaling countries to scientific research and media presentations:

> In consequence they tend to see public expression of the "non-resource" perception as an unacceptable attempt to introduce a "foreign idea". Thus they are led to emphasize "cultural differences", as if such differences were unchanging and the national cultures were not themselves evolving, largely convergently
>
> (Holt 1991:8).

> Whale protectionists have been accused of narrow-mindedly seeking to impose their own values and ethics upon people in other countries who have the right to live by different standards... The fundamental factor, however, is that the issue of whale protection is in fact a global issue which must be resolved in the global arena
>
> (Barstow 1989:12).

The fact that different peoples in different parts of the world develop different patterns of "values and ethics" suited to their own lives and experiences is here overruled by a narrowly conceived notion about cultural change. As "ethics" is taken for granted to change from worse to better according to some evolutionary plan, the change of views about whales in some parts of the Western world is seen as a natural step in the right direction towards a higher form of civilization. Deviant cultures, i.e., whaling cultures, are therefore seen as backward and are expected to give way to change, because they have not yet been exposed to appropriate knowledge and education about whales. This view of cultural change is typical of the notion of non-consumptive utilization of whales and reveals not only a strong ethnocentrical, but also missionary way of thinking. As Payne (1991) points out, stopping the "amoral practice of whaling" by simply insulting the whalers will only destroy the opportunities of a "dialogue" with them, and getting many people to recognize the whales' claim to moral concern will require a . . .

> . . . major change in their intellectual and emotional views towards animals. The difficulty of achieving that is not a reason to delay the process. As John Stuart Mill put it, 'every great movement must experience three stages, ridicule, discussion, adoption'. I say, let us get on with this movement!

> (Payne 1991:22).

In other words, changing the whalers' perception into "adoption" makes a tactical approach necessary. Here, the concept of non-consumptive utilization through whale-watching and education seems to provide an appropriate means to implement the mission of putting an end to whaling.

It is thought that the need for public campaigns and education to save the whales is especially relevant for whaling nations, since the only way to affect the moratorium is to convince these countries to abide by the IWC decision (McCloskey 1986:165). The main educational principle is to provide and translate information about whales and their "universal values" into terms that will enlist support of local people, and in such a form that "people of diverse cultures and religions can relate to them". Educators, scientists, conservationists and non-government organizations on the local level are therefore urged to cooperate in combined educational/political campaigns. A moratorium could then be seen as desirable by local people if they learn about the "mysteries" of whales. An alternative could be properly regulated commercial whale-watching, and "if accompanied by naturalists and printed information, the public can learn a great deal about the animals and their survival needs" (ibid:166–167). This is claimed to demonstrate that whales are worth more alive than dead and, above all, that a carefully developed whale-watching industry will benefit local communities (McCloskey 1983:18). In the interest of assisting the local economy, whaling communities are encouraged to consider changing over to whale-watching because of the moratorium:

> Especially in areas where such whaling has been carried on for a very long time . . . some of the whaling communities may want to convert the stations into museums of whaling, including displays on the nature and habits of whales. Former whaling company employees could work in the museum, on the whale-watching boats, and in the tourist support facilities

> (McCloskey 1983:12).

NORWEGIAN WHALE WATCHING

The town of Andenes, on the very northern tip of Vesterålen in the county of Nordland, is heavily dominated by fishing and NATO activities, but has also some tourism. Nordland is the base of contemporary small-type coastal whaling in Norway. There are, however, no whalers or any whaling activity at Andenes itself. Due to the size and location of the town (population 3500), sufficient infrastructure (airport, hotels, shops, tourist facilities) and an excellent seascape (the edge of the continental shelf is only 10–12 nautical miles from the coast), it has proven to be a very good departure point for whale-watching tourism.

The Entrepreneurs

After an initial assessment in 1987, organized by a mainly Swedish organization, Centre for Studies of Whales and Dolphins (CSWD) and local interests at Andenes, the first organized tours took place in 1988 and the business has grown since then with amazing speed every year. In the 1991 season over 4500 tourists purchased $100 tickets to see (mainly) sperm whales outside Andenes. Today the tours are organized from a special tourist building, the Whale Centre, owned by a company called Whale Safari Ltd. The Whale Centre has a reception area for reservations and general tourist information, a souvenir shop and a cafeteria for visitors. Excursion participants also have access to a whale and whaling exhibition, including a colour slide show about whales. In 1991, the Whale Safari Ltd. was controlled by four joint owners: the municipality (40%), local business (20%), the county council (20%), and CSWD (20%). The two latter partners, however, are not represented at board meetings. The boat tours are operated by an enterprising young whaler and his crew on board a family-owned whaling boat. Being a third-generation whaler from a well-known whaling family in Lofoten, he has acquired the special knowledge and great skill necessary for finding whales. Accompanying the trips are also two marine biologists and a number of mostly foreign students, who work as guides and conduct research.

The Centre for Studies of Whales and Dolphins is built up by a variety of people and professions — scientists, artists, craftsmen and media professionals — but with a common and active interest in whales and dolphins. The core consists of a limited number of marine biologists and other professionals. In 1983, members of the CSWD made a field trip to Lofoten in order to investigate the possibilities for an extensive whale research project, and also to "gain insight into the importance of whaling in the Norwegian society today and in the future" (Anonymous 1984:62). At this early stage of the organization's experience of Norway, the purpose of the project was quite clear:

> How shall the attitudes of the whaling industry be changed, so that all whales and dolphins could live in peace in the future without constantly be-

ing threatened by man? Is it reasonable to expect that people who live off whaling should give up their livelihood and income abruptly? Finding appropriate ways for a cautious and sensible development of the whale-watching tourism as a new and alternative source of income ... is a typical project of CSWD. It is interdisciplinary, and if it turns out to be a success, it may be a positive contribution to the whales in our part of the world. Today people are working all over the world to save the whales, and ... the Centre for Studies of Whales and Dolphins is a part of this whole

(Anonymous 1984:64).

Since then, the outspoken connection among whales, whaling and cultural change has been a characteristic trait of the organization. This view has been repeated and exploited by most of the leading members and marine biologists associated with CSWD, as well as by the Scandinavian branches of WWF.

In order to fulfil the prescription of non-consumptive utilization of whales, the Swedish entrepreneurs have introduced whale-watching at Andenes as a suitable means to achieve this goal. Here, several instruments are used. One is mass media, which is the most important ingredient for promoting the tourist project as such. It has thus gained much attention abroad. Television and press from many parts the world has visited Andenes to cover the tours. On location at Andenes, however, the two most crucial concepts are research and education.

The Use of Research

From the very start of the CSWD involvement in northern Norway, there has been a strong connection among research, whale tourism and whalers. The present leader of the research activities at Andenes emphasizes that "the aim is to combine whale research with whale-watching: located in the last stronghold of Norwegian whaling the Centre plays an important role in changing attitudes towards whales" (Similä and Ugarte 1991:18). Whale-watching is then expected, according to another marine biologist, to "create alternative employment for whalers and at the same time use their expertise on cetacean behaviour" (Arnbom 1988:189). And if whale tourism turns out to work well, it is to "secure a long term platform for whale research and general education about whales, and to create a job alternative to former whalers, now unemployed because of the whaling moratorium" (Lindhard et al. 1988:3).

Even if whalers and whale tourism can provide the marine biologists with local expertise and some financial resources, the major costs for the research must come from elsewhere. The key is WWF, which has supported the marine biologists financially since 1987, mainly by funds derived from large "whale adoption" campaigns. WWF has thus offered the public the opportunity to adopt a whale in northern Norway. With brochures titled "Mayday! Save The Whale!", WWF Denmark has invited the public to be subscription sponsors for up to $100 (WWF 1990b) and business firms for $10 000 per sperm whale

(WWF 1990a). Similarly, WWF Sweden has offered the public the opportunity to adopt a whole killer whale for $1000 or part of a whale for $100 (WWF 1989).

As the public is entitled to receive news about "their" whales, the entrepreneurs provide basic information through photographs taken by the marine biologists (Ostrowski 1989:16; WWF 1990a). The purpose of the campaign is, according to WWF Denmark, to save the last whales of the world from extinction by raising money for lobbying the IWC and promoting whale tourism (Dybbro 1990; WWF 1990a). There is an explicit link between research and whaling: "we are launching a research project in Andenes in Northern Norway, which aim is to replace commercial whaling with whale tourism" (WWF 1990b). And the chairman of WWF Sweden is also quite frank about this connection: "slaughter or whale tourism is a political question of choice" (Hammarström 1988:15).

Several of the CSWD marine biologists also reveal an image of whales as living on the brink of extinction, an idea that directly influences their view about whaling and whalers. One biologist expresses the view that the whaling debate in Norway has moved beyond logic, since they try to find simple solutions to difficult problems: "it is the same sort of problem as with farmers and wolves; when the discussion has gone too far, it does not matter if there are only five wolves left" (Carlson 1991:7). In 1988, another marine biologist expressed this idea by stating that "the hunting of the minke whale must be stopped; there is no doubt about it. Norway has perhaps only 10–15 per cent left of the original stock from 1930" (Emanuelsson 1988:23). Few, if any, members of the IWC Scientific Committee believed this at the time. And a third biologist of CSWD explains that "protection of whales is an economic and ethical question: one thing is that the whale is a resource we cannot afford to lose. . . Another thing is the moral aspect of whether animals shall have the right to exist or not, and I mean that they definitely have such rights" (Myklebust 1988:9). The CSWD entrepreneurs also back up these ideas, stating that if whalers and local people become more interested in the tourist project, they will also have an interest in protecting the whale stocks: "we simply try to show the whalers that there are other ways of making money on whales than killing them" (Seinegård 1987:23).

The tour-operating whaling captain, who tries through this means to save his boat from bankruptcy, has found that the economic reality of tourism is not so simple. He has invested over $160 000 in order to meet safety regulations for taking passengers on board, but he has never received any contributions from WWF. Instead, in January 1992, he was so heavily indebted, that he narrowly escaped bankruptcy and loss of both his boat and home. He can only offer two crew members employment, while during the days of whaling his boat carried a crew of six. At the same time, the local organizer employs (even foreign) students as guides, thus offering them summer jobs at the expense of the laid-off crew (ISG 1992:103-104). The captain's view on the whole matter is very straightforward:

> These so-called "researchers" and their WWF friends were the ones who
> stopped whaling for us. Now they are making money on adopted whales.
> They have robbed us of our livelihood, but make a profit on the whales them-
> selves. They keep telling us that we do not own the whales, but they sell
> adoptions as if *they* own them!

(pers. comm. 1991).

In sum, the marine biologists connected to the whale tourism project at
Andenes are caught in a contradiction. At the same time as they are partly dependent on the expertise of the whalers to perform their research, they promote anti-whaling attitudes themselves and accept financial support from organizations that develop large anti-whaling campaigns. Their desire to help the
whalers with their financial problems because of the moratorium must also be
seen against the fact that some of them have actively contributed to uphold
the same moratorium in the IWC.

The Use of Education

The second crucial concept of the Whale Safari activities, education, is linked
to the research sector, since the marine biologists also provide knowledge
about whales:

> Joining whale friends and whalers for a vivid dialogue is an important part
> of the project. Many locals join the tours out to the whales and learn exciting
> facts about the whales. Slowly but surely is the image of the whale changing
> in northern Norway. *That* is the purpose

(Ostrowski 1989:17).

What are the conditions for such a "dialogue" and what do education and
knowledge mean? These words are not neutral in the dictionary sense, but
rather filled with implicit meanings. The citation above indicates that knowledge about whales is something that is provided by experts, i.e., scientists.
Local people, who are believed to know very little about marine mammals, are
expected to join the tours to achieve the new knowledge. In this respect,
school classes are argued to be of equal importance in the educational process
(Schandy 1991:14).

Education also plays an important role vis-à-vis the tourists. Before every
tour, the participants receive guided information in the Whale Centre. The one
hour preparation includes a guided visit to the whale exhibition and a professional colour slide performance. Then the party leaves (in one body) for the
boat, and once on board the guides provide another round of information.
During the trip the guides make comments on the behaviour of the whales
and answer questions from the tourists.

It seems that their relation to the tourists serves an extra purpose: to provide the tourists with a special kind of knowledge, since the inner meaning of
the tour is not necessarily just looking at whales, but also interpreting them in

a special direction. As Horne (1984) suggested, the social significance of tourism can be clarified if one imagines the traveller/sightseer as a modern pilgrim and what is looked for as a relic. What matters then is what tourists are told they are seeing. The fame of the objects becomes its meaning (Horne 1984:9-10). In the Andenes case, then, the Whale seems to be the "relic", which hopefully can heal some of our urban worries about the deteriorating state of the natural world.

A more clear cut example of the notion of knowledge and education inherent in the whale tourism project is the recent debate on a whaling museum in the Whale Centre. In 1990, CSWD worked out a plan for an extensive permanent whaling exhibition about "the importance of the whale for the people of coastal Norway". The emphasis is centred upon 'the two periods of coastal whaling in Norway', i.e., historical large-type and modern small-type whaling, and "impressions, knowledge, experiences, and artifacts" from them (Ostrowski and Steijner 1990:1). WWF Sweden has contributed $50 000 for its realization, but the CSWD give strong rhetorical emphasis to the idea that the work with the museum must from the very beginning be carried out in close cooperation with whalers and others who defend continuous whaling and that the museum must not be perceived as biased by whale-protection interests. Instead, the basis of the personal experiences offered by the museum shall be the ancient local pride over the hunting: "local visitors must find parts of their own identity exposed, though perhaps the pride will be put in a historical context" (ibid:2).

At first sight, there seems to be a conflict between sponsors and entrepreneurs, since an international anti-whaling organization is funding the museum and CSWD seeks to emphasize the whalers' perception of whaling. This is not the case, however. On the contrary, the two perspectives show a tendency to coincide. By putting two completely different socioeconomic phenomena in Norway — the historical and industrial-scale large-type coastal whaling with the contemporary family-based small-type coastal whaling — together into one category and then emphasizing the historical perspective, the entrepreneurs manage to promote the idea that whaling in Norway today is not only a closed historical chapter, but also involves the same unsustainable traits as old-time industrial whaling operations.

Needless to say, this perspective upsets the whaling captain at Andenes. He wants no less than to go out whaling again. Though operating whale-watching excursions for the moment, his harpoon is mounted on the forecastle head on every trip, since "this *is* a whaling boat, and I *am* a whaler! I am just waiting for the moratorium to end" (pers. comm. 1991).

The historical purpose of the museum and the support from WWF have also upset local organizations. The secretary of the Norwegian Small-Type Whaling Association finds it outrageous that people who are working for a total ban on whaling are touring northern Norway to collect artifacts and memories for, as he phrased it, a "mausoleum" over the whalers' lives, at the same time as the whalers are fighting with all their means to save their occupation. He wishes to see the development of Norwegian coastal culture, with its deep-rooted traditions, but empasizes that the museum must display this culture in the past *as well* as in the future and that the expertise will be found among the whalers themselves (Storhaug 1991:5; Münter 1991:4).

The local organizer at Andenes, Whale Safari Ltd., argues, however, that both they and CSWD have a neutral approach towards resumed commercial whaling and that they for a long time have wished to establish closer contacts between their activities and northern Norway. In this respect, Whale Safari Ltd. emphasizes strongly that money from WWF campaigns must not be used against the whalers (Hagtun 1991a:12). Further, they find no reasons to sever connections with CSWD and WWF, as long as the cooperation works under acceptable conditions and no propaganda is directed against the whalers. The whalers' reluctance to take an active part in the museum is regarded as satisfactory, since they do not contribute financially anyway. CSWD and WWF, on the other hand, has contributed with "money, enthusiasm, and creativity", and without their support the idea of a whaling museum must be given up. The whale tourism project is seen as far too important to be stopped by the whalers' prejudices, distrust and lack of dialogue (Hagtun 1991b:18).

The entrepreneurs have seemingly chosen to present the whaling museum in a thematically "neutral" and "non-controversial" way by emphasizing the historical perspective. But a museum always has a meaning and communicates by means of its character. It has often a silent language — a traditional museum displays authority. The museum's audience is seldom allowed to control the ordering and structure of things but is merely a passive consumer of the meanings being provided. By organizing the display and interpretation of objects, stories, etc., their original cultural importance is lost in favour of a new set of meanings created primarily to serve the goals of the museum authorities.

Variations on this theme are not unusual among museologists. Smith (1989:9), for instance, emphasizes that one of the most insistent problems that museums face is the idea that artefacts can be divorced from their original context and redisplayed in a different context that is regarded as having a superior authority. As Vergo (1989:2-3) points out, museums make choices determined by judgements, which are rooted in our education, upbringing and prejudices. Every arrangement of objects means placing a certain construction upon history, our own culture or someone else's, and beyond all the information there is a subtext of often contradictory strands, woven from wishes, ambitions and political or educational preconceptions of the museum's designers and sponsors. Accordingly, Hudson (1987:114) sympathizes with a museum's wish to appear "objective" and "scientific" but admits that such an attitude is dishonest. After all, as Horne (1984:2) puts it, public culture such as tourism and museums, is a reaffirmation of what life is supposed to be about: sightseeing helps people in modern industrial societies define who they are, and what matters in the world.

The conflict and the problems of the whaling museum at Andenes are not unique. A whaling museum seems to be the best way of communicating the cultural, social and economic nature of whaling to a broader audience, but one can only hope that it will emerge from a closer cooperation between whalers and professional museologists.

CONCLUSIONS

The notion of non-consumptive utilization of whales represents the idea that future whale management regimes should focus much more on issues like ethics, morals, emotions, public amusement and attractions than on biological and statistical sciences. Since this notion is mainly of Anglo-American origin, the vocabulary used to describe it (global standards, for all humankind, ethics, etc.) should be understood in an Anglo-American context. It is most unlikely, for instance, that large segments of African, Asian and Latin American societies, or even significant parts of the industrial world, would share this view about cetaceans (or other animals, for that matter). Since it is naturally the antithesis of non-consumptive utilization, whaling is seen by those promoting non-consumptive utilization goals as something that sooner or later must come to a definite end, and whale tourism is regarded as an instrument for achieving this goal.

The whale-watching project at Andenes has clearly proved to be a part of this current urban-based trend to promote the non-consumptive utilization of whales. The foreign entrepreneurs have tried to transform the whale from within its traditional cultural context in northern Norway by introducing instead a fundamentally alien view of it. In this context, they have tried to introduce an image of the whale as something humans are only supposed to enjoy by non-material means. It has become a matter of *either* watching whales *or* hunting them. But people in rural communities do not necessarily share this view: showing respect for nature *and* harvesting living natural resources do not imply contradictory values. On the contrary, they presuppose each other. There is undoubtedly plenty of cultural space in Norwegian coastal societies to accept and promote *both* sustainable/consumptive *and* recreational/non-consumptive use of whales, but the foreign entrepreneurs do not, figuratively speaking, "allow" local people and tourists this choice.

The entrepreneurs have tried to place tourism in the context of whaling. In a missionary spirit, whale-watching is promoted as an automatic substitute for whaling in order to help the unfortunate whalers. The entrepreneurs, however, have not succeeded in changing either attitudes or the economic situation for the whalers, since whale-watching never can grow to the extent of offering more than a very few whalers alternative employment, and furthermore it constitutes an economy highly dependent on fluctuations abroad. Instead, whaling communities in northern Norway seem to be relatively strong societies with a high degree of cultural resistance to ideas perceived as threats against their traditional economy based on marine resources. The local organizers are then left in the typical position of balancing the expectations of the foreign entrepreneurs and sponsors and the frustrations of the whalers.

Connecting wildlife and leisure is often a case of eco-tourism. The fact that profit comes second to ideology may have inspired a Swedish journalist to state that the Andenes case is an example of "green tourism" as well:

> It is a combination of recreation and ecological and cultural insights. Its contribution is to save endangered animals. It creates alternatives for those who are dependent on environmentally harmful occupations. So apart from exciting experiences, the whale safari tourist can enjoy a clean environmental conscience

> (Frieberg 1991:27).

Contrary to this, current studies indicate that sustainable coastal whale fisheries are among the least environmentally damaging food-protein-producing systems, when the energy costs, habitat disruption and chemical polluting aspects of other food-producing systems are compared (Freeman 1991). In this limited sense, then, whale tourism within the context of the northern Norwegian whaling culture is an example of Lèvi-Strauss's profound insight (1991:89) that totemic animals are indeed "good to think" (but not to eat).

REFERENCES

Anonymous, 1984.
Valar vid Lofoten. In: S. Ostrowski and P. Rundkvist (eds.), *Valar & Delfiner*, pp. 62–67. Centrum för Studier av Valar och Delfiner. Göteborg.

Anonymous, 1991.
Marine Mammal News 17(5):1–4.

Arnbom, T., 1988.
Skåda val i Norge. *Fauna och Flora* 83(4):189.

Asgrimsson, H., 1989.
Developments Leading to the 1982 Decision of the International Whaling Commission for a Zero Catch Quota 1986–1990. In: S. Andresen and W. Østreng (eds.), *International Resource Management: The Role of Science and Politics*, pp. 221–231. Belhaven Press, London.

Barstow, R., 1986.
Non-consumptive Utilization of Whales. *Ambio* 15(3):155–163.

Barstow, R., 1989.
Beyond Whale Species Survival: Peaceful Coexistence and Mutual Enrichment as a Basis for Human/Cetacean Relations. *Sonar* (2):10–13.

Barstow, R., 1991.
Whales are Uniquely Special. In: N. Davies, A. Smith, S. Whyte, and V. Williams, (eds.). *Why Whales?*. pp. 4–7. Whale and Dolphin Conservation Society, Bath.

Carlson, P., 1991.
Annas val i livet. Göteborgs-Posten (12 May):7.
Dybbro, T., 1990.
Hvalrapport. WWF Denmark public flyer, April 1990. World Wide Fund for Nature, Copenhagen.
Emanuelson, S., 1988.
Val i sikte! Golfströmmens jättar ska locka turisterna till havs. Göteborgs-Posten (10 July):23–24.
Foote, D.C., 1975.
Investigation of Small Whale Hunting in Northern Norway 1964. *Journal of the Fisheries Research Board of Canada* 32(7):1163–1189.
Freeman, M.M.R., 1990.
A Commentary on Political Issues with Regard to Contemporary Whaling. *North Atlantic Studies* 2(1–2):106–116.
Freeman, M.M.R., 1991.
Energy, Food Security and A.D. 2040: The Case for Sustainable Utilization of Whale Stocks. *Resource Management and Optimization* 8(3):235–244.
Frieberg, J., 1991.
Nu strömmar turisterna till Nordnorge: Fångststopp gav valfrihet. *SIF-Tidningen* 71(11):24–27.
Fuller, K.S., 1991.
Letter to the Ambassador of Iceland in the United States, Tomas A. Tomasson, 24 May. Available from World Wildlife Fund, 1250 – 24th Street N.W., Washington, D.C. 20037.
Government of New Zealand, 1991.
Statement on Cetaceans, Working Group II: Item 2, Third Session of the Preparatory Committee of United Nations Conference on Environment and Development, Geneva, 12 August–4 September, 1991.
Greenwood, D.J., 1978.
Culture by the Pound: An Anthropological Perspective on Tourism as Cultural Commoditization. In: V. Smith, (ed.) *Hosts and Guests: The Anthropology of Tourism,* pp. 124–138. Basil Blackwell, Oxford.
Gummer, J., 1991.
Letter to the Parliament and Various Interest Parties, 27 March 1991. Available from the Ministry of Agriculture, Fisheries and Food, Whitehall Place, London SW1-2HH.
Hagtun, A., 1991a.
Hvalfangstnæring og hvalsafari. Harstad Tidende (18 April):12.
Hagtun, A., 1991b.
Hvalfangsten og hvalsenteret. Lofotposten (19 April):18.
Hammarstrom, T., 1988.
Här simmar Astrid Lindgrens adoptivson. Expressen (11 January):15.
Hardin, G., 1968.
The Tragedy of the Commons. *Science* 162(3859):1243–1248.
Henke, J.S., 1985.
Seal Wars! An American Viewpoint. Breakwater Books, St. John's.
Herscovici, A., 1985.
Second Nature: The Animal-Rights Controversy. CBC Enterprises, Montreal and Toronto.

Hoel, A.H., 1990.
   Norwegian marine policy and the International Whaling Commission. *North Atlantic Studies* 2(1–2):117–123.
Holt, S., 1991.
   The Un-ethics of Whaling. In: Davies, A. Smith, S. Whyte, and V. Williams, (eds.) *Why Whales?* pp. 8–16 Whale and Dolphin Conservation Society, Bath.
Horne, D., 1984.
   The Great Museum: The Re-presentation of History. Pluto Press, London.
Hudson, K., 1987.
   Museums of Influence. University Press, Cambridge.
IWC, 1993.
   Report of the International Whaling Commission 44. Cambridge: International Whaling Commission. In press.
ISG (International Study Group on Norwegian Small Type Whaling), 1992.
   Norwegian Small Type Whaling in Cultural Perspectives. Norwegian College of Fisheries, Tromsø.
Jonsgard, Å., 1955.
   Development of the Modern Norwegian Small Whale Industry. *The Norwegian Whaling Gazette,* 44(12):409–430.
Kalland, A., 1993.
   Management by Totemization: Whale Symbolism and the Anti-whaling Campaign. *Arctic* 46(2):124–133.
Keith, R.F., and A. Saunders, (eds.) 1989.
   A Question of Rights: Northern Wildlife Management and the Anti-harvest Movement. Canadian Arctic Resources Committee, Ottawa.
Lévi-Strauss, C., 1991.
   Totemism. Merlin Press, London.
Lindhard, M., L.Ø. Knutsen and H. Strager, 1988.
   The Norwegian Whale Safari Project 1988. Unpubl. ms. Available from Centrum för Studier av Valar och Delfiner, Kaponjärgatan 4d, S-41302 Göteborg, Sweden.
Marine Mammal Commission, 1991.
   Issues Facing the International Whaling Commission and the United States Regarding the Resumption of Commercial Whaling and the Future Conservation of Cetaceans. Unpubl. ms. Available from the Marine Mammal Commission, 1825 Connecticut Avenue, N.W. #512, Washington, D.C. 20009.
McCloskey M., 1983.
   Recreational Whale-Watching. Unpubl. ms. Available from Whale Center, 3929 Piedmont Avenue, Oakland, California 94610.
McCloskey, M., 1986.
   Educational Campaigns for Marine Mammals. *Ambio* 15(3):164–167.
Munter, T., 1991.
   Kvalfangerne protesterer mot fangstmausoleum. Fiskaren (29 May):4.
Myklebust, V., 1988.
   Hval og elektronikk. Lofotposten (1 August):9.
Mønnesland, J., S. Johansen, S. Eikeland and K. Hanssen, 1990.
   Whaling in Norwegian waters in the 1980's: The Economic and Social Aspects of the Whaling Industry and the Effects of its Termination. NIBR-Report 1990:14. Norwegian Institute for Urban and Regional Research, Oslo.

Nash, D., 1978.
  Tourism as a Form of Imperialism. In: V. Smith, (ed.) *Hosts and Guests: The Anthropology of Tourism* pp. 33–47. Basil Blackwell, Oxford.
Ortner, S.B., 1973.
  On Key Symbols. *American Anthropologist* 75(5):1338–1346.
Ostrowski, S., 1989.
  Safari säkrar tillgång på val. *WWF Eko* 2(2):16–17.
Ostrowski, S. and T. Steijner, 1990.
  Fångstmuseum i Hvalcentret i Andenes. CSWD memorandum, 12 September. Available from Centrum för Studier av Valar och Delfiner, Kaponjårgatan 4d, S-41302 Göteborg, Sweden. 1–2.
Payne, R., 1991.
  Is Whaling Justifiable on Ethical and Moral Grounds? In: Davies, N., Smith, A., Whyte, S., and Williams, V., eds. *Why Whales?* Bath: Whale and Dolphin Conservation Society, Bath.
Rossel, P.,1988.
  Tourism and Cultural Minorities: Double Marginalisation and Survival Strategies. In: P. Rossel, (ed.) *Tourism: Manufacturing the Exotic.*, pp. 1–20. IWGIA Documents. Vol. 61. Copenhagen: IWGIA.
Schandy, T., 1991.
  WWF satser på kysten. *Verdens Natur* 6(2):14–16.
Scheffer, V., 1991.
  Why should we care about whales? In: N. Davies, A. Smith, S. Whyte and V. Williams, V., (eds.) *Why Whales?* pp. 17–19. Whale and Dolphin Conservation Society, Bath.
Seinegard, M., 1987.
  De kämpar mot ett valfritt hav. *Götheborgske Spionen* (4):22–23.
  Simila, T., and F. UGARTE, 1991. Fjord Feud for Fish. *Sonar* (5):16–18.
Singer, P., 1990.
  Animal Liberation. Revised ed. Avon Books, New York.
Smith, C.S., 1989.
  Museums, Artefacts, and Meanings. In: P. Vergo, (ed.) *The New Museology,* pp. 6–21. London: Reaktion Books.
Smith, V.L., 1978.
  Introduction. In: V. Smith, (ed.) *Hosts and Guests: The Anthropology of Tourism,* pp. 1–14. Basil Blackwell, Oxford.
Storhaug, J.A., 1991.
  Hvalmotstandere vil ha hvalminner. Lofotposten (15 April):5.
Suter, K.D., 1981.
  The International Politics of Saving the Whale. *Australian Outlook* 35(3):283–294.
Vergo, P., 1989.
  Introduction. In: P. Vergo, (ed.) *The New Museology,* pp. 1–5. Reaktion Books, London.
Wenzel, G., 1991.
  Animal Rights, Human Rights: Ecology, Economy and Ideology in the Canadian Arctic. University of Toronto Press, Toronto.

WWF, 1989.
Späckhuggare blir årets julklappsval! Joint Press Release from WWF Sweden and Centrum för Studier av Valar och Delfiner, 8 December 1989. Available from World Wide Fund for Nature, Ulriksdals Slott, S-17171 Solna, Sweden.

WWF, 1990a.
SOS Red Hvalen. WWF Denmark public flyer to business firms, 19 June 1990. Available from World Wide Fund for Nature, Ryesgade 3F, DK-2200 Copenhagen N, Denmark.

WWF, 1990b.
Et nødråb fra havets kæmper, hvalerne. WWF Denmark public flyer, 1990. Available from World Wide Fund for Nature, Ryesgade 3F, DK-2200 Copenhagen N, Denmark.

15. ARCTIC WHALES: SUSTAINING INDIGENOUS PEOPLES AND CONSERVING ARCTIC RESOURCES

NANCY DOUBLEDAY

INTRODUCTION

By reference to the situation facing Canadian Inuit at the present time, this paper explores changing perceptions about whaling and related activities in the context of developing international norms and values associated with human rights, environment and development.

Inuit have much in common with other peoples and their cultures which have long been dependent upon hunting marine mammal renewable resources stocks. This paper addresses Inuit experiences in regard to marine mammal-based renewable resource economies in terms of the importance of historical events to Inuit societies which have relied upon marine mammals for thousands of years, the interaction of contemporary Inuit societies and world ecological, economic and political systems, and the initiatives taken by Inuit societies with respect to the development of appropriate institutions from their position on the political periphery.

THE INTERNATIONAL CONTEXT OF ENVIRONMENTALISM AND THE CHARACTERIZATION OF THE WHALING ISSUE

The present discourse involving whales and whaling contains all the tensions found in a number of other global economic and environmental disputes. The exploitation of natural resources by an industrializing world, where the operative organizing principle has been one of unbridled competition, has come into conflict with widespread and legitimate concerns about finite and threatened environmental resources. A backlash has developed against an era in which growth and progress was an unquestioned virtue, industrialization was a desired and almost universal goal, and the large-scale exploitation of living resources for profit alone was an accepted practice.

For many people the history of the whaling industry provides a classic example of all that was wrong with industrial societies' relationship with the living environment. In reaction to the excesses of industrial whaling in part, an "anti-whaling" movement developed, and according to one commentator, it intensified when:

> in 1979, some environmentalists encountered a new and dangerous element in the industry that operated outside the existing controls on those who killed

the whales. This was the pirate whalers: gangs of modern freebooters armed with harpoon cannons and the whale-butchers' long flensing blades. Usually illegally set up and financed by the Japanese through dummy corporations, they acted very much in the swashbuckling tradition of the rough old sea pirates, violating all controls and regulations on whaling.

(Day 1989:45).

The "pirate whalers" are outside of the law because:

Following the 1982 moratorium, all treaty and legal agreements relating to the whaling industry — from the UN to the IWC — ruled that commercial whaling operations should cease by 1985. Since then, with the exception of aboriginal hunting, all whalers have been considered pirate operations.

(ibid:50).

This definition appears to adopt international legal standards as the basis for the distinction between aboriginal whaling, as being legal, and all other forms of whaling as illegal. Why, the author asks (op. cit), in the face of such widespread opposition to whaling in the late 1980's, does whaling still continue?

According to that author, there is only one reason:

A handful of men are making large profits in virtually the only market for the product remaining in the world: Japan. All commercial whaling (and certainly all pirate whaling) has continued for the past twenty years simply to supply the Japanese market with whalemeat. This market, which buys and sells legal or smuggled meat indiscriminately, only survives itself through government subsidization and artificial pricing. In the end, the whale war continues because of extremely corrupt internal Japanese politics, and the pressure that a few wealthy fishing industrialists can bring to bear on the Japanese government.

However, despite the simplistic analysis and questionable accuracy of these assertions, and lost in the "anti-whaling" reaction to the tales of havoc wrecked by the relentless global pursuit of profit, there are real stories of the people and communities which also use renewable resources, including whales, in a qualitatively (and often quantitatively) different way from that represented in Day's and other anti-whalers' accounts.

Also lost in the anti-whaling rhetoric is an appreciation that the exploitation of natural resources has almost invariably been coupled with the exploitation of non-industrialized peoples and cultures. The characterization which divides whaling interests into only two types, namely "aboriginal" and "pirate", effectively obscures the cases of coastal peoples and cultures dependent upon marine resources, just as the division of whalers into only two categories (namely "aboriginal" and "commercial") did before the term "pirate whaling" was introduced, and as it continues to do in the debates of the International Whaling Commission (Freeman 1993; Gambell 1993).

When thinking of renewable resource-based economies and their importance to those communities which are directly dependent upon the harvesting of wildlife, a common perspective in many urban societies is to disparage the harvesters, to consider them somehow less advanced or even inferior. This perspective is a consequence of a particular intellectual parochialism; now, in the era of animal rights, we are seeing this parochialism repositioned within western liberal democracies. The current restatement made by many of those who have no sympathy for or understanding of the cultures of those who harvest living resources, is made as a choice between faceless and rapacious economic interests on the one hand and individual blameless animals on the other. Even the "aboriginal" distinction is rapidly disappearing, so that resource users are resource users, without significant distinction when compared, as a class, to the individualized animals they use and continue to depend upon. An example of this method of restating and over-simplifying the issue of resource use appears the question posed in relation to the "commercial extinction" of great whales:

> Do we continue to let [whalers] ply their trade until commercial extinction leads to biological extinction? *It must be a clear choice.* Do we extinguish the industry or do we allow the industry to extinguish the whale?

> (Day 1989:50, emphasis added)

There are many ways of characterizing those peoples and cultures for which the harvesting of marine mammals is an activity critical to the continued survival and well-being of their communities. However, there is no room to distinguish them at all in a discussion artificially posed as a choice between mutually exclusive alternatives of "industry" and "whales".

To better appreciate the issues it is more effective to adopt a framework for discussion which accommodates all concerns. In the present paper discussion moves from an examination of Inuit whaling to the shaping of an international context which would provide the necessary flexibility, while remaining firmly rooted in international law and policy.

THE INUIT WAY OF LIFE: THE PAST AND THE IMPORTANCE OF HISTORY

Archaeological sites record waves of early peoples occupying the Western Arctic as early as 30,000 years Before Present (B.P.), with the ancestors of the Eskimo [Inuit] peoples arriving on the Bering Sea coasts of North America after the retreat of the glaciers, almost 10,000 years B.P. Nine thousand years ago dead whales washed up on beaches in the arctic archipelago. Between 5,000 and 4,000 years ago the first Inuit peoples are believed to have appeared in the Arctic and occupied Arctic North America and Greenland. Between 1000 B.C. and 1000 A.D. the Thule Inuit, the direct biological and cultural ancestors of contemporary Inuit, were well established from Alaska, across the Canadian north (including the coast of Labrador) and into Greenland (McGhee 1978).

All early Eskimoan cultures were hunting cultures, sustained by an economy based upon caribou, marine mammals (e.g. polar bear, walrus, seals and small whales), fish and birds. Gradually the hunting technologies changed, and in the later Thule Inuit developed a tradition of harvesting the large bowhead whales in addition to continuing to hunt and fish for the previous inventory of species (McGhee, 1978). While some individual hunters might excel, nevertheless all members of the community benefitted from the hunting success of these individuals due to the widespread existence of obligatory sharing of food and materiel according to need. The many rules for the division of the hunter's kill, starting with the first bird, seal or hare of the young child, still exist widely among Inuit societies today (e.g. Wenzel 1991:98–105; Freeman et al 1992:67–68,72).

The single most important technological change in this transition to the hunting of large whales was the development of the detachable harpoon head to be used in conjunction with a system of floats made from whole inflated sealskins and drags attached to long lines. The drags were circular disks, some two feet in diameter made of driftwood and attached to extremely strong lines made of walrus skin. This technology meant that the hunter could strike the animal, then wait for the float and drag system to tire it out; the floats prevented dead animals from sinking and being lost. A variety of marine mammals could be taken in this way, but it was particularly important for the capture of the large bowhead whale.

Being able to take bowhead whales reliably changed the way Thule Inuit lived, for rather than being restricted to living in small social groups dispersed along the Arctic coast (which were efficient for taking and processing smaller prey), Thule Inuit could live in larger settlements which were necessary for hunting large whales and which could more easily be supported by the larger prey. Thule Inuit also domesticated the dog and developed dogsled travel, which increased their mobility in winter and again changed their patterns of social activities. It was the Thule Inuit who developed the dome-shaped snow house ("igloo"), an essential technique for safe winter travel as well as for comfortable winter settlement living in areas devoid of driftwood or whale bones otherwise commonly used for shelter construction.

The Thule culture developed and spread at a time when the arctic environment was benefitting from climatic warming, when the permanent sea ice retreated far to the north, opening up new opportunities for hunting and fishing. When the climate began to worsen during the "Little Ice Age" of the 1600s, and the summer extent of the permanent sea ice advanced southward, the marine mammals retreated ahead of the advancing ice and consequently the Inuit followed. In Eastern Canada, Inuit moved southward into what is now Quebec and Labrador as far as the north shore of the St. Lawrence River.

To facilitate harvesting a wide variety of land and marine mammals, fish, birds, berries and wild plants, the Inuit developed an ingenious technology. In addition to stone and bone tools, meteoric iron and native copper was also used to make tools. Inuit also traded, sometimes widely, as shown by archaeological studies of the distribution of artifacts whose geographic origins can be definitively identified on the basis of mineralogy. Indeed the author of one study of trade among Inuit in Alaska during the period prior to Euro-American contact goes so far as to say that:

> It is now clear that inter-societal trade — and the accumulation of material wealth and social prestige through trade — was enormously important in northwestern Alaska and northeastern Asia. In fact, the trade system was so important and so strongly integrated into traditional Inuit cultures that trade cannot be meaningfully separated from its larger cultural context.

(Hickey, 1979).

In addition to the importance of trade for immediate value, Hickey (1979) also points out that trade supported a "delayed reciprocity" mechanism common to all hunting and gathering societies, including Inuit groups in other parts of the circumpolar region. Such trade served as a form of mutual insurance against poor harvests and times of need. So important are these trade and other economic relations for northern peoples that in one study it is stated: "indeed, for all practical purposes, every Eskimo, Indian and Chukchi settlement was a trading centre and the normal routes by which people travelled about the country were also trade routes . . . this system . . . spanned most of northern North America and Eastern Eurasia" (Burch 1988:238). It should be mentioned that inter-cultural trade was known to Canadian Inuit prior to European contact, as in the case of the trade in marine mammal and fish oil, sinew-backed bows and Russian trade goods with Indian groups in the Northwest Territories of Canada (McGhee 1974:10, 13; Smith 1979).

In this brief sketch of pre-contact Inuit history is glimpsed the independent, strong, inventive, resourceful Inuit hunting culture. Within Inuit society the trading of renewable resource products obtained from the hunt constitutes an integral and important part of the culture. This material is intended to convey a general portrait of a people and a way of life prior to the arrival in the Arctic of European and American explorers and traders. Clearly the Inuit have a long history of occupation in a vast regions, an occupation which pre-dates the arrival of western civilization in the Americas. The Inuit have a distinct culture based on a renewable resource harvesting economy, and trading of their surplus production is an integral part of their culture and important for maintaining a variety of customary social relationship.

CONTACT BETWEEN INUIT AND NEWCOMERS TO THE ARCTIC REGIONS

When the Europeans and Americans arrived in the Arctic, a wider cultural and economic exchange of goods and technologies occurred. Among the newcomers to the Arctic were the commercial whalers, arriving first in the Eastern Arctic, and later in the Western Arctic. The Inuit advised them where to look for whales and became crew members and harpooners on their whaling boats. This development proved to be highly significant for Inuit whaling and is germane to the present discussion.

As late as the 1980's the conventional wisdom, as represented in scientific and government reports, was that the Inuvialuit (the Inuit of the Western Arctic of Canada) had not hunted bowhead whales in Canadian waters, and that there was no direct connection between the present Inuvialuit population in the western Arctic and earlier Inuit bowhead whaling peoples in the region. From this set of beliefs, concerning the supposed lack of traditional use of bowhead whales in the region, the consequent conclusion was that the Inuvialuit had no rights to harvest bowheads in the future (Raddi and Weeks-Doubleday 1985).

On the basis of written accounts and extensive interviews with Inuvialuit elders, it has since been shown (op. cit) that, within living memory, Inuvialuit had participated in traditional bowhead hunts in the Western Arctic, but more importantly, their traditional harvest had become blended with the commercial whaling conducted by non-Inuit in the area. The Inuvialuit became harpooners and crew members for the whaling captains, at times bringing their entire families on board the whaling vessels, essentially trading their traditional skills and labour for a share of the harvest, plus some other goods and/or money. The whaling captains took the baleen (whalebone), and the blubber which was rendered into oil. The Inuvialuit received the skin (muktuk), which is a prized food item, and the whale meat. The Inuvialuit benefitted from the use of the ship, and the captain benefitted from the skills and knowledge of the Inuvialuit, producing a symbiotic blend of commercial and subsistence whaling which is significant in terms of the issues under consideration here.

Following the arrival of the commercial whalers, the Inuvialuit ceased using their traditional skin boats, the umiaks and kayaks, which are common to Inuit whaling practices throughout the circumpolar region; they also substituted iron harpoons and eventually the harpoon guns of the commercial whalers for the hand-held harpoons, floats and drags used previously. When commercial whaling in the region eventually ceased due to low commodity prices for whale products, the whaling vessels disappeared, leaving the Inuvialuit without the means to pursue subsistence utilization of bowhead whales and thus leading the authorities to the false conclusion that they had no interest in doing so. The Inuvialuit bowhead whalers had become virtually invisible as whalers, because commercial whaling obscured their interests by providing them with the desired whale products through an entirely new means of production and distribution (Freeman et al. 1992:18–20). When commercial whaling disappeared, and with it the new production technology and source for distribution and sharing practices, no provision was made for the Inuvialuit who remained.

This situation, of having needs largely disregarded by outside authorities, is not unfamiliar to other coastal peoples who rely upon whales as well as other marine resources. If accepting the view presented earlier, which suggests that following the 1982 moratorium any non-aboriginal whalers still operating had to be considered as illegal "pirates", then one can see how the removal by regulation of the commercial whaling category has also taken away the "legal" foundation from any other forms of whaling which existed concurrently.

This raises a very real problem related to contemporary discussions of aboriginal subsistence whaling and other forms of non-commercial whaling: the ease with which categorization of the substance of the activity is confused with superficial matters of appearance. This is a subtle point, but one that holds the key to clarifying the commercial/subsistence confusion involved in discussions at the IWC about traditional whaling cultures (discussed in Freeman 1993 and below).

The other impact of commercial whaling upon aboriginal subsistence whaling concerns the population status of the whale resource. Commercial whalers entered the Arctic because by that time they had seriously depleted whale stocks in more accessible whaling grounds. By the time commercial whaling had effectively ceased in the Arctic (immediately preceding World War I) stocks of bowhead whale had also been drastically reduced, lessening harvesting opportunities for Inuit and adversely affecting the biological potential of the stock remaining for years to come.

UNDERSTANDING THE INTERACTION BETWEEN INUIT SOCIETIES AND WORLD ECOLOGICAL, ECONOMIC AND POLITICAL SYSTEMS

Inuit realities have not changed very much in the Arctic: they remain subject to the constraints imposed by global ecological circumstances. However, recent anthropogenic environmental effects, originating outside of their region, are now causing quite different impacts: global change is predicted, and many experts believe the most extreme climate-related effects will occur in the Arctic. Climatic change is not new to Inuit, and indeed the archaeological evidence indicates that Inuit have successfully adapted to earlier periods of climatic change. However, other new threats are emerging, also a result of the interconnectedness of the Arctic to global physical and ecological systems, including, e.g. toxic contaminants from the industrialized world entering the resource stocks important to the Inuit as food, but which are not discussed further in this report (however, see Barrie et al. 1991; Davies 1991).

Connections between Inuit societies and international economic systems have existed since the time of contact, yet despite the changes having taken place since first contact with newcomers, the Inuit still depend on the success of the hunt for a significant portion of their basic food. Although the new technologies may make the process of obtaining food easier, adequate financial resources are required to purchase and operate this imported equipment. Studies consistently show that Inuit, whether full-time hunters or full-time wage workers (who only hunt occasionally) allocate a considerable proportion of their income to purchasing and maintaining hunting equipment (Freeman 1985:254–255; Kruse 1991:320; Langdon 1991:283).

The reason for the persistence of hunting in communities increasingly in-volved with wage employment and an increasingly monetized economy is only partly economic (Freeman 1985, 1988:165–166). However, in a study of full-time and part-time hunters in the community of Holman Island, Smith and Wright (1989) found that full-time hunters produced food from wildlife at a cost of about $1 (Canadian) per kilogram. Hunters each harvested enough produce to support their families, their dogs and four additional people. Based on the local cost of imported frozen meat, Smith and Wright assigned a substi-tution cost of $10.54 per kilogram of meat produced. The mean harvest re-ported was 1428 kilograms for each full-time hunter, which gives a substitu-tion value of $17,117 for this harvest. The substitution value less the hunting costs gave a net benefit of $15,448 to each hunter.

The capital cost of outfitting one hunter in this study was $21,130, with an-nual depreciation costs of $5,922. The annual total operating cost for food hunting was $1669 per hunter (Smith and Wright 1989). It is evident that sub-sistence harvesting in the contemporary Arctic requires some cash outlay; simi-lar amounts have been calculated for Inuit hunters operating elsewhere in the Arctic (e.g. Caulfield 1994: Table 16.5; Wenzel 1991:125).

In the past this cash requirement could be met through sales of sealskins, furs or other by-products of the subsistence hunt for food. However, since the European Economic Community (EEC) sealskin ban and activity by the anti-fur movement, prices for these commodities have declined drastically (e.g. Wenzel 1991:124), and in some instances, community earned cash income has been reduced by 90 percent or more; changes of such magnitude have placed full-time hunters, as well as Inuit communities, in severe jeopardy. Communi-ties still depend on country food, which is more nutritious and culturally ap-propriate than imported store-bought food, and with reduced incomes they cannot afford the cost of replacing the country food with expensive imported foods (Wein and Freeman 1992).

Today Inuit in Canada harvest a wide range of land and sea animals, fish and birds. Marine mammals utilized include the beluga whale (*Delphinapterus leucas*), narwhal (*Monodon monoceras*), bearded seal (*Erignatus barbatus*), ringed seal (*Phoca hispida*), hooded seal (*Cystophora cristata*), harp seal (*Phoca groenlandica*), walrus (*Odobenus rosmarus*), and polar bear (*Ursus maritimus*). In recent times other whales, like the bowhead whale (*Balaena mysticetus*), have occasionally been harvested. There are many other species including plants and marine invertebrates which are used by Inuit in a seasonally-based har-vesting cycle. Harvesting pressures are shifted from less abundant to more abundant or seasonally-available species. In Inuit society there are cultural pro-hibitions on waste, and traditional Inuit practices and skills are geared toward maximizing utilization of harvested animals, though not maximizing the size of the harvest (Freeman 1986:31–32).

Before the EEC trade ban was imposed on sealskins in 1983, the renewable resource economy was healthy and sustainable, and the standard of living in

Inuit communities was considered reasonable and sustainable. Sealskins, produced year-round as a by-products from the food hunt, provided the cash enabling the hunter to continue to harvest food on a full-time basis. While other commodities compensated to some extent for the loss of sealskin revenues, the lobbying goals of some of the anti-harvesting groups who earlier destroyed the sealskin markets are now directed to closing markets for other Inuit products, such as carvings made from ivory and clothing and duvets filled with eider down. Today Inuit are forced to accept more assistance from the government and there is a consequent sense of loss of autonomy and freedom, and, in some quarters, of hope itself. Youth suicides are now commonly reported in many Inuit communities, related, it is believed, to the lack of any meaningful future young people see for themselves.

Despite these gloomy realities, Inuit culture is alive and Inuit are determined to regain control of their lives and resources. In Canada, a land claims policy is effecting transfer of land and resources into Inuit hands. In 1985, the Inuvialuit Final Agreement provided the Inuvialuit of the western Canadian Arctic with $45 million (in 1979 Canadian dollars), and title to a large area of land, various surface and subsurface rights, harvesting rights in perpetuity, and shared management responsibilities. In 1993, the Canadian Parliament passed into law the Nunavut Agreement which guarantees the Inuit of the eastern and central Arctic rights to lands, royalties, resources, harvesting, and participation in decision-making affecting 250 million hectares of Arctic lands and offshore, as well as $1.5 billion in compensation payment over 15 years for lands surrendered.

The Nunavut agreement also provides for the development of institutions of public government, which in the future are likely to provide Inuit with another means of influencing events affecting their society. The example of the Greenland Inuit is instructive in this regard, for the Greenland Home Rule government is a public government dominated by Inuit (Greenlanders) which has played a major role within the Danish government in terms of developing policies and positions on harvesting matters that to some extent at least have held European anti-harvesting influences in check.

CURRENT THREATS TO INUIT SOCIETY AND CULTURE: THE ANTI-HARVEST MOVEMENT

Many people are surprised to discover that Inuit culture remains alive and indeed is undergoing a renaissance in many regions (e.g. Freeman et al. 1992:38–44). Even more surprising to people is the fact that Inuit are actively contributing to many of the important international debates of our time, including *inter alia*, the recognition of the rights of indigenous peoples in international law through the activities of the United Nations Economic and Social Commission, protection of the arctic environment through the Finnish Initiative (now the Rovaniemi Process), and the implementation of sustainable and equitable development through the work of the World Commission on Environment and Development (the Brundtland Commission, 1987), the IUCN-UNEP-WWF Strategy for Sustainable Living (1991) and the United Nations Conference on Environment and Development (1992). Inuit are also engaged in their own initiatives concerning the arctic environment, the renewable resource economy and those international fora and factors which threaten them.

Inuit are engaging in these international concerns precisely because they recognize the impact that global environmental, economic and political systems have upon their daily existence and the implications of this for their future cultural identity and wellbeing. Inuit are not prepared to have their identity or culture taken from them, and they have chosen the difficult role of becoming players in the global debates while at the same time gathering strength through self-government within the jurisdictions which have come to envelop them.

Inuit societies are not frozen in time, and throughout their existence they have been engaged in various forms of adaptation to their immediate environment and to influences beyond, including those rooted in non-Inuit societies and world systems. In consequence of these past and present events, contemporary Inuit societies are heterogeneous and internally differentiated, with the diversity among and within Inuit societies the observed response to a variety of geopolitical influences impinging upon the North. In Alaska, the imprint of corporate America and U.S.-style representative democracy are seen in the corporations resulting from the Alaska Native Claims Settlement Act (Langdon 1986:33–44; Chance 1990:162–175). In Canada, federalism has fostered Inuit initiatives like Nunavut and the creation of the Western Arctic Regional Municipality, public government based upon divisions of power. In Greenland, Inuit have achieved a high degree of political autonomy as a result of the creation of the Greenland Home Rule government.

When considering established ways of life and customary social and cultural norms extending back to before the time of Christ, abruptly confronting the international legal and political realities of our time, it becomes easier to appreciate the vast cultural gulf separating traditional whaling peoples like the Inuit and the recent activities of the International Whaling Commission (IWC) with respect to aboriginal subsistence whaling.

This is not to say that the differences of perspective are solely historical in nature, for there are vast distances between the realities of modern Inuit life in the Arctic and the life of contemporary urban dwellers living outside of the Arctic. These differences are expressed vigorously by those who do not approve of the taking of whales, or of seals, or of fur-bearing animals; the consequences of their disapproval are equally explicit, seen in the disruption and despair of many Arctic communities. Inuit have suffered greatly through the loss of markets for their renewable resource products and the violation of their right to their way of life, a mode of living which is both culturally and environmentally appropriate and sustainable. In response, they have developed strategies and approaches which are relevant to the issues being discussed, including the subsistence/commercial dichotomy, the development of international environmental legal frameworks, and policy and institutional alternatives directed toward the improved management of living marine resources such as whales.

However, Inuit are not alone in seeking new approaches to these questions: those who oppose the taking of whales are also offering alternatives with respect to the policies of the International Whaling Commission, alternatives which deserve scrutiny. This scrutiny is deserved not merely for what such proposals imply concerning the future of whaling or the IWC, but also for what they imply for the future of humankind's evolving relationship with nature, and the future of our species.

The anti-harvest campaigns are based upon concern for animal welfare and animal rights (and often the welfare of the campaigning organizations, e.g. Bonner 1993; Kalland 1994). Animal rights is a philosophy which is gaining strength in urban centres far removed from natural environments. From an ecological perspective, animal rights is an unnatural philosophy that assists in destroying nature rather than saving it, simply because it weakens the connection between human beings and their ecosystems, driving more people into direct dependency upon the industrial economy which represents a greater threat to the environment than regulated local-level use of renewable resources. Actions based upon an animal rights justification often violates the human rights of peoples who are dependent upon renewable resources (Doubleday 1989).

Through its actions, the anti-harvesting lobby is seriously weakening those cultures and societies which are intimately connected to their natural environment, by manipulating international environmental law and policy and through campaigns aimed at consumers. Such actions are deeply troubling from both ethical and ecological standpoints. Inuit, in the aftermath of such activities have lost markets for their products, suffered reduced income and the loss of freedom to trade in the products of their traditional economy. The Brundtland Report has emphasized the importance of sustainability and the fact that those societies which were truly sustainable were those which were directly dependent upon a renewable resource base and used it wisely; a similar understanding is clearly evident in the successor report to the World Conservation Strategy (IUCN-UNEP-WWF 1991:58ff). The realization is now becoming widespread that in coming to a better understanding of our environment, the knowledge of indigenous and other tribal and rural peoples like the Inuit in the contexts of renewable resource research and management is critical (e.g. McCay and Acheson 1987; Berkes 1989; Ruddle and Johannes 1990). Ironically this interest and understanding is coming at a time when the very survival of the people who have this knowledge is most threatened.

In the context of the whaling issue, colonial policies of the past, both domestic and international, restricted the rights of Inuit societies to some extent through their failure to recognize the significance of whaling to Inuit culture.

More recently, the broader attacks directed to additional harvesting compo-
nents of the renewable resource economy have produced far more profound
and serious consequences for Inuit societies.

ABORIGINAL SUBSISTENCE WHALING AND THE INTERNATIONAL
WHALING COMMISSION

In 1931 the Convention for the Regulation of Whaling was established to over-
see the commercial whaling industry; this Convention contained a clause
which provided an exemption for aboriginal peoples. The 1931 Convention
was superseded by the 1946 International Convention for the Regulation of
Whaling, under which the taking of right whales, was prohibited and a specific exception was established in favour of indigenous peoples. The wording
of this exemption clause states: "It is forbidden to take or kill gray whales or
right whales, except when the meat and the products would be used exclusively for local consumption by aborigines".

Clearly this definition stands in violation of Part I, Article 1 of the International Covenant on Economic, Social and Cultural Rights which provides *inter
alia* that:

> All peoples may, for their own ends, freely dispose of their natural wealth
> and resources without prejudice to any obligations arising out of international
> economic cooperation, based on the principle of mutual benefit, and interna-
> tional law. In no case may a people be deprived of their means of subsistence.

That section of the International Convention for the Regulation of Whaling
(ICRW) pertaining to aboriginal whaling also implicitly prohibits all trade in
whale products outside the local community. Given the views expressed by
Hickey (and others) cited above concerning the important role of trade in sustaining Inuit social and economic institutions, this prohibition is an infringement of aboriginal economic and cultural rights, and rights of self-determination under the Covenant according to which:

> All peoples have the right of self-determination. By virtue of that right they
> freely determine their political status and freely pursue their economic, social
> and cultural development.

The exception provided for "aborigines" in the IRCW also clearly contra-
venes the provision to "freely dispose of their natural wealth and resources".
It is also inconsistent with the rights to self-determination.

Another example of the broader consequence of manipulation of categories
of whaling is found in the case of the Alaskan bowhead hunt, where a 1977
proposal by the International Whaling Commission to eliminate the subsistence
harvest (Gambell 1993:102) threatened to eliminate the legal and constitutional
rights of the Alaskan Inupiat.

More recent attempts to define aboriginal subsistence whaling within the International Whaling Commission (IWC) are more progressive than this early attempt. As a result of the reaction by Alaskan Inupiat [Inuit] to the 1977 attempt to eliminate their bowhead subsistence harvest, the IWC convened the three panels of experts in 1979 to address the Alaskan bowhead hunt from the perspectives of wildlife science, nutrition and culture of the aboriginal people concerned. In so doing, IWC appeared to acknowledge implicitly that other, non-conservation considerations/values were relevant to discussion of whaling quotas, including, arguably, recognition that a competing values framework was operative, namely, that of human rights.

While these new criteria were to be considered in setting bowhead quotas, they were not derived from the scientific regime within the IWC; nonetheless, they are quantifiable criteria, and have subsequently been addressed within the scientific paradigm. Also, while these new criteria were not of the scientific conservation regime, they have consistently been given consideration within it.

In 1980, the IWC agreed to establish an ad hoc working group for the "development of management principles and guidelines for subsistence catches of whales by indigenous (aboriginal) peoples". This meeting was held in 1981 and resulted in the following definitions of "aboriginal subsistence whaling", "local aboriginal consumption", and "subsistence catches":

> Aboriginal subsistence whaling means whaling for purposes of local aboriginal consumption carried out by or on behalf of aboriginal, indigenous or native peoples who share strong community, familial, social and cultural ties related to a continuing traditional dependence on whaling and on the use of whales.

> Local aboriginal consumption means the traditional uses of whale products by local aboriginal, indigenous or native communities in meeting their nutritional, subsistence and cultural requirements. The term includes trade in items which are by-products of subsistence catches.

> Subsistence catches are catches of whales by aboriginal subsistence whaling operations.

The IWC Working Group on Aboriginal Subsistence Whaling also agreed upon objectives for management of whaling stocks subject to aboriginal subsistence whaling, confirming implicitly the position that commercial and aboriginal subsistence whaling were two qualitatively different things and that different approaches could reasonably be taken. In 1982, a Resolution establishing an aboriginal subsistence whaling regime was adopted by the IWC (Gambell 1993:104).

With respect to the current IWC definitions of "aboriginal subsistence whaling" and "local aboriginal consumption", it can be stated that aboriginal subsistence harvesting in general necessarily encompasses trade that may include local and non-local trade in the edible whale products as well as the inedible by-products. If the renewable resource economy is to be sustainable, the capacity to trade in renewable resource products provides a potential tool for the management of such economies, rather than an engine of endless growth. This appreciation is explicitly recognized in recent international documents on sustainable development (e.g. IUCN-UNEP-WWF 1991:42, 58). However, fears of over-exploitation of species due to trade in wildlife, particularly in the United States where the memory of passenger pigeons and buffalo linger like ghosts, must be addressed.

In general terms the trend within the IWC has been to recognize that aboriginal whaling is distinct, but to define it according to non-Inuit standards. Participation of indigenous peoples in the technical meetings of IWC has been increasingly encouraged, but they remain continually burdened by the onus of proving their rights and their needs. While the parties to the ICRW have recognized the competence, in legal terms, of the IWC with respect to whalers they have no prerogative with respect to indigenous peoples; and in the face of provisions of international human rights law to the contrary, the prohibition against aboriginal trade of edible whale products deserves reconsideration.

THE FUTURE OF INUIT SUBSISTENCE WHALING AND THE PROPOSED RIGHT TO LIFE OF WHALES

To put the relationship of aboriginal subsistence whaling in historical perspective, 1992 marks the 125th birthday of the country we know as Canada, the 500th anniversary of the "discovery" of America by Columbus, and more than one thousand years of the catching of whales in the Arctic by the direct ancestors of modern Inuit. Inuit whale hunting pre-dates modern concepts of state and the international law which now tries to regulate it.

The success of the anti-harvesting lobby has resulted from a number of strategies, which include acceptance of a narrow and ill-informed definition of subsistence and campaigns against all "commercial" uses of the harvested product. This approach depends upon drawing a rigid distinction between aboriginal-subsistence and commercial harvesting which provides the basis for attacking most traditional harvesting activities, or at a minimum, causes them to be regulated (Freeman 1993). Other strategies involve co-opting the scientific advisory process in order to politically influence both the decision-making procedure and results obtained (Andresen 1993:114-115).

Most recently a surprising development has taken place affecting policy development concerning the international regulation of whaling and the IWC. The U.S. Marine Mammal Commission (a governmental oversight body based in Washington, D.C.) has proposed for the first time that non-scientific criteria be considered as a basis for making decisions about the resumption of commercial whaling, on the grounds that non-consumptive uses of whales have attained higher commercial value than have consumptive uses. This proposal suggests that the U.S. is considering using its political and economic power to force international compliance with the dictates of the anti-harvesting lobby and that it views this ideology as providing a credible basis for making decisions about the utilization of living resources. The proposal suggests renegotiation of the ICRW leading to a convention recognizing the primacy to be afforded these non-consumptive values.

It appears that this new ideology espoused by the Marine Mammal Commission seeks intellectual and legal sustenance from a recent report proposing that whales have a "right to life" with moral equivalence to the human right to life (D'Amato and Chopra 1991). However, if non-consumptive values are accorded equal or higher values than consumptive values within the IWC, and whales enjoy legal protection from untimely death at the hand of whalers as the MMC now proposes, it seems quite unlikely that any consumptive or cultural value which is dependent upon a whale being killed will be accorded appropriate value in management decisions made under such rules.

The D'Amato and Chopra thesis proposes that the human right to life be extended to whales through the recognition that whales possessed this right inherently by virtue of being whales. In regard to the right of aboriginal peoples to take whales, these authors characterized that as a "counter claim" to the right to life of whales. Not surprisingly, given the structure of their argument, D'Amato and Chopra found in favour of the whales. At the heart of their argument, is the idea that the international developments related to whale regulation have moved through a series of conceptual stages involving whales as a free resource, whales as objects of regulated hunting, whales as objects of conservation, whales as objects of protection, whales as objects of preservation, and now, because of the flowering of human consciousness, whales as *de facto* subjects of an entitlement to life. However, it should be remembered that the shifting framework of values and rights, which has given us contemporary human rights law, is not just due to changing awareness.

Chopra (pers. comm.) believes that the arguments of D'Amato and Chopra (1991) have indeed been accepted by the U.S. Marine Mammal Commission and that some within that body agree that the "right to life" argument should be included in U.S. policy with respect to the IWC. If this is the case, proposals for consideration of non-consumptive values by the U.S. in any re-negotiation of the international whaling regulations need to be scrutinized with considerable care and the questionable logical and legal basis of their position well understood.

According to D'Amato and Chopra (op. cit) stage one (whales as a free resource) through stage five (whales as object of preservation) are related to positive international law and the legitimization of capitalist property rights. However, the jump which these authors make from this framework into the human rights framework with whales as subjects of entitlement to life, is a *non sequitur* logically and legally, for despite arguments to the contrary the "right to life" only has meaning legally in the context of human life, and whales are not human.

To propose (or worse, to assert or insist) that a sufficient mass of humanity and a particular intergovernmental organization has achieved some exalted state of mind and from this to conclude that this new perspective has achieved the force of international customary law, raises questions about what effects the views of the unexalted have, or don't have, on *opinion juris*, and on the vetted "entitlement". It appears that the anti-whaling lobby has foreseen the inevitable increase taking place in certain whale stocks currently protected by a whaling ban, and other (for them) unfavourable trends in developing a safe and sustainable whaling management procedure (see Gambell 1993:100–101) and as a consequence is now trying to legislate morality.

In terms of the "counterclaims" of aboriginal whalers posited by D'Amato and Chopra (op. cit), by subjecting the interests of those outside the dominant group (which controls world resources, economies and international law) to intellectual marginalization, they reveal the imperfections of their scheme for an unfolding of higher law. In fact both aboriginal and subsistence interests are pre-existing, and therefore ought to be accommodated in the developing legal framework if it is truly organic. These interests are a direct and immediate descendant of D'Amato and Chopra's Stage 1, the free resource stage; and in some cases offer a legitimate alternative to the state control implicit in their Stages 2 and 3.

In closing discussion on this matter, it is useful to revisit the Preamble to the Charter of the United Nations which sets out as a goal the reaffirmation of:

> faith in fundamental human rights, in the dignity and worth of the human person . . . and for these ends to practice tolerance and to live together as good neighbours.

It appears that there remains much work to be done among the members of our own species before these reasonable goals are reached. Without social, economic and cultural peace among diverse peoples, how can humankind begin to propose a new order in nature?

ALTERNATIVES TO THE EXISTING FRAMEWORKS FOR THE MANAGEMENT OF INUIT WHALING

In developing alternatives to the existing international regime for the management of whale resources and whaling, a position consistent with international law would seem desirable. Such a position would necessarily accord standing to "peoples", in keeping with the language of the United Nations Charter and the International Covenants, as well as to state governments' own laws. Clearly aboriginal peoples are distinct "peoples", and governments have obligations to them. Renewable resource management regimes should also be consistent with the goals of the World Commission on Environment and Development, as expressed in the Brundtland Report, which calls for sustainable development. Inuit prefer the term "sustainable and equitable development" and at the 1992 U.N. Conference on Environment and Development a clear call was made by all indigenous peoples for "sustainable and equitable development" to be promoted. Part of the difficulty which Inuit have faced in their struggle to be recognized has to do with issues of scale: the IWC attempts to perform a global function, but it does not have exclusive or comprehensive jurisdiction over whales, nor can it deal with issues of habitat. Moreover, its membership is not global.

A new approach should be pragmatic and should take into account geographical, ecological, cultural and political realities. In the Arctic for example, the need exists to address marine mammal issues in an integrated fashion, within the context of the arctic marine ecosystem treated as a whole. The distributions of whales, walrus and seals exists within a circumpolar ecumene consisting of many species having continuing social, cultural, economic and dietary importance to Inuit.

If accepting as legitimate the Inuit aspirations for a future which is dynamic, rather than frozen in the past, and their desire for a sustainable renewable resource economy, a new management regime that adequately addresses these needs is required. Existing mechanisms have proven themselves insensitive, unaccountable and too far removed in understanding from the realities of arctic life. A recent example of this distance from circumpolar realities is the proposal made by the US Marine Mammal Commission suggesting a departure from the use of solely scientific criteria for decision-making within the IWC, and advocating the introduction of non-consumptive values as a possible basis for making decisions about whale use. If accepted, is it reasonable to expect that Inuit concerns will ever receive equitable treatment?

One of the options for improving the present regulation of whaling entails renegotiation of the 1946 International Convention for the Regulation of Whaling. However, as the majority of signatories to the Convention appear to be dominated by the anti-harvesting lobby, the likely outcome of any renegotiation will certainly adversely affect all whaling societies. Another option is to be found in the example of the North Atlantic Marine Mammal Management Commission (NAMMCO) formally established in 1992 to contribute through regional cooperation to the conservation, rational management and study of marine mammals in the North Atlantic (Hoel 1993).

A third option is to be found within Inuit attempts to engage the international dimension from the periphery. In 1980, the Inuit Circumpolar Conference (ICC), the international organization representing Inuit in Canada, Greenland and Alaska (with the participation of Siberian Inuit since 1989) founded the ICC Whaling Commission. Although this commission has not been active to date, a proposal to bring it into immediate and effective existence was unanimously adopted at the 1992 ICC General Assembly (Anon 1993:2). A further regional whale management initiative based upon the principles of sustainable and equitable use of whale resources is the bilateral Canada-Greenland Joint Commission on Beluga and Narwhal that came into existence in 1989. The signing of the Declaration of Rovaniemi on the protection of the arctic environment by all eight arctic states, signals the beginning of international cooperation on matters affecting the environment and renewable resources within the Arctic on a circumpolar basis. Such regional, as opposed to putatively global, management regimes appear to offer a more realistic approach to the needs and aspirations of Inuit.

The most prudent strategy is undoubtedly to pursue both options concurrently, and, if necessary, make a choice at an appropriate time in the future. However, it is not impossible to imagine that a hybrid of the two might evolve, with larger-scale commercial whaling regulated within a renewed IWC, and aboriginal subsistence whaling and other community-based small-scale forms of whaling managed by regional bodies with appropriate membership and jurisdiction.

Regardless of the management option which is finally chosen, there is a need to accommodate users of the resource as participants in the management process, in part because it is mandated by the settlement of territorial claims within Canada, but also because from the point of view of the management of common pool resources, it makes good conservation sense to do so (Berkes 1989; Pinkerton 1989; Williams and Baines 1993). It is now increasingly realized that those dependent upon renewable resources frequently have better opportunity to observe the stocks concerned and have longer term knowledge of the environment than do the science-based managers. Most importantly, the local communities may have legitimate institutions and authority to regulate their own harvesting within a management regime.

An appropriate sequence of events with respect to the management of whales and other marine mammals will involve the following steps:

1) establish goals and objectives for management;
2) obtain the best information on the resource stock (e.g. distribution, abundance, harvestable surplus)
3) make management decisions (e.g. establish restrictions on harvesting by quota, season, gear, area etc)
4) allocate quotas, improve efficiency, refine technology, maximize benefits, distribute products.

In this sequence, steps one to three would be conducted with the direct participation of the users of the resource (in the present case, the Inuit), and step four would be left entirely to the resource users.

Undoubtedly other options will arise, and those concerned must be alert to them as they arise; at present it seems to be in the best interests of coastal resource harvesters like the Inuit to look at regional management bodies as providing the greatest assurance that both sustainability and equity are appropriately factored into the management regime.

REFERENCES

Andresen, S., 1993.
   The Effectiveness of the International Whaling Commission. *Arctic* 46:108–115.

Anon, 1993.
    The 6th Inuit Circumpolar Conference. *INWR Digest* No. 2:1–3. International Network for Whaling Research, Edmonton.
Barrie, L.A., D. Gregor, B. Hargrave, R. Lake, D. Muir, R. Shearer, B. Tracey and T. Bidleman, 1992.
    Arctic Contaminants: Sources, Occurrence and Pathways. *The Science of the Total Environment* 122:1–74.
Berkes, F. (ed), 1989.
    Common Property Resources: Ecology and Community-Based Sustainable Development. Belhaven Press, London.
Bonner, R., 1993.
    At the Hand of Man: Peril; and Hope for Africa's Wildlife. Alfred A. Knopf, New York.
Burch, E.S., Jr., 1989.
    War and Trade. In: W.W. Fitzhugh and A. Crowell (eds), *Crossroads of Continents: Cultures of Siberia and Alaska,* pp. 227–240. Smithsonian Institution Press, Washington, D.C.
Caulfield, R.A., 1994.
    Aboriginal Subsistence Whaling in Greenland. In: M.M.R. Freeman and U.P. Kreuter (eds), *Elephants and Whales: Resources for Whom?* pp. 263–292. Gordon and Breach. Science Publishers, Switzerland.
Chance, N.A., 1990.
    The Inupiat and Arctic Alaska: An Ethnography of Development. Holt, Rinehart and Winston, Fort Worth, Texas.
D'Amato, A. and S.K. Chopra, 1991.
    Whales: Their Emerging Right to Life. *American Journal of International Law* 85: 21–62.
Davies, K., 1991.
    Health and Environmental Impact Assessment in Canada. *Canadian Journal of Public Health* 82:19–21.
Day, D., 1989.
    The Eco Wars: True Tales of Environmental Madness. Key Porter Books, Toronto.
Doubleday, N., 1989.
    Aboriginal Subsistence Whaling: The Right of Inuit to Hunt Whales and Implications for International Environmental Law. *Denver Journal of International Law and Policy* 17:373–393.
Freeman, M.M.R., 1985.
    Effects of Petroleum Activities on the Ecology of Arctic Man. In: F.R. Engelhardt (ed), *Petroleum Effects in the Arctic Environment,* pp. 245–273. Elsevier Applied Science Publishers, London and New York.
Freeman, M.M.R., 1986.
    Renewable Resources, Economics and Native Communities. In: J. Green and J. Smith (eds), *Native People and Renewable Resource Management,* pp. 29–37. Alberta Society of Professional Biologists, Edmonton.
Freeman, M.M. R., 1988.
    Tradition and Change: Problems and Persistence in the Inuit Diet. In: I. de Garine and G.A. Harrison (eds), *Coping with Uncertainty in Food Supply,* pp. 150–169. Oxford University Press, Oxford and New York.

Freeman, M.M.R., 1993.
    The International Whaling Commission, Small-Type Whaling, and Coming to Terms
    with Subsistence. *Human Organization* 52:
Freeman, M.M.R., E.E. Wein and D.E. Keith, 1992.
    *Recovering Rights: Bowhead Whales and Inuvialuit Subsistence in the Western Canadian
    Arctic.* Canadian Circumpolar Institute, Edmonton.
Gambell, R., 1993.
    International Management of Whales and Whaling: An Historical Review of the
    Regulation of Commercial and Aboriginal Subsistence Whaling. *Arctic* 46:97–107.
Hickey, C.G., 1979.
    The Historical Beringean Trade Network: Its Nature and Origins. In: A.P. McCartney
    (ed), *Thule Eskimo Culture: An Anthropological Retrospective*, pp. 411–434. National
    Museums of Canada, Ottawa.
Hoel, A.H., 1993.
    Regionalization of International Whale Management: The Case of the North Atlantic
    Marine Mammal Commission. *Arctic* 46:116–123.
IUCN/UNEP/WWF, 1991.
    Caring for the World. A Strategy for Sustainable Living. Gland, Switzerland.
Kalland, A., 1994.
    Whose Whale is That? Diverting the Commodity Path. In: M.M.R. Freeman and U.P.
    Kreuter (eds), *Elephants and Whales: Resources for Whom?* pp. 159–186. Gordon and
    Breach Science Publishers, Switzerland.
Kruse, J.A., 1991.
    Alaskan Inupiat Subsistence and Wage Employment Patterns: Understanding Indi-
    vidual Choice. *Human Organization* 50:317–326.
Langdon, S.J., 1986.
    Contradictions in Alaskan Native Economy and Society. In: S.J. Langdon (ed), *Con-
    temporary Alaskan Native Economies*, pp. 29–46. University Press of America, Lanham,
    Maryland.
Langdon, S.J., 1991.
    The Integration of Cash and Subsistence in Southwest Alaskan Yup'ik Eskimo Com-
    munities. *Senri Ethnological Studies* 30:269–291.
McCay, B.J. and J.A. Acheson (eds), 1987.
    The Question of the Commons: The Cultural Ecology of Communal Resources. The Univer-
    sity of Arizona Press, Tucson.
McGhee, R., 1974.
    *Beluga Hunters: An Archaeological Reconstruction of the History and Culture of the Mac-
    kenzie Delta Kittegarymiut.* Newfoundland Social and Economic Studies No. 13,
    Memorial University of Newfoundland, St. John's.
McGhee, R., 1978.
    Canadian Arctic Prehistory. National Museums of Canada, Ottawa.
Pinkerton, E. (ed), 1989.
    *Co-operative Management of Local Fisheries: New Directions for Improved Management and
    Community Development.* University of British Columbia Press, Vancouver.
Raddi, S. and N. Weeks-Doubleday, 1985.
    *The Prehistoric and Historic Utilization of Bowhead Whales in the Canadian Western Arc-
    tic: A Community-Based Study.* World Wildlife Fund, Toronto.

Ruddle, K. and R.E. Johannes (eds), 1990.
> *The Traditional Knowledge and Management of Coastal Systems in Asia and the Pacific,* (Second Edition). UNESCO, Paris.

Smith, J.G.E. (ed), 1979.
> Indian-Eskimo Relations: Studies in Inter-Ethnic Relations of Small Societies. *Arctic Anthropology* 16, No. 2.

Smith, T.G. and H. Wright, 1989.
> Economic Status and Role of Hunters in a Modern Inuit Village. *Polar Record* 25(153):93–98.

U.S. Marine Mammal Commission, 1991.
> *Issues Facing the International Whaling Commission and the United States Regarding the Resumption of Commercial Whaling and the Future Conservation of Cetaceans.* Washington, D.C.

Wein, E.E. and M.M.R. Freeman, 1992.
> Inuvialuit Food Use and Food Preferences in Aklavik, Northwest Territories, Canada. *Arctic Medical Research* 51:159–172.

Wenzel, G., 1991.
> *Animal Rights, Human Rights: Ecology, Economy and Ideology in the Canadian Arctic.* University of Toronto Press, Toronto and Buffalo.

Williams, N.M. and G. Baines (eds), 1993.
> *Traditional Ecological Knowledge: Wisdom for Sustainable Development.* Centre for Resource and Environmental Studies, Australian National University, Canberra.

16. ABORIGINAL SUBSISTENCE WHALING IN WEST GREENLAND

RICHARD A. CAULFIELD

INTRODUCTION

The International Whaling Commission's (IWC) aboriginal subsistence whaling regime has come under increasing scrutiny following implementation in 1986 of a moratorium on commercial whaling. As the IWC struggles to implement new management procedures for commercial whaling, some suggest that management procedures for aboriginal subsistence whaling should also be revised (Gambell 1993). A central issue in this discussion is the definition of the terms "subsistence" and "commercial," and the distinctions these imply between aboriginal subsistence and other types of whaling, such as small-type coastal whaling (Freeman 1990, 1993).

Discussions about subsistence and cash in rural economies are certainly not new, nor are they limited to whaling. Research throughout the North has revealed that mixed subsistence-market economies provide rural communities and regions with a reliable economic base even when linkages to the larger world economy are tenuous and uncertain (Wenzel 1991; Wolfe and Walker 1987; Asch 1983; Feit 1983; Usher 1981). In these mixed economies, cash and commercial-wage sectors are mutually supportive, and money generated from wage employment or sales of local products is used to capitalize subsistence harvest activities. In Greenland, research by Dahl (1987, 1989, 1990), Nuttall (1992), and Møller and Dybbroe (1981) demonstrates the close interrelationships between subsistence and commercial-wage sectors in local communities and regions. In fact, Dahl (1989) argues that differentiation between commercial and non-commercial harvesting of wild resources in Greenland is meaningless because the two pursuits are inextricably linked.

Aboriginal subsistence whaling in Greenland is an integral part of these local and regional economies (Caulfield 1991; Josefsen 1990; Larsen and Hansen 1990; Greenland Home Rule Government 1989; Helms *et al.* 1984; Donovan 1982; Petersen *et al.* 1981). Greenlanders catch minke whales (*Balaenoptera acutorostrata*) and fin whales (*Balaenoptera physalus*), both of which are subject to IWC jurisdiction. The IWC first adopted quotas for Greenlanders' catch of humpback whales (*Megaptera novaeangliae*) in 1961, and in 1975 placed quotas on catches of minke and fin whales (Gambell 1991a). In 1985, the IWC reduced Greenlanders' minke whale quota by more than half, and eliminated

Table 16.1 International Whaling Commission (IWC) Quotas for
Greenlandic Aboriginal Subsistence Whaling, 1984–1994

Year	Minke Whales W. Greenland	E. Greenland	Fin Whales	Humpback Whales
1984	2-yr. quota total = 588	10	6	9
1985	max. 444 per year	10	8	8
1986	2-yr. quota total = 220	10	10	0
1987	max. 130 per year	10	10	0
1988	110	12	10	0
1989	60	12	23	0
1990	95	12	21	0
1991	100	12	21	0
1992	3 yr. quota	12	21	0
1993	total = 315 max. 105	12	2 yr. quota total = 42	0
1994	per year	not yet determined		not yet determined

(Source: IWC, 1983–1991; Jessen 1992, pers. com.)

humpback catches entirely, ostensibly due to concern about whale stocks (Table 16.1). Minke quotas were reduced even further in the late 1980s to a low of 60 for West Greenland for 1989. However, this was offset somewhat by increased fin whale quotas. In 1991, an increased minke quota was adopted for the three-year period from 1992 and 1994; the three-year quota is 315 whales (including those struck but lost) with a maximum of 115 in any one year. Hunters in West Greenland were allowed to take 21 fin whales in 1992 under a one-year quota (Anon. 1991). Whales caught under these provisions are used only for local consumption within Greenland, and may not be exported.

While there is widespread acceptance within the IWC of Greenland's aboriginal whaling, critics express fears about the development of internal markets for whale products and concern that profit maximization, commoditization, and capital intensification may be developing (IWC, 1985; Lynge 1990). This case study from Qeqertarsuaq (in Danish, Godhavn) Municipality in West Greenland (Fig. 16.1), illustrates the complex and dynamic relationship between subsistence and cash in contemporary Greenlandic whaling. These questions highlight tensions which exist in Greenland between continuity and change in the procurement, distribution, and exchange of whale products. Furthermore, they underscore the difficulties facing indigenous societies in pursuing sustainable development in the Arctic because of conflicting perspectives in

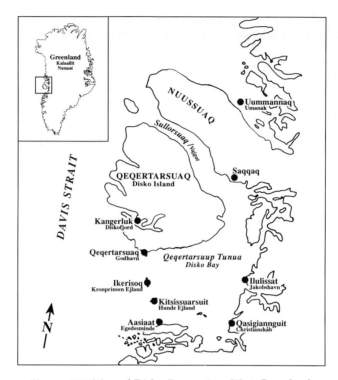

Figure 16.1 Map of Disko Bay region, West Greenland.

Inuit and Euro-american societies about appropriate human–environment relationships.

THE STUDY

The focus of this research is on Qeqertarsuaq Municipality in the Disko Bay region of West Greenland. The municipality encompasses all of Disko Island and a small island group in the bay itself. Qeqertarsuaq, the largest of two communities in the municipality, is located at 69° north latitude, 53° 33′ west longitude (Fig. 16.1). The region has a marine climate, influenced by the adjacent Disko Bay and Davis Strait. The average yearly temperature is –2.5° Celsius. Temperatures range from lows of –25° to –30° Celsius in winter to highs of +15° Celsius or more in summer. During winter, the sun drops below the horizon for a period of six weeks between late November and early January. Sea ice usually covers most of Disko Bay from December until March or April.

Qeqertarsuaq Municipality has a population of 1143 people, 90% of whom are Greenlandic Inuit. The two communities in the municipality are Qeqertarsuaq itself, with a population of 1075, and Kangerluk (Diskofjord), with a population of 68. The language of local Inuit residents is *Kalaallisut*, or Greenlandic Inuit, part of the Eskimo–Aleut linguistic family (Woodbury 1984). The non-Inuit-speaking population of the municipality is almost entirely Danish.

This case study was one of three sponsored by the Greenland Home Rule Government in 1989 and 1990 designed to describe and analyze the changing significance of aboriginal subsistence whaling in West Greenland (cf. Josefsen 1990; Larsen and Hansen 1990). The study communities were selected by the Home Rule government in order to illustrate a cross-section of contemporary whaling practices in Greenland. Hunters in Qeqertarsuaq Municipality participate in both fishing vessel and collective whaling for minke and fin whales. The municipality's mixed economy combines intensive commercial fishing typical of south Greenland with hunting activities (particularly for seal, beluga and narwhal) more typical of municipalities further north.

Research methods employed in the study included a formal survey covering 22% of all households in the municipality (random sample; n = 62), systematic interviews with household members participating in whaling in 1988 and 1989, interviews with key informants, and participant-observation of whaling and other harvest activities. For the formal household survey a survey questionnaire was translated into both West Greenlandic and Danish. For interviews with Greenlandic-speaking households, a local research assistant/translator was employed. Members of the research team recorded responses, which were then coded and analyzed using a SPSS statistical package. In most cases, household interviews were conducted with males, although female household members frequently contributed to responses. Household food consumption data were gathered over a one-year period using a 24-hour recall method. Selected households representing a cross-section of Qeqertarsuaq were asked to report which meat and/or fish products they consumed for one week during the months of October 1989 and January, April and July 1990. In addition to the community-based research, interviews were conducted with officials in the Greenland Home Rule Government in Nuuk and Copenhagen, and in the Greenland Fisheries Research Institute in Copenhagen (See Note 1). Research methods were designed in keeping with ethical principles for social science research in the North (Association of Canadian Universities for Northern Studies, 1982).

THE GREENLANDIC INUIT WHALING COMPLEX: A HISTORICAL OVERVIEW

Contemporary Greenlandic whaling is part of a historical complex of marine resource use dating back at least 4000 years (Grønnow and Meldgaard 1988). Until the end of the eighteenth century, Inuit hunters used skin-covered *umiat* (singl: *umiaq*) in the Thule tradition to catch bowhead and humpback whales

(Petersen 1986). Whales were an important part of Greenlandic diets, provided raw materials for fabricating hunting and fishing equipment, and served as the focus of highly organized collective whaling practices. Prior to contact with Europeans, hunters were governed by customary laws regarding ownership of harpooned whales, distribution of products from flensed whales, and appropriate behavior during whaling and flensing (Glahn 1921). Whales figured prominently in the Greenlandic spiritual world. Like all animals of the sea, whales were a gift from *Sassuma Arnaa*, the 'Mother of the Sea' (Greenland Home Rule Government 1988; Sonne 1986). Hunters demonstrated their respect for these gifts through ritual behavior and right mindfulness; one early-colonial observer noting that "... when they sail out for whale fishing they dress themselves up in their finest clothes ostensibly because the whale demands respect and no filth will tolerate" (Dalager 1915:56). Whale products were also part of a flourishing exchange economy extending the length of Greenland's west coast. Until the seventeenth century, Inuit hunters traveled hundreds of kilometers by kayak and *umiaq* to exchange baleen from whales in Disko Bay for furs and soapstone from South Greenland.

Since the 1700s, Greenlandic whaling practices have changed dramatically due to dynamic ecological conditions, the introduction of new technologies, 270 years of Danish colonial policies, and expanding linkages to the world economy (Fig. 16.2). In the eighteenth and nineteenth centuries, Greenlanders

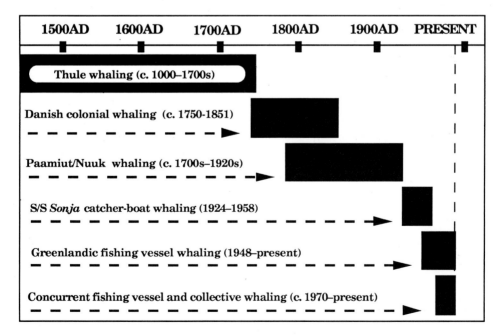

Figure 16.2 Recent Eras in Greenlandic Inuit Whaling.

were employed as whaling crew members by Danish colonial authorities. They gained access to new whaling technology and became increasingly reliant upon goods provided by colonial traders. Yet by the mid- and late-nineteenth century, Greenlandic Inuit whaling had been reduced only to sporadic and isolated catches of larger whales. Humpback whaling persisted in Paamiut and Nuuk into the early twentieth century. However, bowhead stocks were so decimated by Euroamerican whalers that Greenlanders had to shift their efforts to other whale stocks, especially minke and fin whales (Gulløv 1985). As a result, much Inuit knowledge about bowhead whaling was lost, and the spiritual linkages between Inuit hunters and their prey were severely disrupted (Petersen 1986, 1987; Lynge 1990).

This breakdown of the indigenous subsistence whaling regime resulted in major alterations to whaling practices in Greenland. Danish colonial authorities decided to revitalize whaling in Greenland by operating a whaling vessel on behalf of local Greenlanders. From 1924 to 1949, Danish authorities operated the 127-ton catcher boat S/S *Sonja*, which caught fin, blue, sperm, and other large whales. *Sonja*'s Danish crew delivered the whales to local communities where the meat was given to local people in exchange for assistance with flensing. The blubber was shipped to Denmark where it was rendered and sold to offset the ship's operating costs. Between 1950 and 1958 a new vessel, the 250-ton *Sonja Kaligtoq* ('the one that tows [whales]'), was put into service. At first it too delivered whales to local communities, but from 1954 onwards it delivered whales to a shore-based processing plant at Tovqussaq (situated between Nuuk and Maniitsoq). Whale meat processed at the plant was frozen and then shipped to communities along Greenland's west coast (Smidt 1989).

However, Greenlanders themselves revitalized community-based whaling in 1948 when several fishermen began installing small harpoon cannons on their vessels in order to catch minke, fin, and other whales (Kapel 1977, 1979; Fig. 16.3). In the late 1950s and in the 1960s, Greenlandic whaling was largely dominated by a few Greenlandic vessel owners who controlled their own means of production. Meat and *mattak* (whale skin with some blubber attached) from the whales caught were sold in nearby communities. About 1970, fiberglass skiffs and powerful outboard motors became available. Hunters working collectively used these and high-powered rifles to surround, shoot, and then harpoon a minke whale, a practice similar to beluga and narwhal whaling. This collective hunt apparently developed early in the coal-mining community of Qullissat in Disko Bay, where hunters had access to cash for purchasing new technology. Use of this technique enabled Inuit hunters who did not own large fishing vessels to participate in minke whaling. As Kapel (1978) notes:

> the method . . . is in accordance with the collective and co-operative way of life, which was characteristic for the hunting communities, and today needs encouragement and support. In fact, the collective catching could be regarded as a modern version of the traditional Eskimo way of hunting bowheads from umiaks.

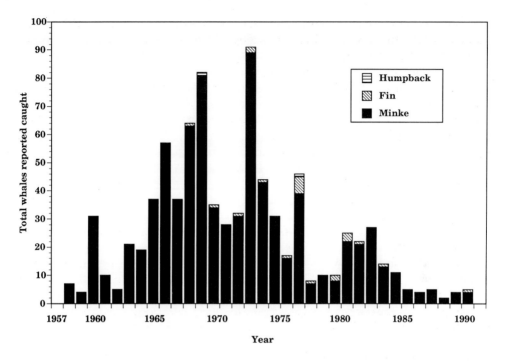

Figure 16.3 Reported Catch of Minke, Fin, and Humpback Whales in Qeqertarsuaq Municipality, 1957 to 1991.

Collective whaling for minkes continues today in West Greenland, concurrent with fishing vessel whaling. However, in part because of criticism of the technique in the IWC, its use is sharply limited by Home Rule government regulations. Currently, only about 25 to 30% of all minke whales are taken by this method (Greenland Home Rule Government 1990a).

QEQERTARSUAQ MUNICIPALITY — A SOCIOECONOMIC PROFILE

Qeqertarsuaq's Whaling History

Whaling was a part of marine-based resource use patterns in Disko Bay long before the arrival of European whalers (Sandgreen 1973). When Dutch and other whalers began frequenting West Greenland in the seventeenth and eighteenth centuries, Qeqertarsuaq became a site of trading activities between Inuit peoples and visiting whalers. The Dutch referred to Qeqertarsuaq's harbor as Liefde Bay, and a Danish missionary visiting the region in 1738 found over 200 people in the vicinity (Sandgreen 1973). Following Denmark's assertion of colonial authority over Greenland in 1721, Danish authorities sought to restrict trade between Greenlanders and others and to initiate their own whaling enterprises. In 1773, the Danes founded the colonial settlement of Godhavn ('good harbor') and, in 1774, began shore-based whaling operations (Gad 1973). Sven Sandgreen, a Swedish-born trader, organized this early colonial whaling for bowheads by hiring local Inuit hunters. However, shore-based whaling produced poor results and faced increasing competition from Dutch, English, and German whaling ships. For example, in 1776 Dutch and German vessels alone caught 132 and 62 bowheads respectively in Disko Bay, while the local shore-based catch in 1777 was only six bowheads (Sandgreen 1973). The Danes strengthened their presence in Disko Bay in 1782 by creating an inspectorate for North Greenland based in Qeqertarsuaq.

Danish colonial whaling in Qeqertarsuaq declined in the early 1800s (Gad 1946). While 20 whales were caught in 1804, the catch declined to 12 in 1816 and to only one or two each year by the 1830s and 1840s (Amdrup et al. 1921; Fisker 1984). Finally, Danish authorities shut down whaling operations in Qeqertarsuaq in 1851, citing economic difficulties (Sveistrup and Dalgaard 1945).

Local Greenlanders continued to catch large whales after 1851 on a sporadic basis, using the meat and *mattak* entirely for local consumption. Oral traditions in Qeqertarsuaq today relate how *Piitarsuaq*, Peter Carl Niels Broberg, a renowned hunter and forebear of families still active in whaling today caught both bowhead and minke whales (Caulfield 1991; Broberg 1945). Born in Qeqertarsuaq in 1825, *Piitarsuaq* obtained a whaling sloop and harpoon cannon from a Scottish whaler whom he assisted in the 1860s. In May of 1864, he led a hunt for two bowheads off Qaqqaliaq (a point of land near Qeqertarsuaq) using this sloop, two *umiat*, and a large number of kayaks (Grønvold 1986). Greenlanders in Qeqertarsuaq continued sporadic whaling into the early 1900s (cf. Bang 1912; Rosendahl 1967).

In the early twentieth century, climate changes throughout West Greenland forced a shift in local economies from marine mammal hunting to cod fishing (Smidt 1989; Vibe 1967). With the introduction of motorboats in Disko Bay in the late 1920s, the use of kayaks began to decline, although not without heated debate. Kayak hunters, objecting to the noise and smell of motorized vessels, would continue to challenge the use of motorboats for hunting through the 1940s and 1950s (Fisker 1984; Rosendahl 1948).

The Greenland Commission's actions in the 1950s to implement a massive modernization program in West Greenland led Qeqertarsuaq to lose its status as an administrative center, and several nearby settlements were abandoned under resettlement policies (Fisker 1984). Qeqertarsuaq's future was itself uncertain, but efforts by local residents in the 1960s to revitalize the economy led to increasing emphasis on shrimp harvesting and processing. In 1962, a shrimp processing ship was stationed in Qeqertarsuaq for the first time by the Royal Greenland Trade (KGH, now Royal Greenland A/S), and in 1966 private investors built a small shrimp processing plant. In 1968, a local Greenlander built a small freezer plant which purchased hunting and fishing products. The plant purchased whale meat and *mattak* for local distribution and also bought salmon for export to Denmark (Berliner 1970). Changes in local fisheries during this period, particularly growth in the shrimp fishery, forced many fishermen to exchange their smaller (20 to 30 foot) multi-purpose vessels for larger vessels (over 40 foot) designed specifically for shrimping.

Whaling was revitalized in Qeqertarsuaq in 1958 when hunters began catching minke whales with fishing vessels equipped with harpoon cannons (Kapel 1978). During the period 1964 to 1977, seven vessels reported minke catches, with the total harvest reaching as high as 89 minke whales in 1973 (Fig. 16.3). The vessels involved typically carried out a multispecies harvest, principally catching shrimp and salmon but taking whales opportunistically. Hunters used minke products both for household consumption or sold them to other individuals. In some cases, minke products were sold to local processing plants or to public institutions such as hospitals in the Disko Bay region.

About 1970, collective hunts for minke whales began in Qeqertarsuaq as skiffs and powerful outboard motors became available. The development of this collective hunt made it possible for many hunters to obtain their own whale meat and *mattak* without having to buy it from fishing vessel owners. During this same period, vessel owners were increasingly involved in the shrimp fishery, which had begun in earnest in the 1950s and had become the major focus of commerical fishing activity in Disko Bay.

Contemporary Mixed Economy in Qeqertarsuaq Municipality

Today Qeqertarsuaq has a mixed subsistence-cash economy based upon commercial shrimping and fishing, seafood processing, public services, and household-based hunting and fishing. This mixed economy is similar that found elsewhere in the North (cf. Asch 1983; Feit 1983; Wenzel 1991; Wolfe and Walker 1987) where household production and reproduction are based upon a mutually-supportive relationship between income from wages and sales of hunting and fishing products on the one hand, and household subsistence production on the other. As is true throughout Greenland, the Home Rule government owns nearly all major infrastructure in the municipality and is the predominant employer (both privately-owned processing plants in the municipality came under Home Rule ownership in the 1980s). The local fishing fleet consists of 23 vessels, four of which are large, privately-owned commercial shrimp trawlers. Most vessels are 20 to 50 feet in length, are family-owned, and are used for multispecies harvests of shrimp, cod, wolffish, halibut, and salmon.

The average household size in the municipality in 1989 was 3.92 persons (Caulfield 1991). Ninety percent of all local households speak *Kalaallisut*, or Greenlandic Inuit, as their principal language (ibid.). Most households own the

Table 16.2 Household Ownership of the Means of Production, Qeqertarsuaq Municipality, 1989 (n = 62)

Means of Production	Percent of Households Owning	Average Number Owned (range)
house/apartment	32	1 (0–1)
skiff	77	2 (0–5)
outboard motor	76	2 (0–4)
sled dogs	73	9 (0–27)
dog sledge	60	2 (0–4)
snowmachine	19	1 (0–2)
car/truck	10	1 (0–1)
fishing cutter	10	1 (0–2)
shrimp trawler	7	1 (0–2)
rifles/shotguns	86	6 (0–20)
fishing or seal nets	66	21 (0–150)
harpoon cannon	5	1 (0–1)
freezer	95	2 (0–4)

(Source: Caulfield, 1991)

means of producing local wild foods for their own consumption (Table 16.2); principally a dog team for winter travel and a skiff with outboard motor for summer use. More than three-quarters of all households surveyed own a skiff with an outboard motor, and nearly the same percentage own sled dogs (the average number owned is nine). Over 80% of all households own rifles or shotguns, with an average of six per household.

Wage employment, particularly in fisheries processing and public service, is a key element in household economic strategies. Fully 87% of all households had at least one wage employee, and the average household had 2.1 persons employed. Public employers provide the great majority of wage jobs in the municipality (Table 16.3), although many of the positions are seasonal or part-time. Figure 16.4, for example, shows the seasonal nature of employment with Royal Greenland A/S (formerly *Kalaallit Tunisassiorfiat* or KTU), the Home Rule-owned shrimp and fish processing plant.

Table 16.3 Full Time and Part Time Wage Employment by Major Public Employers, Qeqertarsuaq Municipality, 1988

Employer	Full Time	Part Time
municipal government	73	36
KTU processing plant	10	246
KNI store	74	0
Nunatek (elec., tele.)	14	40
health service	16	2
police	3	2
church	1	2

(Source: Caulfield, 1991)

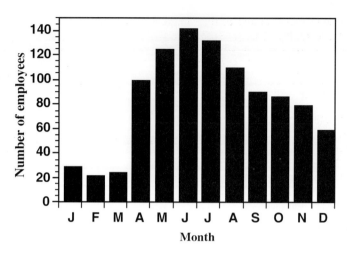

Figure 16.4 Seasonal Nature of Wage Employment with Royal Greenland (formerly KTU) Processing Plant, Qeqertarsuaq, 1989.

Local residents harvest a wide variety of wild resources both for household use and as a source of cash income. Resources providing the greatest economic return are shrimp, cod, wolffish, salmon, beluga, narwhal and seals. In 1989, 39% of all local households received income directly from the sale of wild resources, and 37% of all local residents were classified as either 'full time' or 'part time' hunters and fishers by the Home Rule government (Caulfield 1991). Hunting and fishing products are sold principally through three outlets: (1) the Home Rule owned processing plants, (2) the local *kalaaliaraq*, an outdoor kiosk where unprocessed hunting and fishing products are sold privately, and (3) sales directly to local institutions or other households. Prices paid for locally caught foods are set through negotiations between the hunters and fishers organization and public authorities. Table 16.4 shows selected prices paid for local foods in 1989 (Prices in Table 16.4 and throughout are given in US dollars,

Table 16.4 Selected Prices Paid for Locally-Caught Fish and Wildlife Products at KTU Processing Plant and at *Kalaaliaraq* (outdoor kiosk), Qeqertarsuaq Municipality, 1989 (in U.S. $)

Product	KTU price ($/kg)	Kalaaliaraq Price ($/kg)
seal meat (ringed seal)	.55	6.36
beluga mattak	8.73	14.55
beluga meat (fresh)	2.18	3.64
minke whale meat	1.82	4.00
salmon	4.35	9.09
eider duck	1.82	5.46

(Source: Caulfield, 1991)

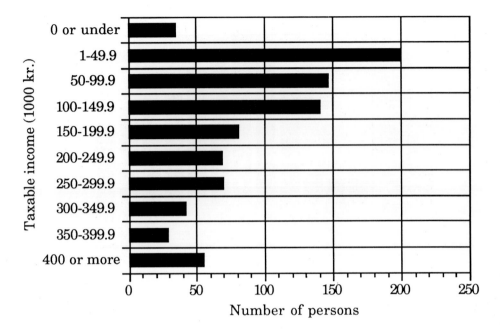

Figure 16.5 Taxable Income in Danish *kroner*, by Income Range, for Residents of Qeqertarsuaq Municipality, 1988.

calculated at an exchange rate of 5.5 Danish kroner per US dollar). Prices shown for sales to KTU are wholesale prices for unprocessed products, while those shown for the *kalaaliaraq* are prices local consumers are charged. When such products are available, KTU processes them and ships them to other communities in Greenland for retail sale. Observations of *kalaaliaraq* sales in Qeqertarsuaq between September 1989 and July 1990 revealed that the most frequently sold foods were fish (26%), seal (22%), beluga or narwhal (14%), and caribou (14%) (Caulfield 1991).

Households in Qeqertarsuaq have some of the highest incomes in all of Greenland because they have access to the shrimp fishery (Greenland Home Rule Government 1991). In 1989, the median taxable income for residents of Qeqertarsuaq Municipality was just over $17 000 (ibid.) and household survey data revealed income ranges from $13 000 to $180 000. Figure 16.5 shows taxable income for residents of Qeqertarsuaq Municipality in 1988, and reveals considerable economic differentiation among households. Households with the highest incomes typically are those owning shrimping and fishing vessels.

Household members in Qeqertarsuaq participate widely in hunting and fishing for local consumption. Fully 90% reported doing so in 1989. Research findings reveal that the average per capita production of wild food by local households in 1989 is 121 kilograms. Nearly three-quarters of all households obtained most or all of their meat or fish from the local environment. The importance of cash to subsistence production is reflected in the high cost of

Table 16.5 Costs of Hunting and Fishing Equipment Typically Used by Household Members to Procure Wild Foods, Qeqertarsuaq Municipality, 1990

Type of equipment	1990 Price (U.S. $)
fiberglass skiff (16′ Pocco 500)	6400.00
outboard motor (40hp Mariner)	4100.00
fuel tanks and hoses (x2)	460.00
shotgun (12 gauge)	450.00
rifle (7.62mm Remington)	675.00
rifle (.222 Sako)	890.00
boat radio, battery, antenna	600.00
plastic floats (x5)	300.00
fish net/arctic char (x2)	160.00
fish net/salmon (x2)	110.00
seal net (x4)	80.00
binoculars	170.00
dog sledge	470.00
dog harnesses and lines (x9)	200.00
dog sledge pad (caribou hide, x2)	65.00
sled dog whip	75.00
harpoon shaft and head (x2)	450.00
ice chisel	29.00
tent	360.00
sleeping bag	130.00
survival kit with flares	110.00
fiberglass skiff (8.5′)	836.00
campstove and tank	38.00
walkie-talkie	282.00
ammunition (12 ga./25 shells, x5)	40.00
ammunition (7.62mm/15 shells, x5)	45.00
ammunition (.222/20 shells, x5)	32.00
TOTAL	$17,557.00

(Source: Caulfield, 1991)

equipment typically used by hunters and fishers. Despite government subsidies of some hunting and fishing equipment, the cost of this equipment if purchased locally can exceed $17 000 (Table 16.5).

CONTEMPORARY MINKE AND FIN WHALING IN QEQERTARSUAQ MUNICIPALITY

Hunters in Qeqertarsuaq Municipality use both fishing vessel and collective whaling techniques. They catch fin whales using fishing vessels equipped with harpoon cannons, and catch minkes using both fishing vessels and collective hunting techniques. Each technique has distinctive technologies, processes, and modes of social organization. Current practices contain elements of both continuity and change, where ancient whaling traditions and knowledge exist side by side with modern technology.

Table 16.6 Characteristics of Older and Newer Style Fishing Vessels Used in Minke and Fin Whaling, Qeqertarsuaq Municipality, 1989

Characteristic	Older Style Vessel	Newer Style Vessel
Year built	1949	1988
Length	37.6 feet	56 feet
Tonnage	19 BRT	46 BRT
Type of hull	wooden	steel
Engine type	69 hp diesel	367 hp diesel
Normal crew	4 persons	5 persons
Cost to present owner	$100,000	$875,000
Est. gross income 1989	$85,000	$550,000
Weeks fishing 1989	12	36+
Weeks whaling 1989	ca. 1	2
Harpoon type	Kongsberg 50mm	Kongsberg 50mm
Principal uses	shrimping, whaling, seal hunting	shrimping, whaling

(Source: Caulfield, 1991)

Contemporary whaling usually takes place between May and October or November, and is highly opportunistic. Participation in whaling in recent years has been widespread, particularly in the collective hunt for minkes where nearly 70% of all households surveyed have participated (Caulfield 1991). Households participating in minke whaling (using both collective and fishing vessels) reported participating an average of one time per year between 1979 and 1989 (ibid.).

Fishing Vessel Whaling for Minke and Fin Whales

In 1989–1990, only two fishing vessels in a fleet of 23 in Qeqertarsuaq were used for catching minke and fin whales (Table 16.6). Both were primarily involved in shrimp trawling, but spent one to two weeks each year catching whales on an opportunistic basis. One vessel is an older style fishing cutter built in 1949 for use in diverse fisheries, and the other is a newer style vessel built in 1988 and designed primarily as a shrimp trawler (Fig. 16.6). The older vessel is owned by a father and son, while the newer is owned by five brothers. In 1989, both types of vessels used a Kongsberg 50mm harpoon cannon with a 'cold' (non-exploding) harpoon. As discussed below, more recent Home Rule regulations now require the use of an exploding penthrite grenade.

Whaling crews on fishing vessels usually number from four to six. Kinship is the most important factor in determining who participates in the hunt. Weather and whaling quotas are the major limiting factors. Typically the time involved in searching for and catching a whale is only a few hours (Caulfield 1991). Once the whale is caught, it is towed to one of several flensing sites near Qeqertarsuaq or Kangerluk. Flensing can take from three to four hours for a minke whale, and as much as six to ten hours for a fin whale. In

Plate 16.1 Older- and Newer-style Fishing Vessels used in Whaling in Qeqertarsuaq.

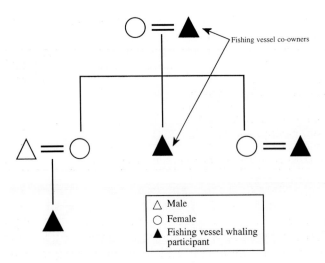

Figure 16.6 Simplified Diagram Showing Kinship Between Fishing Vessel Whaling Participants, Qeqertarsuaq Municipality, 1988.

Qeqertarsuaq, hunters tow the whale at high tide to a rocky point near the harbor where they attach the whale's tail to a hand-operated winch. As the tide falls, flensers use large kitchen knives to remove slabs of meat, blubber, and *mattak*.

The social organization of Greenlanders is typically built upon bilateral kinship systems, with the nuclear family as the most important social unit (Kleivan 1984). This extended family unit continues to be important in fishing vessel whaling today (Fig. 16.6). Whaling crews include both kin members and local elders with considerable whaling experience. Thus, the pattern of social organization of whaling crews differs significantly from that of shrimping, where non-kin members are frequently employed.

Collective Whaling for Minke Whales

In collective whaling for minkes, hunters use skiffs with outboard motors, high-powered rifles, and hand-thrown harpoons. They coordinate their efforts to surround a whale, shoot it with rifles, and then harpoon it. These techniques closely resemble those used for beluga and narwhal (Dahl 1990). Hunters surveyed reported that they typically use 14- to 18-foot fiberglass skiffs with outboard motors of 40 horsepower or more (Table 16.7). The average number of skiffs participating in collective hunts is 16 (range: 5 to 35), with an average of two hunters per skiff. Most hunters use 7.62 mm caliber rifles, and virtually all skiffs have a harpoon, line, and several plastic floats onboard. Walkie-talkie radios are commonly used to communicate during the hunt. The cost of equipment (skiff, outboard, rifles, etc.) used in collective hunting is estimated to be about $12 000 (excluding fuel and ammunition).

Table 16.7 Reported Characteristics of Equipment and Participants in Collective Minke Whaling, Qeqertarsuaq Municipality, 1989 (n = 42)

Characteristic	Description
Most common size of skiff	14 foot
Average number of skiffs participating	16
Average number of hunters participating	30
Average number of hunters per skiff	2
Range of skiffs participating	5 to 35
Most common rifle caliber	.30–06
Average expenses per hunter for fuel and ammunition	$52.91 (292.52 Dkk)

(Source: Caulfield, 1991)

Like fishing vessel whaling, collective whaling is largely opportunistic and success is enhanced by calm winds and seas. Once a minke whale is sighted, hunters communicate by radio about its location. If enough skiffs and hunters are able to participate in the hunt, the whale is pursued. Hunters maneuver their skiffs into position alongside the whale when it surfaces and shoot it with their rifles, aiming for the lungs. Once the whale is slowed by bullets, the hunters hurl a harpoon with line and floats attached into the animal to tire it and to minimize the risk of loss. When the whale is dead, hunters in several skiffs tow it tail-first to the flensing site (typically the same site used in fishing vessel whaling) where it is flensed.

Participation in collective whaling is much more widespread than in fishing vessel whaling because most households own the necessary equipment. As in vessel whaling, kinship is the major factor governing who participates (Fig. 16.7). The cooperative nature of the hunt clearly strengthens kinship networks in the communities.

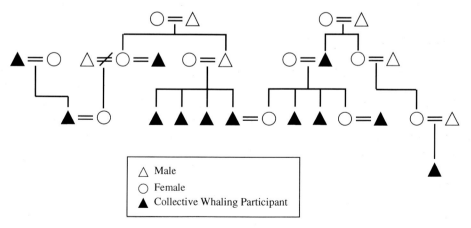

Figure 16.7 Simplified Diagram Showing Kinship Between Collective Whaling Participants, Qeqertarsuaq Municipality, 1988.

Distribution, Exchange, and Utilization of Minke and Fin Whales

Inuit customs determine how minke and fin whales are distributed and utilized within extended families and how they are exchanged in local markets.
While most of the meat, *mattak*, and blubber is used for household consumption, hunters often sell limited amounts for cash. Whale products are distributed in three stages: (1) products are divided among hunters and flensers
themselves, (2) participants in the hunt may distribute their shares to other
households, or sell portions of their shares for cash, and (3) recipient households may share portions with others.

In fishing vessel whaling, the vessel owner(s), crew, and others involved in
flensing share in the first stage of distribution. Typically, 40 to 60% of the catch
is allocated to the vessel in order to cover equipment and fuel costs. For example, a minke whale caught in 1988 and estimated to weigh about 2000 kg was
divided into two shares; about 65% (1300 kg) for the vessel and about 35%
(700 kg) for the owner, his immediate family, and crewmembers, all of whom
were kin-related (Caulfield 1991).

Distribution of whale products in the collective hunt is distinctive because
hunters seek to ensure that all shares are equal. Once the flensing is completed, hunters create equal piles of meat and *mattak* (one for each skiff participating) at the flensing site. Hunters then line up facing two participants
who serve as distributors. One distributor stands with his back to the line of
hunters and points at random to a pile of whale products. The other distributor, who can't see which pile is being pointed to, calls out the name of a
hunter in line who then collects that pile. The distributor calling out the hunter's name has no idea which pile is being pointed to, and thus all hunters are
assured that they are treated equally (see Dahl 1990 regarding use of this technique for distributing shares after beluga hunts).

Prices for minke and fin whale products sold at the KTU processing plant or
at the local *kalaaliaraq* are fixed, generally on an annual basis, through negotiations between the Home Rule government and the national or local hunters
and fishers association (Table 16.8). While avenues exist for selling minke and
fin whale products locally, very few households in Qeqertarsuaq Municipality
sold these products in 1989. Only one household, or two percent of those sur-

Table 16.8 Prices Paid to Hunters (in U.S. $) for Minke Whale Products by KTU
(Disko Laks) and at the *Kalaaliaraq* in Qeqertarsuaq, 1989

Product	KTU Price	Kalaaliaraq Price
minke whale meat (fresh)	1.81/kg	4.00/kg
minke whale *qiporaa* (fluted belly flesh)	2.72/kg	5.46/kg
minke whale *mattak* with blubber	1.14/kg	2.73/kg
minke whale *mattak* without blubber	price not available	9.09/kg
dried minke whale meat (*nikkut*)	7.82/kg	price not available

(Source: Caulfield, 1991)

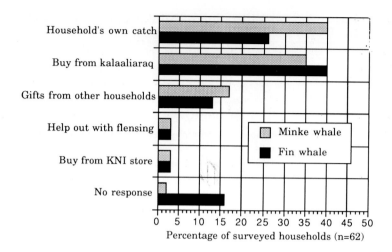

Figure 16.8 Means by Which Households Usually Obtain Minke or Fin Whale Products, Qeqertarsuaq Municipality.

veyed, reported doing so (Caulfield 1991). In that case, the household reported receiving $1 270 for minke meat and *mattak* at the local *kalaaliaraq*. Furthermore, the manager at KTU in Qeqertarsuaq reported that no such products have been purchased by KTU in recent years (N. Bjerregård, pers. comm. 1989). He attributed this to IWC quotas, which sharply limited the number of whales available.

Minke and fin whale meat, *mattak*, and *qiporaq* (ventral grooves on the whale's underside) are highly desired foods in most Greenlandic households. Families prepare whale meat in stews, fried it in butter, or eat it dried (*nikkut*). Greenlanders prefer to eat *mattak* raw or boiled. Household survey data reveals that minke and fin whale products are widely used. Fully 97% of all households use minke whale products, and 73% use fin whale products. Most obtain minke products by participating in hunting and flensing themselves (40%), while 35% usually purchase their products from the *kalaaliaraq* (Fig. 16.8). More households purchase fin whale products because few own a fishing vessel with harpoon cannon.

Hunters also use whale products for sled dog food. Twenty-seven percent of all households reported using minke whale meat for this purpose, while 22% reported using fin whale in this manner. Most households reported a decline in their use of whale meat for sled dog food due to the lower IWC quotas, especially for minke whales. However, hunters stress the importance of obtaining local foods for their sled dog teams. Unlike Inuit in Alaska and Canada, Greenlandic hunters generally do not use snowmachines because they fear that the noise and smell will affect hunting success. In fact, it is illegal to use a snowmachine for hunting on the sea ice in Qeqertarsuaq Municipality (Grønlands Landsting 1987).

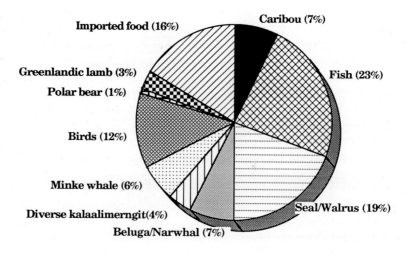

Figure 16.9 Proportion of Meat and Fish Consumed, by Type, for Five Selected Households in Qeqertarsuaq, October 1989–July 1990.

NUTRITIONAL AND SOCIOCULTURAL SIGNIFICANCE OF MINKE AND FIN WHALING

Marine mammals, including whales, contribute to a high calorie diet which is desireable for outdoor activities in an Arctic climate (Helms 1983). Figure 16.9 shows the proportion of wild meats and fish consumed by five selected households in Qeqertarsuaq between October 1989 and July 1990. The largest category of wild food consumed (number of meals where food was present) was fish (23%), followed by seal and walrus (19%), imported meats and fish (16%), birds (12%), beluga and narwhal (7%), caribou (7%), minke whale meat/*mattak* (6%) and other meat/fish (8%).

Greenlanders refer to locally-obtained wild foods as *kalaalimerngit* in the West Greenlandic Inuit language. They differentiate between these foods and *qallunaamerngit*, or "white man's foods." *Kalaalimerngit* comprise a substantial part of household diets, and are integrally linked to Greenlandic identity. As Larsen and Hansen (1990) point out, the distinction between *kalaalimerngit* and other foods is more than simply the origin of the food. In their words, "eating Greenlandic food is of great symbolic weight in determining whether a person is a true Greenlander . . ."(16). Greenlanders emphasize their desire to eat local wild foods for both nutritional and cultural reasons:

> We don't want to eat European food. When we eat European food, we don't feel full. Greenland is a cold place. When we eat European food, we get cold after one or two hours riding in a boat or travelling with a dog team. *Kalaalimerngit* is what we want to eat . . . especially for the old people and for the children, when they get sick. If they eat Greenlandic foods, it's better for them. (author's field notes, 18 February 1990, Qeqertarsuaq).

The procurement, processing, and sharing of *kalaalimerngit* reflects the under-lying systems of reciprocity and community solidarity which continue to be important in Greenlandic life today. While whaling festivals of the type held by Inuit in Alaska do not exist in Greenland, whale products are highly valued during household and community celebrations. These products are often served at a *kaffemik*, a special family celebration held to commemorate birth-days, anniversaries, baptisms, or confirmations, and during community cele-brations like those held on Greenland's National Day (June 21).

Petersen (1989) describes the changing significance of sharing wild foods in Greenlandic communities. Until recently, local residents participated both in generalized sharing and in bartering between households and communities. Sharing thus fostered community solidarity and provided insurance against difficult times. In the twentieth century, these generalized patterns of exchange have declined somewhat, and meat-gifts are increasingly restricted to relatives and close neighbors (Kleivan 1984).

Despite these changes, residents of Qeqertarsuaq Municipality continue to share wild foods with other households. Figure 16.10 shows that 50% of all households surveyed often or always share wild foods with others, while 22% occasionally or rarely do so. Twenty-eight percent do not share at all. For those households which do share (n = 44), virtually all (98%) do so with immediate family members, while 78% also do so with extended family members. How-ever, respondents report that the sharing network is changing over time. Sev-enty-six percent of all households reported that sharing has declined over the past 20 years. The major reason cited was that more households sell hunting and fishing products today in order to obtain cash necessary for equipment and household expenses.

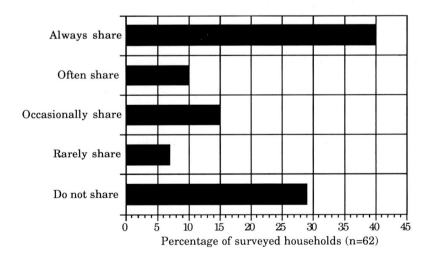

Figure 16.10 Frequency With Which Households Share Wild foods with Others, Qeqertarsuaq Municipality.

In Qeqertarsuaq, sharing practices today are generally limited to members of one's immediate or close extended family. Gifts of meat or fish (called *pajugat*) may be shared with those who household members like or who have helped them in some way (*qujagisaqarneq*). Household members may also give meat to those with whom they have a name relationship (*atsiaqarneq*) (Langgaard 1986). Meat-gifts may be provided to close family members (*ilaquttat*) such as parents, grandparents, or cousins. Gifts may also be given to those who are unable to hunt for themselves (*pilersuissoqanngitsut*) or to those lacking the means to hunt or fish for themselves (*piniuteqanngitsut*). Furthermore, the type of meat given may depend upon the sex of the recipient. For example, boys are typically given the front flipper and claws of a seal in order to give them strength in hunting. Gifts are given without consideration of the material wealth of the recipient. Thus, the owner of a shrimp trawler and the elderly widow both participate fully in contemporary sharing networks.

As Dahl (1989) points out, contemporary marine mammal hunting in Greenland serves complex integrative and cultural functions. The social organization of whaling remains closely tied to kinship relationships between hunters and their families. When hunters speak excitedly over a walkie-talkie about sighting a whale, they share that knowledge with a discrete group because of shared language. Whaling involves sharing of knowledge, values and beliefs about the relationship between animals and hunters. Elders demonstrate culturally appropriate behavior to younger hunters as they work alongside each other in whaling activities. Participation in whaling remains a source of prestige which validates the role of hunting in Greenlandic society. Greenlandic men speak with barely muted pride as they describe their participation in catching a whale. Marine mammal hunting provides Greenlanders with a sense of collective security. For those with limited or seasonal incomes, hunting provides food for the table. As local hunters point out, it is like 'money in the bank'.

Whaling fosters a strong sense of community identity. Use of symbols associated with whaling, such as the presence of a bowhead whale on Qeqertarsuaq Municipality's ceremonial shield, reinforce this. Increasingly, whaling also contributes to a sense of Greenlandic national identity. Greenlanders believe their harvest of marine mammals to be a fundamental human and cultural right, and feel a sense of solidarity with others who share this view. This growing awareness has also led to closer cooperation with fellow Inuit in Canada, Alaska and Chukotka through the Inuit Circumpolar Conference (ICC) and with other nations in the North Atlantic.

GREENLAND, ABORIGINAL WHALING, AND THE IWC

While the IWC clearly recognizes Greenland's minke and fin whale catches as aboriginal subsistence whaling, critics at the international level express fears

that profit maximization, commoditization, and capital intensification could develop in the whaling regime (Lynge 1990). Debate has also focused on infractions of IWC whaling quotas by some Greenlandic hunters, particularly during
the mid-1980s. Greenlanders, in contrast, have voiced strong dissatisfaction
with minke whale quota reductions first imposed in 1985. While more recent
quotas have been increased somewhat, Greenlanders argue that quotas are still
far from providing the estimated 670 tons of whale products needed each year
(Anon. 1988).

Profit Maximization and Commoditization in Whaling

This research demonstrates that neither fishing vessel nor collective whaling in
Qeqertarsuaq Municipality is carried out primarily on a profit-maximizing basis. Fishermen use their vessels largely for shrimp trawling, and spend only
one or two weeks whaling out of about five months of activity. Whaling is
economically marginal for vessel owners, with incomes from whaling in recent
years amounting to no more than 10% of total gross. For example, the older
style trawler described above had a gross income of about $54 000 in 1988,
only $4000 of which came from sales of whale products. The newer style
trawler grossed about $550 000 in 1989, and whale accounted for only about
$5500. Expenses associated with whaling for this vessel were estimated to be
about $900, providing a net return of about $4600.

   Cash plays a smaller role in collective whaling. While customary distribution
practices often leave each hunter with only small amounts of whale, participation can be beneficial. As an example, in a typical hunt with 16 skiffs and 30
hunters participants might divide a minke whale weighing 2000 kg into equal
shares of 60 kg. At current prices, each share would be worth about $340.
Thus even with operating costs of about $50 per skiff, participation in the hunt
clearly brings a positive economic return.

   While avenues exist in Qeqertarsuaq Municipality for selling minke and fin
whale, hunters have rarely used them in recent years. Even in the 1960s and
1970s when more whale products were sold, cultural factors restrained any interest in maximizing profits. Hunters were reluctant to treat whales as a simple commodity because of the prestige involved in carrying out all aspects of
whaling. Hunters refused to develop specialized hunting and processing systems, choosing instead to remain personally involved in all aspects of the
hunt.

Capital Intensification and New Technology in Whaling

Vessel owners who also whale have kept pace with the demands of the shrimp
and fishing industry by improving their vessels' technology and efficiency
(Berthelsen et al. 1989). However, whaling technology used today is virtually
the same as that used 40 years ago. Interestingly, the only significant change in

technology has been made to meet international requirements for the humane killing of whales. As of 1991, the Home Rule government required all fishing vessel whalers to use the Norwegian-designed penthrite grenade, the so-called "hot harpoon." Vessel owners in Qeqertarsuaq seem favorably inclined toward this development, even though additional cost and specialized training are required.

Greenlandic Whaling and External Regulatory Regimes

The Greenland Parliament (in Greenlandic, *Naalakkersuisut*) enacts laws to regulate whaling practices, and the Home Rule administration establishes regulations to carry out these laws. According to Home Rule regulations, hunters involved in minke or fin whaling must (1) have a full time hunting license, (2) reside in Greenland, and (3) have a "close affiliation" with Greenlandic society. Hunters must also have a special whaling permit, issued for each whale taken. Furthermore, hunters using fishing vessels must use a 50 mm harpoon cannon capable of firing the penthrite grenade. Regulations also specify that whales with young may not be caught, and that only the larger and more mature fin whales may be taken.

Hunters wishing to participate in a collective hunt must first receive special dispensation from the Home Rule government before a hunt can take place (Greenland Home Rule Government 1990b). Typically, requests for dispensation are submitted to the Home Rule offices in Nuuk by municipal authorities. This is given if the authorities determine that the hunt has major significance for the local community, and if there are insufficient quantities of whale products from fishing vessel whaling. Home Rule regulations also require that a minimum of five skiffs participate, that all skiffs have a harpoon and floats on board, and that rifles of 7.62 mm caliber or larger be used.

The Home Rule government allocates IWC quotas for minke and fin whales to local municipalities in consultation with the national hunters and fishers association (KNAPK), and the nationwide municipal government organization (KANUKOKA). In making the allocations, the Home Rule considers population size, availability of suitable vessels with harpoon cannons, and the availability of economic alternatives. Once municipalities receive their annual quota, they allocate specific whale quotas to vessels and collective hunters after consultation with the local hunters and fishers association. Local authorities have become increasingly strict in monitoring whaling in recent years.

DISCUSSION AND CONCLUSIONS

Greenlandic Inuit whaling in Qeqertarsuaq Municipality has undergone dramatic transformation because of ecological dynamics, colonial policies, and growing interaction with the world economy. Despite this transformation, whaling remains closely linked to Inuit customs and traditions of great depth. Contemporary whaling practices in Greenland include the sale of whale products for local consumption. However, the data reveal that this exchange is limited and not carried out to maximize profits. Instead, the purpose is to obtain cash necessary for continuation of hunting pursuits and to sustain flows of hunting products to other Greenlanders who are unable to obtain their own whale products. Market forces do not dominate contemporary whaling practices. Rather, as Freeman (1993) notes, the goal is to sustain local social, cultural and economic activity intergenerationally, a goal which encompasses changing technologies designed to improve efficiency and safety.

Greenland's whaling management regime remains in a state of flux, and is subject to both external and internal forces. Within the IWC, scientific uncertainty about the status of whale stocks and whale population dynamics makes management decisions difficult. Debates about "humane" killing of whales and the IWC's jurisdiction over small cetaceans continue to dominate IWC meetings. As improved data about the status of whale stocks becomes available, the political debate could increasingly turn more to culturally-based arguments about the ethics and appropriateness of whaling. Anticipating this shift, the governments of Greenland, Iceland, the Faroe Islands and Norway have formed a North Atlantic Marine Mammal Commission (NAMMCO), which may well serve as a forum for developing regionally-based research and management strategies for marine mammals (Anon. 1992; Hoel 1991).

Within Greenlandic society, whaling regulation continues to be a source of some controversy. In recent years, increasingly strict IWC quotas have reduced significantly the number of whales caught, placed a strain on culturally based hunting practices, and fostered increasing alienation of hunters from whaling management regimes. For example, hunters in Qeqertarsuaq believe that quotas are far too low to meet household requirements. In 1989, 86% of households surveyed reported that they could not obtain enough minke whale products, citing IWC quotas as the primary reason. Furthermore, Home Rule regulations limiting whaling only to full time hunters exclude many who have participated in whaling in the past, particularly in the collective hunt. Many part time wage-earners are also active hunters who would like to participate. The effect of these regulations has been to deepen social differentiation in local communities and to undermine egalitarian hunting practices. Recent quota infractions in some municipalities may be a reflection of the alienation that many hunters feel toward these quotas and regulations. Modest increases in IWC quotas for minke whales enacted in 1991 may reduce these tensions somewhat, but many Greenlandic hunters continue to believe that decisions about whaling made at the international level are motivated more by political than biological considerations.

NOTE

1. I am grateful for the approval and support given this research by residents of
   Qeqertarsuaq Municipality, West Greenland. In particular, I would like to thank
   Borgmester Jens Johan Broberg, members of the *kommunalbestyrelse*, and the leadership of KNAPP, the local hunters and fishers association. A special note of apprecia-
   tion goes to Rev. Lars Pele Berthelsen and his family. *Ilissinnut tamassi, qujanarsuaq.*
   This article is based upon a more extensive report about whaling in Qeqertarsuaq
   Municipality presented to the IWC in 1991 (Caulfield 1991). The research was ap-
   proved by the Greenland Home Rule Government and the Commission for Scien-
   tific Research in Greenland. Funding was provided in part by the Greenland Home
   Rule Government, the Wenner-Gren Foundation for Anthropological Research, and
   the Sir Phillip Reckitt Foundation. Special thanks to the Hon. K. Egede, B. Rosing,
   A. Jessen, Dr. H. Thing, E. Lemche, J. Paulsen, J. Jervin, Professor R. Petersen, H.C.
   Petersen, A. Jakobsen, Dr. O. Bennike, P. Bennike, F. Lynge, and professors J. Dahl,
   I. Kleivan, M. Fortescue, P. Blaikie and N. Abel. Drs. F. Larsen and F. Kapel at the
   Greenland Fisheries Research Institute in Copenhagen, and Drs. R. Gambell and G.
   Donovan at the International Whaling Commission in Cambridge, England pro-
   vided valuable information regarding whale stocks in the North Atlantic Ocean. S.
   Mitchell at the University of Alaska Fairbanks and two anonymous reviewers pro-
   vided valuable editorial suggestions.

REFERENCES

Amdrup, G.C., L. Bobe, A.S. Jensen and H.P. Steensby, 1921.
   Grønland i Tohundredaaret for Hans Egedes Landing. *Meddelelser om Grønland*
   60–61:272–341.
Anon., 1955–1963.
   Almindelig beretning. Ministeriet for Grønland.
Anon., 1991.
   Flere vågehvaler til Grønland. Sermitsiak' 22 (31 May 1991):3. Nuuk.
Anon., 1992.
   Miluumasut imarmiut pillugit suleqatigiinneq annertusarneqarpoq. Sermitsiak' 16
   (15 April 1992):7. Nuuk.
Asch, M, 1983.
   Dene Self-Determination and the Study of Hunter-Gatherers in the Modern World.
   In: E. Leacock and R. Lee,. (eds) *Politics and History in Band Societies.* Cambridge
   University Press, Cambridge.
Association of Canadian Universities for Northern Studies, 1982.
   Ethical Principles for Northern Research. Ottawa: ACUNS, Ottawa.
Bang, H.V., 1912.
   Om Rationel Hvalfangst i Grønland. *Det Grønlandske Selskabs Aarsskrifter* 1912:19–41.
Berliner, F., 1970.
   Vestgrønland: Fra Narssarssuak til Upernavik. Carit Andersens Forlag, Kobenhavn.
Berthelsen, C., I. Mortensen and E. Mortensen, 1989.
   Kalaallit Nunaat/Grønland Atlas. Pilersuiffik, Nuuk.

Broberg, A., 1945.
K'ekertarssuarme piniartorssusimasok. *Avangnamiok'* 1:3–5.

Caulfield, R., 1991.
Qeqertarsuarmi arfanniarneq: Greenlandic Inuit Whaling in Qeqertarsuaq Kommune, West Greenland. Document TC/43/AS4. International Whaling Commission, Cambridge.

Dahl, J., 1987.
Tradition og Kultur i Den Grønlandske Naturudnyttelse. *Grønland* 35(10):295–304.

Dahl, J., 1989.
The Integrative and Cultural Role of Hunting and Subsistence in Greenland. *Etudes/Inuit/Studies*, 13(1):23–42.

Dahl, J., 1990.
Hvidhvalerne i Saqqaq. *Naturens Verden* 1990:149–160.

Dalager, L., 1915.
Grønlandske Relationer indeholdende Grønlændernes Liv og Levnet, Deres Skikke og Vedtægter samt Temperament og Superstitioner tillige nogle Korte Reflectioner over Missionen Sammenskrevet ved Friderichshaabs Colonie i Grønland Anno 1752 af Lars Dalager Kjøbman. Det Grønlandske Selskabs Skrifter 2.

Donovan, G. (ed), 1982.
Aboriginal Subsistence Whaling. International Whaling Commission. Special Issue 4, Cambridge.

Feit, H.A., 1983.
The Future of Hunters within Nation-States: Anthropology and the James Bay Cree. In: E. Leacock and R. Lee (eds), *Politics and History in Band Societies.* pp. 363–411. Cambridge University Press, Cambridge.

Fisker, J., 1984.
K'ek'ertarssuak'/Godhavn. Nordiske Landes Bogforlag, Uummannaq.

Freeman, M.M.R., 1990.
A Commentary on Political Issues with regard to Contemporary Whaling. *North Atlantic Studies* 2(1–2):106–116.

Freeman, M.M.R., 1993.
Introduction: Community-based Whaling in the North. *Arctic* 46(2):iii–iv.

Gad, F., 1946.
Grønlands Historie; En Oversigt fra ca. 1500 til 1945. *Det Grønlandske Selskabs Skrifter*, 14 (1946).

Gad, F., 1973.
The History of Greenland: II, 1700–1782. C. Hurst & Co., London.

Gambell, R., 1993.
International Management of Whales and Whaling: An Historical Review of the Regulation of Commercial and Aboriginal Subsistence Whaling. *Arctic* 46(2):97–107.

Glahn, H.C., 1921.
Dagbøger 1767–68. *Det Grønlandske Selskabs Skrifter* 4 (1921).

Greenland Home Rule Government, 1988.
Our Way of Whaling — Arfanniariaaserput. Nuuk.

Greenland Home Rule Government, 1989.
Greenland subsistence hunting. Document TC/41/22. International Whaling Commission, Cambridge.

Greenland Home Rule Government, 1990a.
   "Notat . . . vedr. dispensationsansøgning om riffelfangst af hvaler 1988/89." Unpubl.
   memorandum, Erhvervsdirektoratet, Box 269, DK-3900 Nuuk. 3 August 1990.
Greenland Home Rule Government, 1990b.
   Hjemmestyrets bekendtgørelse nr. 49 af 18. december 1990 om fangst af store hvaler.
   Namminersornerullutik Oqartussat, Nuuk.
Greenland Home Rule Government, 1991.
   Grønland/Kalaallit Nunaat, 1990 Statistisk Årbog. Atuakkiorfik, Nuuk.
Grønsland Landsting, 1987.
   Qamutit savequtillit pillugit Qeqertarsuarsuup kommuniani ileqqoreqqusaq/Vedtægt
   for Godhavn kommune om kørsel med snescootere, 9. november 1987. In:
   Nalunaerutit–Grønlandsk Lovsamling, Serie B. Nuuk.
Grønnow, B. and M. Meldgaard, 1988.
   Boplads i Dybfrost: Fra Christianshåb Museums Udgravninger på Vestgrønlands
   Ældste Boplads. *Naturens Verden* 1988:409–440.
Grønvold, H.P., 1986.
   Qeqertarsuarmiu arfanniaq. [the whaler from Qeqertarsuaq]. Unpubl. manuscript in
   possession of the author.
Gulløv, H.C., 1985.
   The Impact of European Whaling on Eskimo Society in West Greenland. In: W.W.
   Fitzhugh (ed) *Cultures in Contact: The Impact of European Contacts on Native American
   Institutions, A.D. 1000–1800.* Smithsonian Institution Press, Washington, D.C.
Helms, P., 1983.
   Nutritional Needs relating to Aboriginal/Subsistence Whaling among the Inuit in
   Greenland. Document TC/36/AS/2. International Whaling Commission, Cambridge.
Helms, P., O. Hertz and F. Kapel, 1984.
   The Greenland Aboriginal Whale Hunt. Document TC/36/AS2. International Whal-
   ing Commission, Cambridge.
Hoel, A.H., 1993.
   Regionalization of International Whale Management: The Case of the North Atlantic
   Marine Mammals Commission. *Arctic* 46(2):116–123.
(IWC), 1983-91.
   International Convention for the Regulation of Whaling, 1946; Schedule. Cambridge.
(IWC), 1985.
   Definition of Aboriginal Subsistence Whaling Proposed by India. Document IWC/
   37/22. Cambridge.
(IWC), 1988.
   Aboriginal Subsistence Working Group — Danish Statement. Document TC/40/AS3.
   International Whaling Commission, Cambridge.
Josefsen, E., 1990.
   Cutter Harvests of Minke Whales in Qaqortoq (Greenland). Document TC/42/
   SEST5. International Whaling Commission, Cambridge.
Kapel, F.O., 1977.
   Catch Statistics for Minke Whales, West Greenland, 1954-74. *Reports of the Interna-
   tional Whaling Commision* 27:456–459.
Kapel, F.O., 1978.
   Catch of minke whales by fishing vessels in West Greenland. *Reports of the Interna-
   tional Whaling Commission* 28:217–226.

Kapel, F.O., 1979.
Exploitation of Large Whales in West Greenland in the Twentieth Century. *Reports of the International Whaling Commission* 29:197–214.

Kapel, F.O. and R. Petersen, 1984.
Subsistence Hunting — The Greenland Case. *Reports of the International Whaling Commission,* Special Issues 4:51–73.

Kleivan, I., 1984.
West Greenland Before 1950. In: D. Damas (ed) *Handbook of North American Indians: Arctic.* pp. 595–621. Smithsonian Institution, Washington, D.C.

Langgaard, P., 1986.
Modernization and Tradtional Interpersonal Relations in a Small Greenlandic Community: A Case Study from Southern Greenland. *Arctic Anthropology* 23(1 and 2): 299–314.

Larsen, S.E. and K.G. Hansen, 1990.
Inuit and Whales at Sarfaq (Greenland): Case Study. Document TC/42/SEST4. International Whaling Commission, Cambridge.

Lynge, F., 1990.
Kampen om de vilde dyr — en arktisk vinkel. Akademisk Forlag, København.

Møller, P. and S. Dybroe, 1981.
Fanger/Fisker-Lønarbejder? En Undersøgelse af Sammenhaengene Mellem Fangst og Lønarbejde i Godhavn, 1976 og 1977. Institute for Forhistorisk Arkaeologi, Middelalder-Arkaeologi, Etnografi og Socialantropologi, Arhus Universitet, Arhus.

Nuttal, M.A., 1992.
Arctic Homeland: Kinship, Community and Development in Northwest Greenland. Belhaven Press, London.

Petersen, H.C., 1986.
Skinboats of Greenland. The National Museum of Denmark, the Museum of Greenland and the Viking Ship Museum in Roskilde, Roskilde.

Petersen, H.C., 1989.
Traditional and Present Distribution Channels in Subsistence Hunting in Greenland. In: Greenland Subsistence Hunting. Document TC/41/22. International Whaling Commission, Cambridge.

Petersen, R., E. Lemche and F.O. Kapel, 1981.
Subsistence Whaling in Greenland. Document TC/33/WG/S3. International Whaling Commission, Cambridge.

Rosendahl, P., 1948.
Grønlandske dyrefredningsproblemer. *Jagtvennen* 24:225–230.

Rosendahl, P., 1967.
Jakob Danielsen, a Greenlandic Painter. Rhodos, Copenhagen.

Sandgreen, O., 1973.
Kekertarssuak'. Kalatdlit Nunane Nakiterisitsissarfik, Nuuk.

Smidt, E.L.B., 1989.
Min tid i Grønland; Grønland i min tid. Nyt Nordisk Forlag-Arnold Busck, København.

Sonne, B., 1986.
Toornaarsuk, an Historical Proteus. *Arctic Anthropology* 23(1–2):199–219.

Statsministeriet, S., 1977–81.
Grønland/Kalaallit Nunaat Arbog. København.

Sveistrup, P.P. and S. Dalgaard, 1945.
   Det Danske Styre af Grønland, 1825–50. *Meddelelser om Grønland* 145(10).
Usher, P., 1981.
   Sustenance or Recreation? The Future of Native Wildlife Harvesting in Northern
   Canada. In: M.M.R. Freeman, (ed), *Proceedings, First International Symposium on Re-
   newable Resources and the Economy of the North*, pp. 56–71. Association of Canadian
   Universities for Northern Studies, Ottawa.
Vibe, C., 1967.
   Arctic Animals in Relation to Climatic Fluctuations. *Meddelelser om Grønland* 170(5).
Wenzel, G., 1991.
   Animal Rights, Human Rights: Ecology, Economy and Ideology in the Canadian Arctic.
   Belhaven Press, London.
Wolfe, R.J. and R.J. Waler, 1987.
   Subsistence Economies in Alaska: Productivity, Geography and Development Im-
   pacts. *Arctic Anthropology* 24(2):56–81.
Woodbury, A.C., 1984.
   Eskimo and Aleut languages. In: D. Damas (ed), *Handbook of North American Indians:
   Arctic*, pp. 89–63. Smithsonian Institution, Washington, D.C.

17. INTERNATIONAL ATTITUDES TO WHALES, WHALING AND THE USE OF WHALE PRODUCTS: A SIX-COUNTRY SURVEY

MILTON M.R. FREEMAN and STEPHEN R. KELLERT

INTRODUCTION

This study has been designed to determine how people in six countries view issues related to the management of whales and whaling. A further goal of the research is to assess how informed people in selected countries are about whales and whaling.

A questionnaire (Appendix 1) was designed by the authors and administered to a representative random sample of about 500 adults in each of the following countries: Australia, England, Germany, Japan, and Norway. A larger-sized sample, numbering about 1000 adults was questioned in the United States. The actual sample sizes and the selected characteristics used to describe these national samples and the level of accuracy in polling results from these representative samples are shown in Appendix 2.

Survey companies in these six countries were selected by Gallup Canada, which administered the technical aspects of the project. In nearly all cases the polling companies selected were affiliated with the Gallup organization. Letters were sent to the Whaling Commissioners in each of the 36 member governments of the International Whaling Commission, inviting those governments' financial support and advice in the design and conduct of the study. Thirteen replies to these letters were received, and subsequent correspondence with those who indicated interest in the study resulted in most of the funding being obtained from these sources. No government agency requested changes be made to the preliminary survey questionnaire provided by the researchers. Additional funds needed to complete the study were obtained from a number of sources (see Note 1).

DESIGN OF THE SURVEY

The purpose of the study was to determine public attitudes toward and knowledge concerning the conservation of whales and the management of whaling. Six questions were asked in order to determine peoples' views about whaling (Appendix 3). A further fifteen questions were asked to determine the importance of various policy goals likely to be considered important in whaling management (Appendices 4 and 5).

At the present time, the main purpose for hunting whales is to obtain human food. Therefore, to provide one particular context for understanding prevailing attitudes toward killing animals for food, nine questions asked people the degree to which they approved or disapproved of the production and sale of a number of domesticated and wild animals (including whales) as food (Appendix 6). Additional analytical context is provided by ten questions that seek to assess how much factual information people possess about whales, whaling and the use of whale products (Appendices 7 and 8).

For most questions respondents were asked to answer on a scale of 1 to 5, where 1 indicated their strong disapproval/disagreement, and 5 their strong approval/agreement with a particular statement; a response rated 3 indicates the respondent held no strong opinion for or against the proposition. Respondents could also answer that they held no opinion or did not know the answer. Some of the knowledge-testing questions asked respondents if they believed the statement was either "true" or "false". Respondents could also answer "don't know" to these particular knowledge-testing questions.

RESULTS

In discussing the results of this survey, an "approve"/"agree" response represents the summing of responses obtained in the "strongly" or "moderately" approve/agree categories (i.e. 1 + 2 on the five-point scale). In similar manner, the rates obtained for "moderately" or "strongly" disapprove/disagree (4 + 5) responses are summed to obtain the "disapprove"/"disagree" response rate (expressed as a percentage).

Questions Concerning the Acceptability of Whaling

The results presented in Table 17.1 indicate that respondents in Australia, England, Germany and the U.S. held opinions markedly different from those expressed by Japanese and Norwegians when each was questioned about whaling. For example, when asked whether they "opposed the hunting of whales under any circumstances" a majority of respondents in Australia (60%) and Germany (59%) agreed. However, approximately equal-sized majorities in Norway (61%) and Japan (57%) disagreed with the statement that whales should not be hunted under any circumstances. Opposition to whaling under any circumstances was more moderated in the U.S. where 48% opposed whaling, and even more evenly divided in England with 43% opposed to whaling, 37% not opposed, and a further 19% expressing no strong opinion one way or the other.

However, in a related questions ("I can't imagine why anyone would want to kill anything as intelligent as a whale") respondents in England followed the same trend as those in Australia, Germany and the U.S. in opposing the

Table 17.1 Public Attitudes Toward Whaling, Percent Response Rate by Country (from Appendix 3)

		Australia	England	Germany	Japan	Norway	U.S.A.
Can't imagine why anyone would kill intelligent whales	Agree	64	64	59	25	22	57
	No particular opinion	13	13	18	24	16	20
	Disagree	22	20	24	50	57	25
Opposed to hunting of whales under any circumstances	Agree	60	43	54	24	21	48
	No particular opinion	11	19	18	18	18	17
	Disagree	29	37	27	57	61	34
There is nothing wrong with whaling if it is properly regulated	Agree	21	19	26	64	74	27
	No particular opinion	13	17	14	13	9	17
	Disagree	66	64	60	22	15	55

killing of whales. Again, clear majorities in Japan and Norway disagreed with that particular statement. In response to the statement that "there is nothing wrong with whaling if it is properly regulated", about two-thirds of respondents in Australia and England disagreed, whereas between two-thirds and three-quarters of Japanese and Norwegians respectively agreed that regulated whaling was an acceptable practice. The U.S. position (55% disagreeing with the statement) appeared intermediate between these extremes (Table 17.1).

Policy Questions to be Addressed by a Whaling Authority

The answers to these questions showed varying priorities in different countries for fifteen listed policy objectives (Appendices 4 and 5). The highest policy priority for respondents in Australia, England, Germany, Norway and the U.S. was that the most humane methods of killing be utilized and that strict international controls be put into place. In Japan respondents placed highest priority on the sustainability of the whale fishery and minimizing wasteful practices. Respondents in all six countries indicated high levels of support for the requirement that harvests should be based upon the best scientific advice.

In further questions about broad areas of policy to be followed in future management initiatives, all respondents placed protection of the whales' environment (against pollution or industrial disturbance) as the highest goal. There was also high priority accorded in each country to the importance of managing whales in the context of marine ecosystem considerations. In each country, strictly economic goals (i.e. the profitability of the whaling industry) ranked as the lowest policy priority.

Policies in support of social and cultural goals (i.e. maintaining the wellbeing of traditional whaling communities) were afforded relatively high priority in both Japan and Norway compared to the opinions expressed in Australia,

Table 17.2 Ranking of Countries According to Support for Selected Whaling Policy Options
(**Scale: 5** = strongly support, **3** = neither support nor oppose, **1** = strongly oppose)

	Most Supportive					*Least Supportive*
Whaling policy should support cultural goals (traditional whaling communities' way of life)	Japan (3.71)	United States (3.42)	Norway (3.35)	England (3.31)	Australia (3.12)	Germany (3.01)
Whaling policy should support social goals (jobs and local residents' wellbeing in traditional whaling communities)	Norway (3.62)	Japan (3.61)	United States (3.14)	Germany (3.04)	England (2.87)	Australia (2.80)
If whales species are not endangered, economic and cultural needs of traditional whalers justifies continued hunting	Japan (3.92)	Norway (3.70)	United States (2.79)	Australia (2.63)	England (2.44)	???
If whales become plentiful again, harvesting them for useful products is acceptable	Norway (4.04)	Japan (3.87)	United States (2.59)	England (2.44)	Germany (2.31)	Australia (2.19)
Non-endangered whales can be killed to provide food for humans	Norway (4.14)	Japan (3.40)	United States (2.44)	Germany (2.41)	Australia (2.10)	England (2.00)
Nothing wrong with harvesting whales if properly regulated	Norway (4.10)	Japan (3.79)	United States (2.46)	Germany (???)	Australia (2.15)	England (2.11)

England, Germany, and the U.S. However, in respect to maintaining the cultural traditions of whaling communities, responses in England and the U.S. appeared intermediate between the higher levels of support in Japan and Norway and the lower levels of support indicated for this policy goal in both Australia and Germany (Tables 17.2 and 17.3). Norwegian respondents indicated strong support (60% in favour, 15% opposed) for a policy that views whales as a protein food source for human use (Appendix 4). This policy enjoyed decreasing levels of support among respondents in Japan (33% opposed), Germany and the U.S. (44%), England (50%) and Australia (60% opposed).

Questions Concerning Peoples' Attitudes Toward Different Sources of Meat

It is apparent that answers to questions about the acceptability of producing and consuming various food animals reflect different national, as well as respondents' personal, food habits. Thus, as might be expected, Australians (at 28%) exhibited levels of acceptance of kangaroo meat twice as high as found among Germans, Japanese and Norwegians. Factors other than familiarity in-

Table 17.3 Response Rates (%) in Regard to Whaling in Support of Traditional Communities (See Appendix 3, 4, and 5). "**Agree**" based on summing "4" and "5" responses on five-point scale; "**Neutral**" is "3", "**Disagree**" is sum of "1" and "2" responses; see Appendix 3 for basis of scaling

		Norway	Japan	United States	England	Australia	Germany
Small-scale whaling	Agree	47	53	57	77	72	77
to primarily benefit	Disagree	25	21	22	9	13	10
local communities	Neutral	23	23	19	11	12	8
	Mean	3.39	3.58	3.64	4.29	4.13	4.34
Whaling policy	Agree	44	59	49	46	40	37
to support cultural	Disagree	25	16	23	27	33	37
traditions	Neutral	26	22	26	26	25	23
	Mean	3.35	3.71	3.42	3.31	3.12	3.01
Whaling policy to	Agree	56	59	38	33	30	33
support social goals	Disagree	18	20	30	39	40	39
(jobs and whaling	Neutral	21	20	32	25	29	25
communities)	Mean	3.62	3.61	3.14	2.87	2.80	2.87
Harvesting non-	Agree	56	67	34	23	30	28
endangered whales	Disagree	19	19	44	53	49	52
justified to meet	Neutral	18	11	21	23	21	19
economic and	Mean	3.70	3.92	2.79	2.44	2.63	2.54
cultural needs of							
traditional whalers							

fluence food preferences or acceptability, and these various factors likely account for the very low rates of acceptance (around 7%) of kangaroo among respondents in England and the U.S., compared to higher rates from other countries where kangaroo meat is equally likely to be encountered rarely (Appendix 6). Similarly, whereas very strong disapproval was expressed for eating horseflesh in England and the U.S. (only 4% and 10% approving), a more tolerant attitude toward consumption of horseflesh was expressed by Germans (23%), Japanese (26%) and Norwegians (27%). In the case of eating lamb, Norwegians indicated the highest approval ratings (81%), with respondents in Australia (67%) and England (61%) registering lower approval rates and German and U.S. respondents (both at 45%) and Japanese (40%) even lower levels of approval.

In regard to eating whale meat, very small numbers of Australian, English, German, and U.S. respondents indicated approval (in the 2–8% range). Considerably higher levels of support for eating whale meat were registered in both Japan (33%) and Norway (37%). However, in both Japan and Norway those disapproving the use of whale meat were 38.5% and 41% respectively, with a further 26% and 21% respectively registering no strong opinion for or against the production and sale of whale meat. These response rates are quite similar to those registered by U.S. respondents in regard to approval and disapproval toward the production and sale of deer meat and wildfowl. The German respondents' disapproval rate for whale meat production and sale (at

79%) was about equal to Germans' equally high disapproval rate for eating wildfowl (78%), a food source that Australian, English, Japanese, Norwegian, and U.S. respondents find varyingly more acceptable (disapproval ratings ranging from 32% in Norway, to 49% in Australia).

Peoples' Knowledge about Whales and Whaling

One set of ten questions assessed respondents' general knowledge about: (a) the natural history of whales, (b) whaling, (c) the use of whale products (Appendix 7). Another set of questions was directed toward obtaining some measure of public perceptions of whale species' scarcity or abundance (Appendix 8). As these representative national surveys question non-specialists (i.e. the public-at-large), and the questions asked tend to be of a "technical" or specialized nature, it is reasonable to anticipate that many people would answer "Don't know" to many questions, and indeed, this was often the case. For example, in answer to the question "Is the sperm whale the only great whale to use teeth to feed?" in the six countries an average of 40% answered: "Don't know".

In comparison to respondents in Japan and Norway, a much higher number of people in Australia, England, Germany, and the U.S. incorrectly believe that "all large whale species are currently in danger of extinction" (ranging from 70% incorrect answers in Australia, to 41% incorrect in Norway). A somewhat similar pattern of incorrect answers was obtained in response to the statement "Some whale species have become extinct in modern industrial times", where about 80% of Australian, English, German, and U.S. respondents believed, incorrectly, that statement to be true compared to a 55% incorrect response rate in Norway. In answer to the statement "Some countries continue to kill more than a thousand whales every year for scientific research" around 70% of Australian, English, and German respondents provided the incorrect answer, as did 67% of U.S. respondents, compared with 40% of Japanese and 34% of Norwegian answering incorrectly.

These examples indicate the extent to which people fail to provide factually correct answers to these questions. This same pattern of correct and incorrect answers was to a great extent repeated throughout the questionnaire sequence dealing with matters of fact. Thus, the proportion of Australian (at 44%), German (50%), U.S. (54%) and English (57%) respondents believing, again incorrectly, that the main reason for whaling is to supply cheap sources of edible oils, is much higher than the 22% of Norwegians and 30% of Japanese sharing that belief.

A somewhat similar pattern of incorrect responses is obtained to the statement that "most whale meat today is being consumed in expensive Japanese restaurants" (where only 14% and 20% of English and German respondents recognized that to be an incorrect statement). However, about half of the U.S. respondents recognized that statement to be factually untrue, which was the highest correct response rate obtained for that particular question. More Japanese (62%) and U.S. (60%) respondents knew that the blue whale has been protected for a quarter century by international agreement (compared to only 39% of Germans) whereas correct answers to the question about echo-locating abilities of the great whales ranged from a low of 4% (U.K.) to a high of 18% (Japan).

In summary, it appears that in each country the general public tends to know very little about whales, whaling and the use of whale products as gauged by answers to these general and topical questions. The average correct score for the six countries was only about 40%, with a low score of 31% (in the U.K.) to a high score of 52% (in Norway).

Knowledge Concerning Whale Population Levels

One idea yet to be statistically explored from the data collected in this survey, is that peoples' beliefs concerning population status of whales might influence their level of approval or disapproval of whaling. Stated differently, that opposition to whaling may be influenced by a widespread belief that whales can be considered as endangered species. In order to obtain data needed to further consider this question, respondents were asked to indicate where, over a broad range of population numbers, they would place each of seven species of whale (i.e. some of the better known baleen whales and the sperm whale).

A high level of knowledge about whale population numbers was not expected. However, in order to allow respondents in different geographical locations to have an opportunity to find a known species of whale on the list, the species listed in the questionnaire included those most often mentioned (e.g. blue, humpback, minke, etc) in the mass media and the "environmental" or popular literature that receives wide distribution. The level of correct answers was, as expected, low (see Appendix 8). Thus less than 1% of Germans, about 2% of Australian and U.K. respondents and 8% and 9.5% of U.S. and Norwegian respondents respectively knew that sperm whales numbers exceed 1 million. In Germany about half the respondents believed the sperm whale population numbers fewer than 10,000. In Germany and Australia about half the respondents (60% in the U.S.) believe there are less than 10,000 minke whales in the world, and only about 5% thought that the number was greater than 100,000 (the correct answer being about one million). Respondents in Japan and Norway were three to four times as likely to select a correct answer for minke whale population levels compared to those in Australia, England, Germany, or the U.S.

CONCLUSIONS

From the information obtained from these surveys some general conclusions can be offered at this time.

1. Substantial differences appear to exist between public attitudes toward whale management and use in the four non-whaling countries (see Note 2) where widespread opposition to whaling is found, and the two whaling countries where only a minority opposes whaling.
2. There appears to be a shared perspective in all six countries regarding the great importance of (i) protecting whale habitat from pollution and disturbance, (ii) maintaining an ecosystem perspective in whale management, (iii) basing harvest levels upon the best scientific advice.
3. Though there exists widespread opposition to whaling in the non-whaling countries, in the event that whaling is to continue in the future, respondents in most countries place high priority on ensuring that whales be killed in as humane a fashion as is technologically possible and that strict international inspection of whaling is in effect.
4. In regard to whale-management policy, the public in all countries appears to place least importance upon strictly economic objectives in comparison to those of an environmental nature.
5. In regard to the acceptability of producing and selling whale meat for human consumption, considerable differences exist between the two whaling countries and the four non-whaling countries. Thus whale meat received the highest disapproval ratings of any meats in the four non-whaling countries; in the two whaling countries whale meat enjoyed neither high approval nor high disapproval ratings.
6. The public in each country appears to have only a limited amount of correct information concerning whales and issues related to whaling. However, somewhat greater knowledge was possessed by respondents in the two countries (Japan and Norway) where public support for whaling is strongest.

NOTE

1. The authors of this report wish to thank the individuals and agencies who provided generous support for this study: Canadian Circumpolar Institute (Edmonton), Farøya Fiskasøla (Torshavn), Home Rule Government of the Faroe Islands (Torshavn), Department of Fisheries of the Home Rule Government of Greenland (Nuuk), Hvalfur H.F. (Iceland), The Institute of Cetacean Research (Tokyo), National Science Foundation (Washington, D.C.), Norwegian Fisheries Research Council (Trondheim), Social Sciences and Humanities Research Council of Canada (Ottawa), University of Alberta Central Research Fund (Edmonton). Thanks also are due to Dr. Stephen Popiel, Gallup Canada, and Ms. Elaine Maloney, Canadian Circumpolar Institute, for management and technical support provided with skill and understanding.
2. For the purpose of this study, Alaskan whaling is ignored when characterizing the U.S. as a "non-whaling country".

APPENDIX 1

SAMPLE QUESTIONNAIRE (AS USED IN THE AUSTRALIAN SURVEY)

Hello. My name is <SAY NAME>. I'm from the Roy Morgan Research Centre, the people who conduct the Morgan Gallup Poll. Today we're conducting a survey about people's attitudes and knowledge about whales and whaling.

The purpose of this survey is to learn more about how people regard marine animals, whales and other environmental problems. We are asking these questions in various countries throughout the world. Your opinions will help governments in these countries to develop better policies for managing and conserving marine mammals, especially whales. There are no right or wrong answers to most of these questions and this survey will require no more than 15 minutes of your time. We very much appreciate your help.

On a scale that goes from 1 to 5, where 1 means strongly disapprove and 5 means strongly approve, how do you view the production and sale of the following animals for human consumption?

Q1A. How do you view the production and sale of CHICKEN for human consumption?

Q1B. How do you view the production and sale of DEER for human consumption?

Q1C. How do you view the production and sale of HORSE for human consumption?

Q1D. How do you view the production and sale of KANGAROO for human consumption?

Q1E. How do you view the production and sale of LAMB for human consumption?

Q1F. How do you view the production and sale of LOBSTER for human consumption?

Q1G. How do you view the production and sale of SEAL for human consumption?

Q1H. How do you view the production and sale of WHALES for human consumption?

Q1I. How do you view the production and sale of WILDFOWL (e.g. ducks, geese, pheasant) for human consumption?

I'm now going to read a series of statements to you and I would like you to tell me how strongly you agree or disagree with each one.

Q2A. Non-endangered whales can be killed in order to provide food for humans.

Q2B. You see nothing wrong with harvesting whales if it is properly regulated.

Q2C. If a whale species is not endangered, the economic and cultural needs of people who traditionally hunt these animals justifies their continued hunting.

Q2D. You cannot imagine anyone would want to kill anything as intelligent as a whale.

Q2E. If whale populations become plentiful again, you think people should be allowed to harvest them for useful products.

Q2F. You are opposed to the hunting of any kind of whale under any circumstances.

Now I'd like to know how important you believe the following goals should be when the International Whaling Commission establishes whaling policies.

Q3A. Ecological goals, such as the role whales play in ocean ecosystem management.

Q3B. Ethical goals such as the right of whales to exist without harm from human interference.

Q3C. Economic goals, such as the profitability of the whaling industry.

Q3D. Resource use goals, to ensure a continued supply of protein or meat for human consumption.

Q3E. Animal welfare concerns, such as the possible pain and suffering inflicted on whales by whaling.

Q3F. Social goals, such as maintaining jobs and the local residents' wellbeing in traditional whaling communities.

Q3G. Environmental goals, such as protection of whales' habitat from the threat of marine pollution or industrial activity.

Q3H. Cultural goals, such as maintaining traditional whaling communities and their way of life.

A possibility exsts that whaling may be resumed in the future. If whaling is resumed, please indicate your support for the following conditions that should be observed.

Q4A. Only hunting a limited number from abundant, non-endangered species.

Q4B. The number of whales to be caught should be based upon the best scientific information.

Q4C. Regular and strict international inspection and regulation of whaling operations be put in place.

Q4D. Killing of whales should be carried out in as humane a fashion as is technically possible.

Q4E. Whaling should be carried out on a small scale primarily to provide benefits to local communities.

Q4F. All edible parts of the whale to be utilized for human food only.

Q4G. Distribution of all edible products be required to prevent wastage.

Now I'm going to read a list of statements and would like you to tell me, to the best of your knowledge, which of them are true or false.

Q5A. All large whale species are currently in danger of extinction.

Q5B. Some whale species have become extinct in modern industrial times.

Q5C. The largest whale species for example the blue and the humpback, mostly feed on moderate-sized fish like cod and salmon.

Q5D. Today it's possible to kill whales quickly using modern technology.

Q5E. Some countries continue to kill more than a thousand whales every year for scientific research.

Q5F. All whales can navigate by echo-location.

Q5G. The sperm whale is the only great whale using teeth to feed.

Q5H. The main justification for commercial whaling is to provide cheap oil for various industrial uses.

Q5I. Nearly all the whale meat sold today is consumed in expensive Japanese restaurants.

Q5J. The blue whale has been protected from whaling for a quarter century by international agreement.

Now I am going to read a list of different species of whales and would like you to tell me what you think the global or worldwide population of each is.

Q6A. Do you think that the global population of BLUE WHALES is less than 1,000,

between 1,000 and 9,999
between 10,000 and 99,999
between 100,000 and 999,999
or one million or more?
(IF DOESN'T KNOW: Well, your best guess?)

Q6B. Do you think that the global population of GRAY WHALES is less than 1,000,

between 1,000 and 9,999
between 10,000 and 99,999
between 100,000 and 999,999
or one million or more?
(IF DOESN'T KNOW: Well, your best guess?)

Q6C. Do you think that the global population of SPERM WHALES is less than 1,000,

between 1,000 and 9,999
between 10,000 and 99,999
between 100,000 and 999,999
or one million or more?
(IF DOESN'T KNOW: Well, your best guess?)

Q6D   Do you think that the global population of MINKE WHALES is less than 1,000,

between 1,000 and 9,999
between 10,000 and 99,999
between 100,000 and 999,999
or one million or more?
(IF DOESN'T KNOW: Well, your best guess?)

Q6E. Do you think that the global population of HUMPBACK WHALES is less than 1,000,

between 1,000 and 9,999
between 10,000 and 99,999
between 100,000 and 999,999
or one million or more?
(IF DOESN'T KNOW: Well, your best guess?)

Q6F   Do you think that the global population of FIN WHALES is less than 1,000,

between 1,000 and 9,999
between 10,000 and 99,999
between 100,000 and 999,999
or one million or more?
(IF DOESN'T KNOW: Well, your best guess?)

Q6G.   Do you think that the global population of RIGHT WHALES is less than 1,000,

between 1,000 and 9,999
between 10,000 and 99,999
between 100,000 and 999,999
or one million or more?
(IF DOESN'T KNOW: Well, your best guess?)

Q6H.   And do you think that the global population of ALL SPECIES OF WHALES is less than 1,000

between 1,000 and 9,999
between 10,000 and 99,999
between 100,000 and 999,999
or one million or more?
(IF DOESN'T KNOW: Well, your best guess?)

To make sure we have a true cross-section of people, I'd like to ask you a few questions about yourself.

QAGE.   Would you mind telling me your approximate age please?

IF REFUSES READ OUT.

IF STILL REFUSES ESTIMATE

18–19
20–24
25–29
30–34
35–39
40–44
45–49
50–54
55–59
60–64
65+

QEDUC.   What is the highest level of education you have reached?
(IF OTHER, HIGHLIGHT OTHER AND TYPE IN RESPONSE).

PRIMARY SCHOOL
SOME SECONDARY SCHOOL
SOME TECHNICAL OR COMMERCIAL
PASSED 4TH FORM/YEAR 10
PASSED 5TH FORM/YEAR 11/LEAVING
FINISHED TECHNICAL SCHOOL
COMMERCIAL COLLEGE OR TAFE
FINISHED/NOW STUDYING H.S.C./V.C.E/YEAR 12
DIPLOMA FROM C.A.E.
SOME UNIVERSITY/C.A.E.
DEGREE FROM UNIVERSITY OR CAE
OTHER
CAN'T SAY

QWORK.   Are you now in paid employment?

IF YES ASK: Is that full-time or 35 hours or more a week, or part-time?

IF NOT EMPLOYED:

QNONW.   Are you now looking for a paid job?

IF NOT LOOKING ASK: Are you retired, a student, a non-worker or home duties?

QOCC.   What is your occupation — the position and industry?

1: Professional
2: Owners of Executives
3: Owners of Small Businesses

1: Sales
2: Semi-professional
4: Other White Collar
5: Skilled
6: Semi-Skilled
7: Unskilled
8: Farm Owners
Farm
Workers (Number)
Occupation

QINC.   Would you mind telling me your approximate annual income from all sources before tax?

IF CAN'T SAY: Well, what's your best guess?

LESS THAN $10,000
$10,000–$14,999
$15,000–$19,999
$20,000–$24,999
$25,000–$29,999
$30,000–$34,999
$35,000–$39,999
$40,000–$44,999
$45,000–$49,999
$50,000 OR MORE
CAN'T SAY
REFUSED

Thank you for your time and assistance.

RECORD SEX OF RESPONDENT.
RECORD YOUR OWN NAME FOR A TRUE AND HONEST IN-TERVIEW.

APPENDIX 2

	Australia	England	Germany	Japan	Norway	U.S.
Sample Size	517	517	507	517	536	1006
Urban/Rural	+	+	+	+	+	+
Age	+	+	+	+	+	+
Gender	+	+	+	+	+	+
Marital Status		+				
Household Composition		+	+			
Education	+	+	+	+	+	+
Occupation	+	+	+		+	+
Income	+		+	+	+	+
Race						+
Date of polling	January 9–12 1992	February 1992	January 23–26 1992	January 15–18 1992	Early Feb. 1992	Early June 1992

Sample size and data characteristics

	Sample Size				
	1,000	600	500	400	200
Percentages near 10	2	3	3	3	4
Percentages near 20	2	3	4	4	6
Percentages near 30	3	4	4	4	6
Percentages near 40	3	4	4	5	7
Percentages near 50	3	4	4	5	7
Percentages near 60	3	4	4	5	7
Percentages near 70	3	4	4	4	6
Percentages near 80	2	3	4	4	6
Percentages near 90	2	3	3	3	4

Percentage variation (error) contained in survey results at 95% confidence level (i.e., results will occur with stated variation 19 times out of 20).

APPENDIX 3

Public Attitudes Toward Whaling, by Country. Response on a Five-Point Scale: **1** = Strongly Disapprove, **2** = Moderately Disapprove, **3** = No Position for or Against, **4** = Moderately Agree, **5** = Strongly Agree.

Questions	Response	Australia	England	Germany	Japan	Norway	U.S.A.
Non-endangered	1 + 2	65.9	66.9	61.5	29.2	12.7	55.0
whales can be	3	14.3	15.5	15.4	14.9	11.9	18.0
killed for human	4 + 5	19.2	16.6	21.9	54.7	62.7	26.3
food	Can't say	0.6	1.0	1.2	0.8	2.6	0.6
	Mean	2.10	2.00	2.21	3.40	4.13	2.44
Harvesting whales	1 + 2	65.8	63.5	59.6	21.8	14.6	55.2
is acceptable if	3	12.6	17.0	14.2	13.2	8.6	17.0
properly regulated	4 + 5	20.9	19.0	26.1	64.2	74.1	27.2
	Can't say	0.8	0.6	0.2	0.8	2.8	0.5
	Mean	2.15	2.11	???	3.79	4.10	2.46
Harvesting non-	1 + 2	49.1	52.6	52.0	18.9	19.2	44.0
endangered whales	3	20.5	22.6	18.7	11.2	18.3	21.4
justified for economic	4 + 5	29.6	23.2	28.2	68.6	55.6	34.2
and cultural needs of	Can't say	0.8	1.5	1.0	1.0	6.9	0.4
traditional whalers	Mean	2.63	2.44	2.54	3.92	3.70	2.79
Can't imagine why	1 + 2	21.7	20.1	23.5	47.9	57.1	24.8
anyone would kill	3	13.3	13.0	17.6	23.8	15.5	19.6
intelligent whales	4 + 5	63.9	64.2	55.8	24.6	21.8	57.0
	Can't say	1.2	2.7	2.4	3.7	5.6	0.7
	Mean	3.89	3.85	3.67	2.63	2.36	3.61
Harvesting plentiful	1 + 2	62.9	54.6	55.9	17.2	15.9	49.8
whales for useful	3	15.3	16.1	20.3	15.9	9.0	21.7
products is O.K.	4 + 5	21.2	27.3	31.3	65.6	71.6	28.0
	Can't say	0.6	2.1	2.6	1.4	3.5	0.5
	Mean	2.19	2.44	2.31	3.87	4.04	2.59
Opposed to whale	1 + 2	29.0	37.1	26.5	56.6	61.0	34.4
hunting under any	3	11.0	19.3	18.1	18.4	15.7	17.1
circumstances	4 + 5	60.0	43.1	54.3	23.5	21.1	48.4
	Can't say	0	0.4	1.2	1.6	2.2	0.2
	Mean	3.66	3.18	3.57	2.41	2.27	3.30

APPENDIX 4

Public Attitudes Toward Selected Whaling Policies, by Country.
(Response scale as in Appendix 3).

Questions	Response	Australia	England	Germany	Japan	Norway	U.S.A.
Environmental	1 + 2	3.5	5.4	3.6	9.8	5.0	7.0
(protection against	3	4.8	5.0	3.9	12.4	13.4	9.2
pollution, industrial	4 + 5	91.1	88.0	91.3	75.6	77.3	83.0
activity, etc.)	Can't say	0.6	1.5	1.2	2.1	4.3	0.8
	Mean	4.68	4.54	4.70	4.19	4.35	4.44
Ecological	1 + 2	3.6	6.0	2.8	12.0	7.4	6.7
(part played by whales	3	6.0	8.7	3.7	16.1	14.4	12.9
in marine ecosystems)	4 + 5	87.4	80.1	90.9	65.3	71.1	78.2
	Can't say	2.9	5.2	2.6	6.6	7.1	2.2
	Mean	4.62	4.43	4.74	4.06	4.21	4.33
Cultural	1 + 2	33.3	26.9	37.3	15.8	24.9	24.3
(maintaining traditions	3	24.6	25.9	22.7	22.1	25.6	25.8
in whaling	4 + 5	40.0	45.5	36.9	58.9	44.2	48.7
communities)	Can't say	2.1	1.7	3.2	3.1	5.4	1.1
	Mean	3.12	3.31	3.01	3.71	3.35	3.42
Social	1 + 2	39.6	39.3	34.1	19.7	18.3	29.7
(jobs, social wellbeing	3	29.2	25.3	28.2	19.7	20.9	31.5
in traditional whaling	4 + 5	30.0	32.9	35.5	58.6	56.1	37.5
communities)	Can't say	1.2	2.5	2.2	1.9	4.7	1.3
	Mean	2.80	2.87	3.04	3.61	3.62	3.14
Animal Welfare	1 + 2	9.7	8.1	6.0	23.6	10.6	16.6
(avoid whales' pain	3	6.6	5.2	6.7	29.8	11.2	14.0
and suffering)	4 + 5	83.2	85.5	85.4	42.5	74.4	68.5
	Can't say	0.6	1.2	2.0	4.1	3.7	0.9
	Mean	4.42	4.48	4.56	3.32	4.20	3.97
Ethical	1 + 2	5.2	6.9	6.3	16.3	26.5	12.1
(whales' right to exist)	3	12.4	13.7	10.8	25.5	27.1	17.9
	4 + 5	81.4	78.5	81.0	53.4	41.0	69.4
	Can't say	1.0	0.8	1.8	4.9	5.4	0.7
	Mean	4.45	4.31	4.43	3.67	3.32	4.07
Resource Use	1 + 2	59.2	49.2	43.9	32.9	15.4	43.6
(protein/food supply	3	15.3	22.6	17.2	24.2	19.4	24.2
for humans)	4 + 5	24.5	26.1	34.7	40.6	60.2	41.1
	Can't say	1.0	2.1	4.2	2.3	4.9	1.1
	Mean	2.33	2.56	2.80	3.12	3.76	2.74
Economic	1 + 2	69.6	50.3	66.9	45.4	33.2	51.8
(profitability of	3	15.5	20.7	13.8	25.7	26.1	23.8
the whaling industry)	4 + 5	14.1	25.2	17.1	24.8	36.3	22.9
	Can't say	0.8	3.9	2.2	4.1	4.3	1.6
	Mean	2.02	2.52	2.09	2.64	2.97	2.48

APPENDIX 5

Response Rates to Selected Whaling Policies, by Country.
(Response scale as in Appendix 3).

Policy	Response Rate	Australia	England	Germany	Japan	Norway	U.S.A.
Only limited harvest from abundant non-endangered species	1 + 2	7.4	9.2	11.9	6.4	13.6	9.5
	3	4.1	10.4	11.2	10.4	16.8	8.7
	4 + 5	86.1	76.0	73.6	81.1	65.9	90.1
	Can't say	2.5	4.3	3.4	2.1	3.7	1.7
	Mean	4.55	4.29	4.22	4.40	4.01	4.30
Harvest level based on best scientific advice	1 + 2	4.1	8.7	10.3	8.1	5.8	8.1
	3	2.3	6.2	5.5	8.9	10.1	8.4
	4 + 5	90.7	80.1	80.7	80.2	80.2	81.7
	Can't say	2.9	5.0	3.6	2.7	3.9	1.8
	Mean	4.72	4.39	4.42	4.34	4.41	4.40
Distribution in order to ensure complete utilization and minimize waste	1 + 2	9.4	12.0	7.7	9.1	9.4	8.8
	3	7.9	13.5	10.7	7.7	15.7	11.0
	4 + 5	78.0	64.6	75.0	80.5	70.6	77.6
	Can't say	4.6	9.9	6.7	2.7	4.5	2.6
	Mean	4.33	4.08	4.35	4.34	4.14	4.31
Use most humane killing method technologically possible	1 + 2	1.4	5.1	4.0	12.1	3.9	6.9
	3	1.5	3.1	1.8	15.5	4.5	5.7
	4 + 5	95.0	88.6	91.0	66.9	89.8	85.0
	Can't say	2.1	3.3	3.5	5.4	1.9	2.5
	Mean	4.88	4.68	4.78	4.03	4.68	4.54
Whaling on a small-scale to primarily benefit local communities	1 + 2	12.9	8.7	10.1	20.7	25.4	21.9
	3	12.4	11.0	7.5	22.6	23.3	19.2
	4 + 5	72.2	76.8	77.0	53.2	47.2	57.1
	Can't say	2.5	3.5	5.5	3.5	4.1	1.8
	Mean	4.13	4.29	4.34	3.58	3.39	3.64
All edible whale products for human consumption only	1 + 2	24.0	22.8	15.6	31.9	28.0	28.1
	3	19.0	18.8	15.6	24.0	17.2	23.2
	4 + 5	52.1	46.6	62.5	39.7	49.1	46.3
	Can't say	5.0	11.8	6.3	4.4	5.8	2.5
	Mean	3.57	3.51	3.97	3.18	3.41	3.34
Regular and strict international inspection of whaling to be in place	1 + 2	1.5	4.5	5.5	9.6	5.3	4.7
	3	1.2	3.7	3.0	11.6	10.1	6.4
	4 + 5	95.7	89.8	88.3	77.0	82.1	87.3
	Can't say	1.5	2.1	3.2	2.7	2.6	1.6
	Mean	4.89	4.66	4.66	4.26	4.47	4.57

APPENDIX 6

Public Attitudes Toward Production and Sale of Selected Animal Meats.
(Response scale as in Appendix 3).

Species	Response Rate	Australia	England	Germany	Japan	Norway	U.S.A.
Chicken	1 + 2	8.5	10.7	16.4	10.3	6.5	5.8
	3	14.9	22.8	22.3	16.1	12.7	10.0
	4 + 5	76.2	66.1	60.4	71.5	79.3	84.1
	Mean	4.25	3.93	3.85	4.09	4.31	4.48
Deer	1 + 2	49.7	62.6	35.7	61.3	23.9	39.2
	3	21.5	18.2	22.5	19.3	22.8	21.9
	4 + 5	27.5	17.6	40.2	17.2	51.5	38.2
	Mean	2.59	2.13	3.07	2.21	3.47	2.96
Horse	1 + 2	72.3	84.9	60.2	52.2	51.4	80.6
	3	13.7	10.1	16.0	20.1	19.2	8.8
	4 + 5	13.4	4.4	23.1	26.3	27.1	10.0
	Mean	1.86	1.44	2.28	2.51	2.46	1.65
Kangaroo	1 + 2	47.7	80.9	64.1	67.7	55.2	79.6
	3	23.6	9.7	14.8	14.7	15.5	11.0
	4 + 5	28.2	6.8	15.9	13.2	12.9	7.5
	Mean	2.61	1.54	2.01	1.95	2.02	1.60
Lamb	1 + 2	9.8	16.1	29.0	33.0	7.2	30.9
	3	13.3	22.4	24.7	25.3	10.4	23.4
	4 + 5	66.6	61.3	45.3	40.1	80.6	45.2
	Mean	4.22	3.75	3.29	3.07	4.32	3.27
Lobster	1 + 2	16.6	26.3	48.2	21.2	10.4	13.3
	3	18.6	21.5	17.8	17.6	18.1	18.7
	4 + 5	63.9	49.9	30.4	57.1	68.6	67.5
	Mean	3.88	3.36	2.65	3.58	4.01	3.99
Seal	1 + 2	89.0	91.1	86.6	65.6	44.2	87.8
	3	5.2	5.0	4.9	16.4	20.7	5.3
	4 + 5	5.2	2.9	6.3	14.4	31.7	6.3
	Mean	1.38	1.28	1.40	2.06	2.71	1.42
Whale	1 + 2	93.1	92.8	79.1	40.7	38.5	87.7
	3	4.8	4.4	10.7	25.9	21.3	4.9
	4 + 5	2.0	2.3	8.5	32.5	37.4	6.7
	Mean	1.23	1.23	1.63	2.81	2.96	1.41
Wildfowl	1 + 2	48.8	44.7	78.1	42.2	32.2	38.2
	3	26.3	24.0	10.1	25.1	30.8	22.7
	4 + 5	24.1	30.6	10.7	31.1	34.1	38.5
	Mean	2.53	2.68	1.70	2.75	3.00	2.99

APPENDIX 7

Response Rate for Knowledge Questions About Whales and Whaling;
(**C** = Correct Answer, **I** = Incorrect Answer, **DK** = Don't Know).

Question	Answer	Australia	England	Germany	Japan	Norway	U.S.A.
All large species	C	24	20	18	38	41	31
of whales in danger	I	70	68	69	50	41	65
of extinction	DK	5	12	13	11	19	4
Some whale species	C	15	7	6	22	22	11
have become extinct	I	80	81	81	65	55	84
in modern times	DK	5	12	13	12	22	5
Largest whales	C	30	35	42	42	39	42
largely feed on	I	57	37	39	40	36	48
moderately-sized fish	DK	13	28	18	18	24	9
Whales can be quickly	C	50	50	64	60	72	66
killed using modern	I	42	33	25	23	12	29
technology	DK	8	17	11	16	17	5
Some countries kill	C	22	12	16	37	31	24
> 1,000 whales/year	I	69	70	71	40	34	67
for scientific research	DK	9	18	13	22	35	9
All whales can	C	9	4	6	18	6	17
navigate by	I	82	82	75	56	71	77
echolocation	DK	10	14	18	25	23	6
Sperm whales only	C	28	20	31	27	22	30
great whale using	I	45	20	21	39	22	53
teeth for feeding	DK	27	60	48	34	56	17
Main reason for	C	50	22	32	30	49	40
commercial whaling	I	44	57	50	52	22	54
is to provide cheap	DK	6	21	18	18	29	6
industrial oils							
Nearly all whale meat	C	30	14	20	41	36	49
today sold to expensive	I	57	64	54	49	32	40
Japanese restaurants	DK	13	22	25	10	32	11
Blue whale protected	C	47	52	39	62	50	60
for a quarter century	1	31	14	24	20	17	26
	DK	22	34	37	18	32	14

APPENDIX 8

Response Rate to Knowledge Questions About Whale Population Size.
(Rates in percent; may not total 100 due to wording and small numbers not answering).
Population Data from W. Aron, *Coastal Management* 16:99–110 (1988).

	Australia	England	Germany	Japan	Norway	U.S.A.
Blue Whale (7,500–15,000)						
< 1,000	19	13	21	9	14	21
1,000–9,999	39	26	43	28	24	37
10,000–99,999	23	12	17	29	20	25
100,000–999,999	4	3	2	7	5	6
> 1 Million	1	1	< 1	2	2	1
Don't know	15	45	17	25	36	10
Fin Whale (105,000–122,000)						
< 1,000	16	8	13	8	7	24
1,000–9,999	31	10	21	19	10	30
10,000–99,999	20	10	12	22	14	22
100,000–999,999	5	3	2	10	7	4
> 1 Million	1	< 1	< 1	3	2	1
Don't know	27	69	51	37	61	19
Gray Whale (ca. 19,000)						
< 1,000	13	5	15	10	6	14
1,000–9,999	37	18	29	18	18	36
10,000–99,999	26	15	15	19	18	30
100,000–999,999	4	5	4	9	6	8
> 1 Million	2	2	1	2	2	2
Don't know	19	55	35	43	50	11
Humpback (ca. 10,000)						
< 1,000	16	11	18	9	7	18
1,000–9,999	43	22	20	16	13	37
10,000–99,999	22	12	12	23	14	28
100,000–999,999	4	4	2	12	7	6
> 1 Million	< 1	1	< 1	3	2	2
Don't know	14	50	47	37	58	10
Minke Whale (ca. 800,000)						
< 1,000	24	13	10	10	6	32
1,000–9,999	30	10	19	14	9	29
10,000–99,999	16	9	14	19	19	14
100,000–999,999	3	4	6	12	13	4
> 1 Million	< 1	1	< 1	5	5	1
Don't know	25	64	50	39	48	21

Appendix table (*continue*)

	Australia	England	Germany	Japan	Norway	U.S.A.
Right Whale (ca. 3,000)						
< 1,000	17	10	13	16	8	26
1,000–9,999	36	10	19	20	9	33
10,000–99,999	22	5	8	17	8	20
100,000–999,999	4	2	1	7	5	3
> 1 Million	< 1	< 1	< 1	2	1	1
Don't know	21	72	58	39	70	18
Sperm Whale (ca. 1,500,000)						
< 1,000	13	7	22	8	5	17
1,000–9,999	36	21	27	15	15	36
10,000–99,999	30	14	12	27	19	28
100,000–999,999	5	5	3	15	7	7
> 1 Million	2	2	< 1	3	2	1
Don't know	14	51	40	33	52	11
All Whale Species (more than 1 Million)						
< 1,000	< 1	< 1	1	2	1	1
1,000–9,999	7	4	9	6	3	6
10,000–99,999	32	15	36	18	8	31
100,000–999,999	36	22	26	28	30	36
> 1 Million	15	20	7	29	28	19
Don't know	10	40	22	16	38	8

INDEX